A Course on Large Deviations with an Introduction to Gibbs Measures

A Course on Large Deviations with an Introduction to Gibbs Measures

Firas Rassoul-Agha
Timo Seppäläinen

Graduate Studies in Mathematics
Volume 162

American Mathematical Society
Providence, Rhode Island

2010 *Mathematics Subject Classification.* Primary 60-01, 60F10, 60J10, 60K35, 60K37, 82B05, 82B20.

For additional information and updates on this book, visit
www.ams.org/bookpages/gsm-162

Library of Congress Cataloging-in-Publication Data

Rassoul-Agha, Firas.
A course on large deviations with an introduction to Gibbs measures / Firas Rassoul-Agha, Timo Seppäläinen.
pages cm. — (Graduate studies in mathematics ; volume 162)
Includes bibliographical references and indexes.
ISBN 978-0-8218-7578-0 (alk. paper)
1. Large deviations. 2. Probabilities. 3. Measure theory. I. Seppäläinen, Timo O. II. Title.

QA273.67.R375 2015
519.2—dc23

2014035275

∞ The paper used in this book is acid-free and falls within the guidelines
established to ensure permanence and durability.
Visit the AMS home page at `http://www.ams.org/`

10 9 8 7 6 5 4 3 2 1   20 19 18 17 16 15

To Alla, Maxim, and Kirill

To Celeste, David, Ansa, and Timo

Contents

Part III. Additional large deviation topics

Appendixes

Preface

This book arose from courses on large deviations and related topics given by the authors in the Departments of Mathematics at the Ohio State University (1993), at the University of Wisconsin-Madison (2006 and 2013), and at the University of Utah (2008 and 2013).

Our goal has been to create an attractive collection of material for a semester's course which would also serve the broader needs of students from different fields. This goal has had two implications for the book.

(1) We have not aimed at anything like an encyclopedic coverage of different techniques for proving large deviation principles (LDPs). Part I of the book focuses on one classic line of reasoning: (i) upper bound by an exponential Markov-Chebyshev inequality, (ii) lower bound by a change of measure, and (iii) an argument to match the rates from (i) and (ii). Beyond this technique Part I covers Bryc's theorem and proves Cramér's theorem with the subadditive method. Part III of the book covers the Gärtner-Ellis theorem and an approach based on the convexity of a local rate function due to Baxter and Jain.

(2) We have not felt obligated to stay within the boundaries of large deviation theory but instead follow the trail of some interesting material. Large deviation theory is a natural gateway to statistical mechanics. A discussion of statistical mechanics would be incomplete without some study of phase transitions. We prove the phase transition of the Ising model in two different ways: (i) first with classical techniques: the Peierls argument, Dobrushin's uniqueness condition, and correlation inequalities and (ii) the second time with random cluster measures. This means leaving large deviation theory completely behind. Along the way we have the opportunity to learn coupling methods which are central to modern probability theory

but do not get serious application in the typical first graduate course in probability.

We give now a brief overview of the contents of the book.

Part I covers core general large deviation theory, the relevant convex analysis, and the large deviations of independent and identically distributed (i.i.d.) processes on three levels: Cramér's theorem, Sanov's theorem, and the process level LDP for i.i.d. variables indexed by a multidimensional square lattice.

Part II introduces Gibbs measures and proves the Dobrushin-Lanford-Ruelle variational principle that characterizes translation-invariant Gibbs measures. After this we study the phase transition of the Ising model. Part II ends with a chapter on the Fortuin-Kasteleyn random cluster model and the percolation approach to Ising phase transition.

Part III develops the large deviation themes of Part I in several directions. Large deviations of i.i.d. variables are complemented with moderate deviations and with more precise large deviation asymptotics. The Gärtner-Ellis theorem is developed carefully, together with the necessary additional convex analysis beyond the basics covered in Part I. From large deviations of i.i.d. processes we move on to Markov chains, to nonstationary independent random variables, and finally to random walk in a dynamical random environment. The last two topics give us an opportunity to apply another approach to proving large deviation principles, namely the Baxter-Jain theorem. The Baxter-Jain theorem has not previously appeared in textbooks, and its application to random walk in random environment is new.

Here is a guide to the dependencies between the parts of the book. Sections 2.1-2.3 and 3.1-3.2 are foundational for all discussions of large deviations. In addition, we have the following links. Chapter 5 relies on Sections 4.1-4.2, and Chapter 6 relies on Chapter 5. Chapter 8 relies on Chapters 6 and 7. Chapter 9 can be read independently of large deviations after Sections 7.1-7.3 and 7.6. Section 10.2 makes sense only in the context of Chapter 9. Chapters 12 and 14 are independent of each other and both rely on Sections 4.1-4.2. Chapter 13 relies on Chapter 5. Chapter 15 relies on Section 13.1 and Chapter 14. Chapter 16 relies on Chapter 14.

The ideal background for reading this book would be some familiarity with the language of measure-theoretic probability. Large deviation theory does also require a little analysis, point set topology, and functional analysis. For example, readers should be comfortable with lower semicontinuity and the weak topology on probability measures. It should be possible for an instructor to accommodate students with quick lectures on technical prerequisites whenever needed. It is also possible to consider everything in the framework of concrete finite spaces, in which case probability measures become simply probability vectors.

In practice our courses have been populated by students with very diverse backgrounds, many with less than ideal knowledge of analysis and probability. This has turned out less problematic than one might initially fear. Mathematics students are typically fully satisfied only after every theoretical point is rigorously justified. But engineering students are content to set aside much of the theory and focus on the essentials of the phenomenon in question. There is great interest in probability theory among students of economics, engineering, and the sciences. This interest should be encouraged and nurtured with accessible courses.

The appendixes in the back of the book serve two purposes. There is a quick overview of some basic results of analysis and probability without proofs, for the reader who wants a quick refresher. In particular, here the reader can look up textbook tools such as convergence theorems and inequalities that are referenced in the text. The other material in the appendixes consists of specialized results used in the text, such as a minimax theorem and inequalities from statistical mechanics. These are proved.

Since this book evolved in courses where we tried to actively engage the students, the development of the material relies on frequent exercises. We realize that this feature may not appeal to some readers. On the other hand, spelling out all the technical details left as exercises might make for tedious reading. Hopefully an instructor can fill in those details fairly easily if he or she wants to present full details in class. Exercises that are referred to in the text are marked with an asterisk.

One of us (Timo Seppäläinen) first learned large deviations from a course taught by Steven Orey in 1988–1989 at the University of Minnesota. We are greatly indebted to the existing books on the subject, especially those by Amir Dembo and Ofer Zeitouni [**15**], Frank den Hollander [**16**], Jean-Dominique Deuschel and Daniel Stroock [**18**], Richard Ellis [**32**], and Srinivasa Varadhan [**79**].

As a text that combines large deviations with equilibrium statistical mechanics, [**32**] is a predecessor of ours. There is obviously a good degree of overlap but the books are different. Ours is a textbook with a lighter touch while [**32**] is closer to a research monograph, covers more models in detail, and explains much of the physics. We recommend [**32**] to our readers and students for further study. Our phase transition discussion covers the nearest-neighbor Ising model while [**32**] also covers long-range Ising models. On the other hand, [**32**] does not cover Dobrushin's uniqueness theorem, random cluster models, general lattice systems, or their large deviations.

Our literature references are sparse and sometimes do not assign credit to the originators of the ideas. We encourage the reader to consult the superb historical notes and references in the monographs of Dembo-Zeitouni, Ellis, and Georgii.

We thank Jeff Steif for lecture notes that helped shape the proof of Theorem 9.2, Jim Kuelbs for material for Chapter 11, and Chuck Newman for helpful discussions on the liquid-gas phase transition for Chapter 7. We also thank Davar Khoshnevisan for several valuable suggestions. We thank the team at the AMS and especially Ed Dunne for his patience in the face of serial breaches of agreed deadlines, and we thank the several reviewers for valuable suggestions.

Support from the National Science Foundation, the Simons Foundation, and the Wisconsin Alumni Research Foundation is gratefully acknowledged.

Firas Rassoul-Agha
Timo Seppäläinen

Part I

Large deviations: General theory and i.i.d. processes

Chapter 1

Introductory discussion

Toss a fair coin n times. When n is small there is nothing to say beyond enumerating all the outcomes and their probabilities. With a large number of tosses patterns and order emerge from the randomness: heads appear about 50% of the time and the histogram approaches a bell curve. As the number of tosses increases, these patterns become more and more pronounced. But from time to time a random fluctuation might break the pattern: perhaps 10,000 tosses of a fair coin give 6,000 heads. In fact, we know that there is a chance of $(1/2)^{10,000}$ that all tosses yield heads. The point is that to understand the system well one cannot be satisfied with understanding only the most likely outcomes. One also needs to understand rare events. But why care about an event that has a chance of $(1/2)^{10,000}$?

Here is a simplified example to illustrate the importance of probabilities of rare events. Imagine that an insurance company collects premiums at a steady rate of c per month. Let X_k be the random amount that the insurance company pays out in month k to cover customer claims. Let $S_n = X_1 + \cdots + X_n$ be the total payout in n months. Naturally the premiums must cover the average outlays, so $c > E[X_k]$. The company stays solvent as long as $S_n \le cn$. Quantifying the chances of the rare event $S_n > cn$ is then of obvious interest.

This is an introductory book on the methods of computing asymptotics of probabilities of rare events: the theory of large deviations. Let us start with a basic computation.

Example 1.1. Let $\{X_k\}_{k\in\mathbb{N}}$ be a sequence of independent and identically distributed (i.i.d.) Bernoulli random variables with success probability p (each $X_k = 1$ with probability p and 0 with probability $1-p$). Denote the partial sum by $S_n = X_1 + \cdots + X_n$. The strong law of large numbers says that, as $n \to \infty$, the sample mean S_n/n converges to p almost surely. But at any given n there is a chance p^n for all heads ($S_n = n$) and also a chance $(1-p)^n$ for all tails ($S_n = 0$). In fact, for any $s \in (0,1)$ there is a positive chance of a fraction of heads close to s. Let us compute the asymptotics of this probability.

Denote the integer part of $x \in \mathbb{R}$ by $\lfloor x \rfloor$; that is, $\lfloor x \rfloor$ is the largest integer less than or equal to x. From binomial probabilities

$$\begin{aligned} P\{S_n = \lfloor ns \rfloor\} &= \frac{n!}{\lfloor ns \rfloor!(n-\lfloor ns \rfloor)!}\, p^{\lfloor ns \rfloor}(1-p)^{n-\lfloor ns \rfloor} \\ &\sim \frac{n^n p^{\lfloor ns \rfloor}(1-p)^{n-\lfloor ns \rfloor}}{\lfloor ns \rfloor^{\lfloor ns \rfloor}(n-\lfloor ns \rfloor)^{n-\lfloor ns \rfloor}} \sqrt{\frac{n}{2\pi\lfloor ns \rfloor(n-\lfloor ns \rfloor)}}\,. \end{aligned}$$

We used Stirling's formula (Exercise 3.5)

$$n! \sim e^{-n} n^n \sqrt{2\pi n}\,. \tag{1.1}$$

The notation $a_n \sim b_n$ means that $a_n/b_n \to 1$. Abbreviate

$$\beta_n = \sqrt{\frac{n}{2\pi\lfloor ns \rfloor(n-\lfloor ns \rfloor)}}$$

and to get rid of integer parts, let also

$$\gamma_n = \frac{(ns)^{ns}(n-ns)^{n-ns}}{\lfloor ns \rfloor^{\lfloor ns \rfloor}(n-\lfloor ns \rfloor)^{n-\lfloor ns \rfloor}} \cdot \frac{p^{\lfloor ns \rfloor}(1-p)^{n-\lfloor ns \rfloor}}{p^{ns}(1-p)^{n-ns}}\,.$$

Then

$$P\{S_n = \lfloor ns \rfloor\} \;\sim\; \beta_n \gamma_n \exp\Bigl(ns \log\frac{p}{s} + n(1-s)\log\frac{1-p}{1-s}\Bigr).$$

***Exercise 1.2.** Show that there exists a constant C such that

$$\frac{1}{C\sqrt{n}} \le \beta_n \le \frac{C}{\sqrt{n}} \quad \text{and} \quad \frac{1}{Cn} \le \gamma_n \le Cn$$

for large enough n. By being a little more careful you can improve the second statement to $C^{-1} \le \gamma_n \le C$.

The asymptotics above give the limit

$$\begin{aligned} &\lim_{n\to\infty} \frac{1}{n} \log P\{S_n = \lfloor ns \rfloor\} = -I_p(s) \\ &\text{with} \quad I_p(s) = s\log\frac{s}{p} + (1-s)\log\frac{1-s}{1-p}. \end{aligned} \tag{1.2}$$

Note the minus sign introduced in front of $I_p(s)$. This is a convention of large deviation theory.

It is instructive to look at the graph of I_p (Figure 1.1). I_p extends continuously to $[0,1]$ with values $I_p(0) = \log\frac{1}{1-p}$ and $I_p(1) = \log\frac{1}{p}$ that

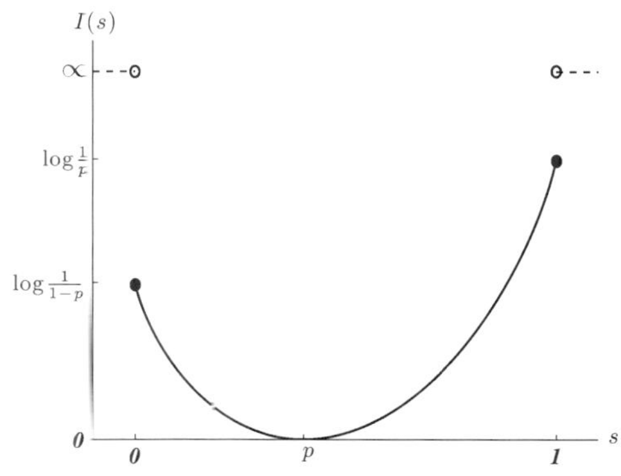

Figure 1.1. The rate function for coin tosses.

match the exponential decay of the probabilities of the events $\{S_n = 0\}$ and $\{S_n = n\}$. The unique zero of I_p is at the law of large numbers limit p which we would regard as the "typical" behavior of S_n/n. Increasing values of I_p correspond to less likely outcomes. For $s \notin [0,1]$ it is natural to set $I_p(s) = \infty$.

The function I_p in (1.2) is a *large deviation rate function.* We shall understand later that $I_p(s)$ is also the *relative entropy* of the coin with success probability s relative to the one with success probability p. The choice of terminology is not a coincidence. This quantity is related to both information-theoretic and thermodynamic entropy.

For this reason we take a brief detour to discuss these well-known notions of entropy and to point out the link with the large deviation rate function I_p. The relative entropy that appears in large deviation theory will take center stage in Chapters 5–6, and again in Chapter 8 when we discuss statistical mechanics of lattice systems.

Limit (1.2) is our first large deviation result. One of the very last ones in the book is limit (16.12), which is the analogue of (1.2) for a random walk in a dynamical random environment, that is, in a setting where the success probability of the coin also fluctuates randomly.

1.1. Information-theoretic entropy

A coin that always comes up heads is not random at all, and the same of course for a coin that always comes up tails. On the other hand, we should probably regard a fair coin as the "most random" coin because we cannot predict whether we see more heads or tails in a sequence of tosses with better than even odds. We discuss here briefly the quantification of the degree of randomness of a sequence of coin flips. We take the point of view that the degree of randomness of the coin is reflected in the average number of bits needed to encode a sequence of tosses. This section is inspired by Chapter 2 of Ash's book [**4**].

Let $\Omega = \{0,1\}^n$ be the space of *words* $\omega \in \Omega$ of length n. A *message* is a concatenation of words. The message made of words $\omega_1, \omega_2, \ldots, \omega_m$ is written $\omega_1\omega_2\cdots\omega_m$. A *code* is a map $C : \Omega \to \bigcup_{\ell \geq 1}\{0,1\}^\ell$ that assigns to each word $\omega \in \Omega$ a *code word* $C(\omega)$ which is a finite sequence of 0's and 1's. $|C(\omega)|$ denotes the length of code word $C(\omega)$. A concatenation of code words is a *code message*. Thus, a message is encoded by concatenating the code words of its individual words to make a code message: $C(\omega_1\cdots\omega_m) = C(\omega_1)\cdots C(\omega_m)$. A code should be *uniquely decipherable*. That is, for every finite sequence $c_1\cdots c_\ell$ of 0's and 1's there exists at most one message $\omega_1\cdots\omega_m$ such that $C(\omega_1)\cdots C(\omega_m) = c_1\cdots c_\ell$.

Now sample words at random under a probability distribution P on the space Ω. In this discussion we employ the base 2 logarithm $\log_2 x = \log x/\log 2$.

Noiseless coding theorem. *If C is a uniquely decipherable code, then its average length satisfies*

$$\sum_{\omega\in\Omega} P(\omega)\,|C(\omega)| \geq -\sum_{\omega\in\Omega} P(\omega)\log_2 P(\omega) \tag{1.3}$$

with equality if and only if $P(\omega) = 2^{-|C(\omega)|}$.

In information theory the quantity on the right of (1.3) is called the *Shannon entropy* of the probability distribution P. For a simple proof of the theorem see [**4**, Theorem 2.5.1, page 37].

Consider the case where the n *characters* of the word ω are chosen independently, and let $s \in [0,1]$ be the probability that a character is a 1. Then $P(\omega) = s^{N(\omega)}(1-s)^{n-N(\omega)}$, where $N(\omega)$ is the number of ones in ω. (As usual, $0^0 = 1$.) By the noiseless coding theorem, the average length of a decipherable code C satisfies

$$\sum_{\omega\in\Omega} |C(\omega)|\, s^{N(\omega)}(1-s)^{n-N(\omega)}$$
$$\geq -\sum_{\omega\in\Omega} s^{N(\omega)}(1-s)^{n-N(\omega)}\log_2 s^{N(\omega)}(1-s)^{n-N(\omega)}.$$

Since $\sum_\omega s^{N(\omega)}(1-s)^{n-N(\omega)} = 1$ and $\sum_\omega N(\omega)s^{N(\omega)}(1-s)^{n-N(\omega)} = ns$, the right-hand side equals $nh(s)$ where

$$h(s) = -s\log_2 s - (1-s)\log_2(1-s) = 1 - \frac{I_{1/2}(s)}{\log 2},$$

and we see the large deviations rate function from (1.2) appear. Thus we have the lower bound

$$\sum_{\omega\in\Omega} |C(\omega)|\, s^{N(\omega)}(1-s)^{n-N(\omega)} \geq nh(s). \tag{1.4}$$

In other words, any uniquely decipherable code for independent and identically distributed characters with probability s for a 1 must use, on average, at least $h(s)$ bits per character. In this case the Shannon entropy and the rate function $I_{1/2}$ are related by

$$-\sum_{\omega\in\Omega} P(\omega)\log_2 P(\omega) = 1 - \frac{I_{1/2}(s)}{\log 2}.$$

Here is a simplistic way to see the lower bound $nh(s)$ on the number of bits needed that makes an indirect appeal to large deviations in the sense that deviant words are ignored. With probability s for symbol 1, the typical word of length n has about ns ones. Suppose we use code words of length L to code these typical words. Then

$$2^L \geq \binom{n}{\lfloor ns \rfloor}$$

and the lower bound $L \geq nh(s) + \mathcal{O}(\log n)$ follows from Stirling's formula.

The values $h(0) = h(1) = 0$ make asymptotic sense. For example, if $s = 0$, then a word of any length n is all zeroes and can be encoded by a single bit, which in the $n \to \infty$ limit gives 0 bits per character. This is the case of complete order. At the other extreme of complete disorder is the case $s = 1/2$ of fair coin tosses where all n bits are needed because all words of a given length are equally likely. For $s \neq 1/2$ a 1 is either more or less likely than a 0 and by exploiting this bias one can encode with less than 1 bit per character on average.

David A. Huffman [**48**], while a Ph.D. student at MIT, developed an optimal decipherable code, that is, a code C whose average length cannot be improved upon. As $n \to \infty$, the average length of the code generated by this algorithm is exactly $h(s)$ per character and so the lower bound (1.4) is achieved asymptotically. We illustrate the algorithm through an example. For a proof of its optimality and asymptotic average length see page 42 of [**4**].

Example 1.3 (Huffman's algorithm)**.** Consider the case $n = 3$ and $s = 1/4$. There are 8 words. Word 111 comes with probability $1/4^3$, words 110, 101, and 011 come each with probability $3/4^3$, words 100, 010, and 001 come with probability $3^2/4^3$ each, and word 000 comes with probability $(3/4)^3$. These 8 words are the terminal leaves of a binary tree that we build.

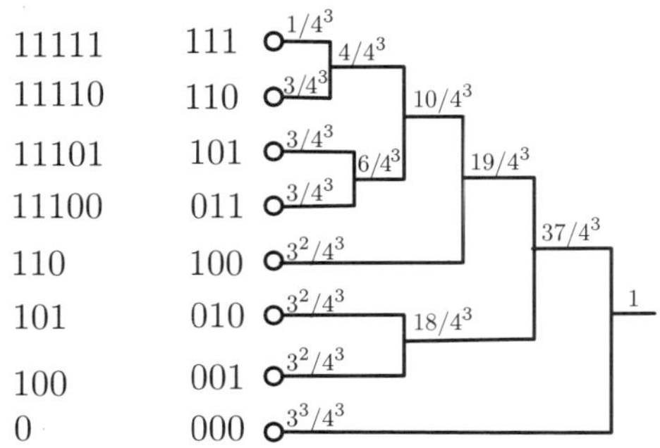

Figure 1.2. The tree for Huffman's algorithm in the case $n = 3$ and $s = 1/4$. The leftmost column shows the resulting codes.

First, find the two leaves with the smallest probabilities. Ties can be resolved arbitrarily. Give these two leaves a and b a common ancestor labeled with a probability that is the sum of the probabilities of a and b. In our example, leaves 111 and 110 are given a common parent labeled with probability $4/4^3$.

Now leaves a and b are done with, and their parent is regarded as a new leaf. Repeat the step. Continue until there is one leaf left. In our example, the second step gives a common ancestor to leaves 101 and 011. This new node is lab, and so on. Figure 1.2 presents the final tree.

To produce the code of a word, start at the root and follow the tree to the leaf of that word. At each fork encode a down step with a 0 and an up step with a 1 (in our figure). For instance, word 101 is reached from the root by three successive up steps followed by a single down step and then another up step. Thus word 101 is encoded as 11101.

The average length of the code is

$$\frac{5 \times 1 + 5 \times 3 + 5 \times 3 + 5 \times 3 + 3 \times 3^2 + 3 \times 3^2 + 3 \times 3^2 + 1 \times 3^3}{4^3} = \frac{158}{64}.$$

This is $158/192 \approx 0.8229$ bits per character. As the number of characters n grows, the average length of the encoding per character will converge to the information-theoretic entropy $h(1/4) \approx 0.811$.

1.2. Thermodynamic entropy

The next discussion of thermodynamics is inspired by Schrödinger's lectures [**70**]. After some preliminary computations we use the first and second laws of thermodynamics to derive an expression for entropy. In the simplest case of a system with two energy levels this expression can be related to the rate function (1.2). The reader should be aware that this section is not mathematically rigorous.

Let A denote a physical system whose possible energy levels are $\{\varepsilon_\ell : \ell \in \mathbb{N}\}$. Then consider a larger system A_n made up of n identical physically independent copies of A. By physically independent we mean that these components of A_n do not communicate with each other. Each component can be at any energy level ε_ℓ. Immerse the whole system in a large heat bath at fixed absolute temperature T which gives the system total energy $\mathcal{E} = nU$.

Let a_ℓ be the number of components in state ε_ℓ. These numbers must satisfy the constraints

$$(1.5) \qquad \sum_\ell a_\ell = n \quad \text{and} \quad \sum_\ell a_\ell \varepsilon_\ell = \mathcal{E}.$$

For given values a_ℓ, the total number of possible arrangements of the components at different energy levels is $\frac{n!}{a_1! \cdots a_\ell! \cdots}$. When n is large, it is reasonable to assume that the values a_ℓ that appear are the ones that maximize the number of arrangements, subject to the constraints (1.5).

To find these optimal a_ℓ values, maximize the logarithm of the number of arrangements and introduce Lagrange multipliers α and β. Thus we wish to differentiate

$$(1.6) \qquad \log \frac{n!}{a_1! \cdots a_\ell! \cdots} - \alpha\Big(\sum_\ell a_\ell - n\Big) - \beta\Big(\sum_\ell a_\ell \varepsilon_\ell - \mathcal{E}\Big)$$

with respect to a_ℓ and set the derivative equal to zero. To use calculus, pretend that the unknowns a_ℓ are continuous variables and use Stirling's formula (1.1) in the form $\log n \sim n(\log n - 1)$. We arrive at

$$\log a_\ell + \alpha + \beta \varepsilon_\ell = 0 \quad \text{for all } \ell.$$

Thus $a_\ell = Ce^{-\beta \varepsilon_\ell}$. Since the total number of components is $n = \sum a_\ell$,

$$(1.7) \qquad a_\ell = \frac{n e^{-\beta \varepsilon_\ell}}{\sum_j e^{-\beta \varepsilon_j}}.$$

The second constraint gives

$$\mathcal{E} = \frac{n \sum_\ell \varepsilon_\ell e^{-\beta \varepsilon_\ell}}{\sum_\ell e^{-\beta \varepsilon_\ell}}.$$

These equations should be understood to hold only asymptotically. Divide both equations by n and take $n \to \infty$. We interpret the limit as saying that when a typical system A is immersed in a heat bath at temperature T the system takes energy ε_ℓ with probability

$$(1.8) \qquad p_\ell = \frac{e^{-\beta \varepsilon_\ell}}{\sum_j e^{-\beta \varepsilon_j}}$$

and then has average energy

$$U = \frac{\sum_\ell \varepsilon_\ell \, e^{-\beta\varepsilon_\ell}}{\sum_\ell e^{-\beta\varepsilon_\ell}}. \tag{1.9}$$

Expression (1.6) suggests that β is a function of $\{\varepsilon_\ell\}$ and U. We argue with physical reasoning that β is in fact a universal function of T alone.

Consider another system B with energy levels $\{\bar{\varepsilon}_m\}$. Let B_n denote a composite system of n identical and independent copies of B, also physically independent of A_n. Immersing A_n in a heat bath with temperature T specifies a value of β for it. Since β can a priori depend on $\{\varepsilon_\ell\}$, which is a characteristic of system A, we denote this value by β_A. Similarly, immersing B_n in the same heat bath leads to value β_B.

We can also immerse A_n and B_n together in the heat bath and consider them together as consisting of n independent and identical copies of a system AB. This system acquires its own value β_{AB} which depends on the temperature T and on the energies a system AB can take. Since A and B are physically independent, AB can take energies in the set $\{\varepsilon_\ell + \bar{\varepsilon}_m : \ell, m \in \mathbb{N}\}$.

Let $a_{\ell,m}$ be the number of AB-components whose A-part is at energy level ε_ℓ and whose B-part is at energy level $\bar{\varepsilon}_m$, when A_n and B_n are immersed together in the heat bath. Solving the Lagrange multipliers problem for the AB-system gives

$$a_{\ell,m} = \frac{n e^{-\beta_{AB}(\varepsilon_\ell + \bar{\varepsilon}_m)}}{\sum_{i,j} e^{-\beta_{AB}(\varepsilon_j + \bar{\varepsilon}_i)}} = n \cdot \frac{e^{-\beta_{AB}\varepsilon_\ell}}{\sum_j e^{-\beta_{AB}\varepsilon_j}} \cdot \frac{e^{-\beta_{AB}\bar{\varepsilon}_m}}{\sum_i e^{-\beta_{AB}\bar{\varepsilon}_i}}.$$

To obtain a_ℓ, the number of A-components at energy ε_ℓ, sum over m:

$$a_\ell = \sum_m a_{\ell,m} = \frac{n e^{-\beta_{AB}\varepsilon_\ell}}{\sum_j e^{-\beta_{AB}\varepsilon_j}}.$$

Since A_n and B_n do not interact, this must agree with the earlier outcome (1.7):

$$a_\ell = \frac{n e^{-\beta_A \varepsilon_\ell}}{\sum_j e^{-\beta_A \varepsilon_j}} = \frac{n e^{-\beta_{AB}\varepsilon_\ell}}{\sum_j e^{-\beta_{AB}\varepsilon_j}} \quad \text{for all } \ell \in \mathbb{N}.$$

It is reasonable to assume that system A can take at least two different energies $\varepsilon_\ell \neq \varepsilon_{\ell'}$ for otherwise the discussion is trivial. Then the above gives $e^{-\beta_A(\varepsilon_\ell - \varepsilon_{\ell'})} = e^{-\beta_{AB}(\varepsilon_\ell - \varepsilon_{\ell'})}$ and so $\beta_A = \beta_{AB}$. Switching the roles of A and B leads to $\beta_B = \beta_{AB} = \beta_A$. Since system B was arbitrary, we conclude that β is a universal function of T.

We regard β as the more fundamental quantity and view T as a universal function of β. The state of the system is then determined by the energy levels $\{\varepsilon_\ell\}$ and β by equations (1.8) and (1.9).

Next we derive the precise formula for the dependence of T on β. Working with fixed energies ε_ℓ and considering β to be the only variable will not help since we can replace β by any monotone function of it and nothing in the above reasoning changes. We need to make energies ε_ℓ vary, which leads to the notion of work done by the system.

The first law of thermodynamics states that if the parameters of the system (i.e. its energies ε_ℓ) change, it will absorb an average amount of heat $dQ = d\mathcal{E} + dW$, where dW is the work done by the system. If the energies change by $d\varepsilon_\ell$, then $dW = -\sum_\ell a_\ell \, d\varepsilon_\ell$ and

$$dQ = d\mathcal{E} - \sum_\ell a_\ell \, d\varepsilon_\ell.$$

Let nS be the *entropy* of the system A_n. By the second law of thermodynamics

$$dQ = nT \, dS.$$

Define the *free energy* $F = \log \sum e^{-\beta\varepsilon_j}$. Divide the two displays above by n to obtain

$$\begin{aligned}
dS &= \frac{1}{T}\Big(dU - \sum p_\ell \, d\varepsilon_\ell\Big) \\
&= \frac{1}{T\beta}\Big(d(\beta U) - U\,d\beta - \beta \sum p_\ell \, d\varepsilon_\ell\Big) \\
&= \frac{1}{T\beta}\Big(d(\beta U) + \frac{\partial F}{\partial \beta}\,d\beta + \sum \frac{\partial F}{\partial \varepsilon_\ell}\,d\varepsilon_\ell\Big) \qquad (1.10) \\
&= \frac{1}{T\beta}\,d(\beta U + F).
\end{aligned}$$

Abbreviate $G = \beta U + F$, which, by the display above, has to be a function $f(S)$ such that $f'(S) = T\beta$.

Recall that the three systems A, B, and AB acquire the same β when immersed in the heat bath. Consequently $F_A + F_B = F_{AB}$. Since $U = -\frac{\partial F}{\partial \beta}$, the same additivity holds for the function G, and so

$$f(S_A) + f(S_B) = f(S_{AB}).$$

Then by (1.10), since T is a universal function of β, $dS_{AB} = dS_A + dS_B$, which implies $S_{AB} = S_A + S_B + c$. Now we have

$$f(S_A) + f(S_B) = f(S_A + S_B + c).$$

Differentiate in S_A and S_B to see that $f'(S_A) = f'(S_B)$. Since the system B was chosen arbitrarily, entropy S_B can be made equal to any number regardless of the value of temperature T. Therefore $f'(S)$ must be a universal constant, which we call $1/k$. (This constant cannot be zero because T and β vary with each other.) This implies $\beta = \frac{1}{kT}$ and $G = k^{-1}S$. The constant k is called *Boltzmann's constant*. If $k < 0$, (1.8) would imply that as $T \to 0$

the system chooses the highest energy state, which goes against physical sense. Hence $k > 0$.

Let us compute S for a system with two energy levels ε_0 and ε_1. By symmetry, recentering, and a change of units, we can assume that $\varepsilon_0 = 0$ and $\varepsilon_1 = 1$. The system takes energy 0 with probability p_0 and energy 1 with probability p_1. The average energy $U = p_1$ and from (1.8) $p_1 = e^{-\beta}/(1+e^{-\beta})$. Then

$$\begin{aligned} S = kG = k(\beta U + F) &= k\big(p_1(\beta + F) + (1-p_1)F\big) \\ &= -k\big(p_1 \log p_1 + (1-p_1)\log(1-p_1)\big) \\ &= k\log 2 - kI_{1/2}(p_1). \end{aligned}$$

Thus rate function $I_{1/2}$ of Example 1.1 is, up to a universal positive multiplicative factor and an additive constant, the negative thermodynamic entropy of a two-energy system. In the previous section we saw that $-I_{1/2}$ is a linear function (with positive slope) of information-theoretic entropy. Together these observations imply that the thermodynamic entropy of a physical system represents the amount of information needed to describe the system or, equivalently, the amount of uncertainty remaining in it.

The identity $(k\beta)^{-1}S = U + \beta^{-1}F$ expresses an energy-entropy balance and reappears several times later in various guises. It can be found in Exercise 5.19, as equation (7.8) for the Curie-Weiss model, and in Section 8.3 as part (c) of the Dobrushin-Lanford-Ruelle variational principle for lattice systems.

1.3. Large deviations as useful estimates

The subject of large deviations is about controlling probabilities of atypical events. There are two somewhat different forms of this activity.

(i) Proofs of limit theorems in probability require estimates to rule out atypical behavior. Such estimates could be called "ad-hoc large deviations".

(ii) Precise limits of vanishing probabilities on an exponential scale are stated as large deviation principles.

The subject of this book is the second kind of large deviations. The next chapter begins a systematic development of large deviation principles. Before that, let us look at two textbook examples to illustrate the use of independence in the derivation of estimates to prove limit theorems.

Example 1.4. Let $\{X_n\}$ be an i.i.d. sequence with $E[X] = 0$. (Common device: X is a random variable that has the same distribution as all the X_n's.) We wish to show that, under a suitable hypothesis,

$$S_n/n^p \to 0 \quad P\text{-almost surely (a.s.), for any } p > 1/2. \tag{1.11}$$

In order to illustrate a method, we make a strong assumption. Assume the existence of $\delta > 0$ such that $E[e^{\theta X}] < \infty$ for $|\theta| \leq \delta$. When $p \geq 1$ limit (1.11) follows from the strong law of large numbers. So let us assume $p \in (1/2, 1)$.

For $t \geq 0$ Chebyshev's inequality gives

$$P\{S_n \geq \varepsilon n^p\} \leq E[e^{tS_n - \varepsilon t n^p}] = \exp\{-\varepsilon t n^p + n \log E[e^{tX}]\}.$$

The exponential moment assumption implies that $E[|X|^k]t^k/k!$ is summable for $t \in [0, \delta]$. Recalling that $E[X] = 0$,

$$\begin{aligned} E[e^{tX}] &= E[e^{tX} - tX] \leq 1 + \sum_{k=2}^{\infty} \frac{t^k}{k!} E[|X|^k] \\ &\leq 1 + t^2\delta^{-2} \sum_{k=2}^{\infty} \frac{\delta^k}{k!} E[|X|^k] \leq 1 + ct^2 \quad \text{for } t \in [0, \delta]. \end{aligned}$$

Then, taking $t = \frac{\varepsilon n^p}{2nc}$ and n large enough,

$$\begin{aligned} P\{S_n \geq \varepsilon n^p\} &\leq \exp\{-\varepsilon t n^p + n \log(1 + ct^2)\} \\ &\leq \exp\{-\varepsilon t n^p + nct^2\} = \exp\Big\{ -\frac{\varepsilon^2}{4c} n^{2p-1} \Big\}. \end{aligned} \tag{1.12}$$

Applying this to the sequence $\{-X_n\}$ gives the matching bound on the left:

$$P\{S_n \leq -\varepsilon n^p\} \leq \exp\Big\{ -\frac{\varepsilon^2}{4c} n^{2p-1} \Big\}. \tag{1.13}$$

Inequalities (1.12)–(1.13) can be regarded as large deviation estimates. (Although later we see that since the scale is n^p for $1/2 < p < 1$, technically these are called moderate deviations. But that distinction is not relevant here.) These estimates imply the summability

$$\sum_n P\{|S_n| \geq \varepsilon n^p\} < \infty.$$

The Borel-Cantelli lemma implies that for any $\varepsilon > 0$

$$P\{\exists n_0 : n \geq n_0 \Rightarrow |S_n/n^p| \leq \varepsilon\} = 1.$$

A countable intersection over $\varepsilon = 1/k$ for $k \in \mathbb{N}$ gives

$$P\{\forall k \ \exists n_0 : n \geq n_0 \Rightarrow |S_n/n^p| \leq 1/k\} = 1,$$

which says that $S_n/n^p \to 0$, P-a.s.

We used an unnecessarily strong assumption to illustrate the exponential Chebyshev method. We can achieve the same result with martingales under the assumption $E[|X|^2] < \infty$. Since S_n is a martingale (relative to the filtration $\sigma(X_1, \dots, X_n)$), Doob's inequality (Theorem 5.4.2 of [**27**] or (8.26) of [**54**]) gives

$$P\Big\{ \max_{k \le n} |S_k| \ge \varepsilon n^p \Big\} \le \frac{1}{\varepsilon^2 n^{2p}} E[|S_n|^2] = \frac{nE[|X|^2]}{\varepsilon^2 n^{2p}} = \frac{c}{\varepsilon^2} n^{-(2p-1)}.$$

Pick $r > 0$ such that $r(2p-1) > 1$. Then,

$$P\Big\{ \max_{k \le m^r} |S_k| \ge \varepsilon m^{pr} \Big\} \le \frac{c_1}{m^{r(2p-1)}}.$$

Hence, $P\{\max_{k \le m^r} |S_k| \ge \varepsilon m^{pr}\}$ is summable over m and the Borel-Cantelli lemma implies that $m^{-rp} \max_{k \le m^r} |S_k| \to 0$ P-a.s. as $m \to \infty$.

To get the result for the full sequence pick m_n such that $(m_n - 1)^r \le n < m_n^r$. Then,

$$n^{-p} \max_{k \le n} |S_k| \le \Big(\frac{m_n^r}{n} \Big)^p m_n^{-rp} \max_{k \le m_n^r} |S_k| \longrightarrow 0 \quad \text{as } n \to \infty$$

because $m_n^r / n \to 1$.

Example 1.5 (Longest run of heads). Let $\{X_n\}$ be an i.i.d. sequence of Bernoulli random variables with success probability p. For each $n \ge 1$ let R_n be the length of the longest success run among $(X_1, \dots, X_n)$. We derive estimates to prove a result of Rényi [**66**] that

$$P\Big\{ \lim_{n \to \infty} \frac{R_n}{\log n} = -\frac{1}{\log p} \Big\} = 1. \tag{1.14}$$

Fix $b > 1$ and r such that $r(b-1) > 1$. Let $\ell_m = \lceil -br \log m / \log p \rceil$. ($\lceil x \rceil$ is the smallest integer larger than or equal to x.) If $R_{m^r} \ge \ell_m$, then there is an $i \le m^r$ such that $X_i = X_{i+1} = \cdots = X_{i+\ell_m-1} = 1$. Therefore

$$P\{R_{m^r} \ge \ell_m\} \le m^r p^{\ell_m} \le 1/m^{r(b-1)}. \tag{1.15}$$

By the Borel-Cantelli lemma, with probability one, $R_{m^r} \le \ell_m$ for large enough m (though how large m needs to be is random). Consequently, with probability one,

$$\overline{\lim_{m \to \infty}} \frac{R_{m^r}}{\log m^r} \le -\frac{b}{\log p}.$$

Given n, let m_n be such that $m_n^r \le n < (m_n+1)^r$. Then $R_n \le R_{(m_n+1)^r}$ and

$$\overline{\lim_{n \to \infty}} \frac{R_n}{\log n} \le \overline{\lim_{n \to \infty}} \frac{\log(m_n+1)^r}{\log m_n^r} \cdot \frac{R_{(m_n+1)^r}}{\log(m_n+1)^r} \le -\frac{b}{\log p}.$$

Taking $b \searrow 1$ along a sequence shows that

$$P\Big\{ \overline{\lim_{n \to \infty}} \frac{R_n}{\log n} \le -\frac{1}{\log p} \Big\} = 1.$$

We have the upper bound for the goal (1.14).

Fix $a \in (0,1)$ and let $\ell_n = \lfloor -a \log n / \log p \rfloor$. Let A_i be the event that $X_{i\ell_n+1} = \cdots = X_{(i+1)\ell_n} = 1$. Then

$$\{R_n < \ell_n\} \subset \bigcap_{i=0}^{\lfloor n/\ell_n \rfloor - 1} A_i^c.$$

By the independence of the A_i's

$$P\{R_n < \ell_n\} \le (1 - p^{\ell_n})^{\frac{n}{\ell_n} - 1} \le e^{-p^{\ell_n}(\frac{n}{\ell_n} - 1)} \le e^{1 - n^{1-a}/\ell_n}. \tag{1.16}$$

Once again, by the Borel-Cantelli lemma, $R_n < \ell_n$ happens only finitely often, with probability one, and thus $\underline{\lim}_{n\to\infty} R_n / \log n \ge -a/\log p$. Taking $a \nearrow 1$ proves that

$$P\Big\{ \underline{\lim}_{n\to\infty} \frac{R_n}{\log n} \ge -\frac{1}{\log p} \Big\} = 1.$$

Looking back, the proof relied again on a right-tail estimate (1.15) and a left-tail estimate (1.16). It might be a stretch to call (1.15) a large deviation bound since it is not exponential, but (1.16) can be viewed as a large deviation bound.

Remark 1.6. Combining the limit theorem above with the fact that the variance of R_n remains bounded as $n \to \infty$ (see [**10**], [**42**]) provides a very accurate test of the hypothesis that the sequence $\{X_n\}$ is i.i.d. Bernoulli with probability of success p.

Chapter 2

The large deviation principle

2.1. Precise asymptotics on an exponential scale

Since the 1960s a standard formalism has been employed to express limits of probabilities of rare events on an exponential scale. The term for these statements is *large deviation principle* (LDP). We introduce this in a fairly abstract setting and then return to the Bernoulli example.

There is a sequence $\{\mu_n\}$ of probability measures whose asymptotics we are interested in. These measures exist on some measurable space $(\mathcal{X}, \mathscr{B})$. Throughout our general discussion we take $\mathcal{X}$ to be a Hausdorff topological space, unless further assumptions are placed on it. $\mathscr{B} = \mathscr{B}_{\mathcal{X}}$ is the Borel σ-algebra of $\mathcal{X}$, and $\mathcal{M}_1(\mathcal{X})$ is the space of probability measures on the measurable space $(\mathcal{X}, \mathscr{B}_{\mathcal{X}})$. Thus $\{\mu_n\}$ is a sequence in the space $\mathcal{M}_1(\mathcal{X})$. In Example 1.1 $\mathcal{X} = \mathbb{R}$ and μ_n is the probability distribution of S_n/n: $\mu_n(A) = P\{S_n/n \in A\}$ for Borel subsets $A \subset \mathbb{R}$.

Remark on mathematical generality. A reader not familiar with point-set topology can assume without any harm that $\mathcal{X}$ is a metric space. Even taking $\mathcal{X} = \mathbb{R}$ or $\mathbb{R}^d$ will do for a while. However, later we will study large deviations on spaces of probability measures, and the more abstract point of view becomes a necessity. If the notion of a Borel set is not familiar, it is safe to think of Borel sets as "all the reasonable sets for which a probability can be defined".

To formulate a general large deviation statement, let us look at result (1.2) of Example 1.1 for guidance. The first ingredient of interest in (1.2) is the normalization n^{-1} in front of the logarithm. Obviously this can change in a different example. Thus we should consider probabilities $\mu_n(A)$ that decay roughly like $e^{-r_n C(A)}$ for some normalization $r_n \nearrow \infty$ and a constant $C(A) \in [0, \infty]$ that depends on the event A.

In (1.2) we identified a rate function. How should the constant $C(A)$ relate to a rate function? Consider a finite set $A = \{x_1, \dots, x_n\}$. Then asymptotically

$$r_n^{-1} \log \mu_n(A) = r_n^{-1} \log \sum_i \mu_n\{x_i\} \approx \max_i r_n^{-1} \log \mu_n\{x_i\}$$

so that $C(A) = \min_i C(x_i)$. This suggests that in general $C(A)$ should be the infimum of a rate function I over A.

The final technical point is that it is in general unrealistic to expect $r_n^{-1} \log \mu_n(A)$ to actually converge on account of boundary effects, even if A is a nice set. A reasonable goal is to expect statements in terms of limsup and liminf.

From these considerations we arrive at the following tentative formulation of a large deviation principle: for Borel subsets A of the space $\mathcal{X}$,

$$(2.1) \quad -\inf_{x \in A^\circ} I(x) \le \varliminf_{n\to\infty} \frac{1}{r_n} \log \mu_n(A) \le \varlimsup_{n\to\infty} \frac{1}{r_n} \log \mu_n(A) \le -\inf_{x \in \overline{A}} I(x),$$

where A° and $\overline{A}$ are, respectively, the topological interior and closure of A. This statement is basically what we want, except that we need to address the uniqueness of the rate function.

Example 2.1. Let us return to the i.i.d. Bernoulli sequence $\{X_n\}$ of Example 1.1. We claim that probability measures $\mu_n(A) = P\{S_n/n \in A\}$ satisfy (2.1) with normalization $r_n = n$ and rate I_p of (1.2). This follows from (1.2) with a small argument.

For an open set G and $s \in G \cap [0, 1]$, $\lfloor ns \rfloor / n \in G$ for large enough n. So

$$\varliminf_{n\to\infty} \frac{1}{n} \log P\{S_n/n \in G\} \ge \lim_{n\to\infty} \frac{1}{n} \log P\{S_n = \lfloor ns \rfloor\} = -I_p(s).$$

This holds also for $s \in G \smallsetminus [0, 1]$ because $I_p(s) = \infty$. Taking the supremum over $s \in G$ on the right gives the inequality

$$\varliminf_{n\to\infty} \frac{1}{n} \log P\{S_n/n \in G\} \ge \sup_{s\in G}\bigl(-I_p(s)\bigr) = -\inf_{s \in G} I_p(s).$$

With $G = A^\circ$ this gives the lower bound in (2.1).

Split a closed set F into $F_1 = F \cap (-\infty, p]$ and $F_2 = F \cap [p, \infty)$. First prove the upper bound in (2.1) for F_1 and F_2 separately. Let $a = \sup F_1 \le p$ and $b = \inf F_2 \ge p$. (If F_1 is empty, then $a = -\infty$, and if F_2 is empty, then

$b = \infty$.) Assume first that $a \geq 0$. Then

$$\frac{1}{n}\log P\{S_n/n \in F_1\} \leq \frac{1}{n}\log P\{S_n/n \in [0,a]\} = \frac{1}{n}\log \sum_{k=0}^{\lfloor na \rfloor} P\{S_n = k\}.$$

***Exercise 2.2.** Prove that $P\{S_n = k\}$ increases with $k \leq \lfloor na \rfloor$.

By the exercise above,

$$\overline{\lim_{n\to\infty}} \frac{1}{n}\log P\{S_n/n \in F_1\} \leq \lim_{n\to\infty} \frac{1}{n}\log(\lfloor na \rfloor + 1)P\{S_n = \lfloor na \rfloor\} = -I_p(a).$$

This formula is still valid even when $a < 0$ because the probability vanishes. A similar upper bound works for F_2. Next write

$$\begin{aligned}\frac{1}{n}\log P\{S_n/n \in F\} &\leq \frac{1}{n}\log\Big(P\{S_n/n \in F_1\} + P\{S_n/n \in F_2\}\Big)\\ &\leq \frac{1}{n}\log 2 + \max\Big(\frac{1}{n}\log P\{S_n/n \in F_1\}, \frac{1}{n}\log P\{S_n/n \in F_2\}\Big).\end{aligned}$$

I_p is decreasing on $[0,p]$ and increasing on $[p,1]$. Hence, $\inf_{F_1} I_p = I_p(a)$, $\inf_{F_2} I_p = I_p(b)$, and $\inf_F I_p = \min(I_p(a), I_p(b))$. Finally,

$$\overline{\lim_{n\to\infty}} \frac{1}{n}\log P\{S_n/n \in F\} \leq -\min(I_p(a), I_p(b)) = -\inf_F I_p.$$

If we now take $F = \overline{A}$, the upper bound in (2.1) follows.

We have shown that (2.1) holds with I_p defined in (1.2). This is our first example of a full-fledged large deviation principle.

Remark 2.3. The limsup for closed sets and liminf for open sets in (2.1) remind us of weak convergence of probability measures where the same boundary issue arises. Section B.4 gives the definition of weak convergence.

The following exercises contain other instances where the rate function can be derived by hand.

Exercise 2.4. Prove (2.1) for the distribution of the sample mean of an i.i.d. sequence of real-valued normal random variables. Identifying I is part of the task.

Hint: The density of S_n/n can be written down explicitly. This suggests $I(x) = (x-\mu)^2/(2\sigma^2)$, where μ is the mean and σ^2 is the variance of X_1.

Exercise 2.5. Prove (2.1) for the distribution of the sample mean of an i.i.d. sequence of exponential random variables and compute the rate function explicitly.

Hint: Use Stirling's formula.

2.2. Lower semicontinuous and tight rate functions

We continue with some general facts and then in Definition 2.12 state precisely what is meant by a large deviation principle. We recall the definition of a *lower semicontinuous* function.

Definition 2.6. A function $f : \mathcal{X} \to [-\infty, \infty]$ is lower semicontinuous if $\{f \le c\} = \{x \in \mathcal{X} : f(x) \le c\}$ is a closed subset of $\mathcal{X}$ for all $c \in \mathbb{R}$.

***Exercise 2.7.** Prove that if $\mathcal{X}$ is a metric space, then f is lower semicontinuous if and only if $\underline{\lim}_{y \to x} f(y) \ge f(x)$ for all x.

An important transformation produces a lower semicontinuous function f_{lsc} from an arbitrary function $f : \mathcal{X} \to [-\infty, \infty]$. This *lower semicontinuous regularization* of f is defined by

$$f_{\text{lsc}}(x) = \sup\Big\{ \inf_{y \in G} f(y) : G \ni x \text{ and } G \text{ is open}\Big\}. \tag{2.2}$$

This turns out to be the maximal lower semicontinuous minorant of f.

Lemma 2.8. *f_{lsc} is lower semicontinuous and $f_{\text{lsc}}(x) \le f(x)$ for all x. If g is lower semicontinuous and satisfies $g(x) \le f(x)$ for all x, then $g(x) \le f_{\text{lsc}}(x)$ for all x. In particular, if f is lower semicontinuous, then $f = f_{\text{lsc}}$.*

Proof. $f_{\text{lsc}} \le f$ is clear. To show f_{lsc} is lower semicontinuous, let $x \in \{f_{\text{lsc}} > c\}$. Then there is an open set G containing x such that $\inf_G f > c$. Hence by the supremum in the definition of f_{lsc}, $f_{\text{lsc}}(y) \ge \inf_G f > c$ for all $y \in G$. Thus G is an open neighborhood of x contained in $\{f_{\text{lsc}} > c\}$. So $\{f_{\text{lsc}} > c\}$ is open.

To show $g \le f_{\text{lsc}}$ one just needs to show that $g_{\text{lsc}} = g$. For then

$$\begin{aligned} g(x) &= \sup\Big\{ \inf_G g : x \in G \text{ and } G \text{ is open}\Big\} \\ &\le \sup\Big\{ \inf_G f : x \in G \text{ and } G \text{ is open}\Big\} = f_{\text{lsc}}(x). \end{aligned}$$

We already know that $g_{\text{lsc}} \le g$. To show the other direction let c be such that $g(x) > c$. Then, $G = \{g > c\}$ is an open set containing x and $\inf_G g \ge c$. Thus $g_{\text{lsc}}(x) \ge c$. Now increase c to $g(x)$. □

The above can be reinterpreted in terms of *epigraphs*. The epigraph of a function f is the set $\operatorname{epi} f = \{(x,t) \in \mathcal{X} \times \mathbb{R} : f(x) \le t\}$. For the next lemma we endow $\mathcal{X} \times \mathbb{R}$ with its product topology.

Lemma 2.9. *The epigraph of f_{lsc} is the closure of* $\operatorname{epi} f$.

Proof. Note that the epigraph of f_{lsc} is closed. That it contains the epigraph of f (and thus also the closure of the epigraph of f) is immediate because

$f_{\text{lsc}} \le f$. For the other inclusion we need to show that any open set outside the epigraph of f is also outside the epigraph of f_{lsc}. Let A be such a set and let $(x,t) \in A$. By the definition of the product topology, there is an open neighborhood G of x and an $\varepsilon > 0$ such that $G \times (t-\varepsilon, t+\varepsilon) \subset A$. So for any $y \in G$ and any $s \in (t-\varepsilon, t+\varepsilon)$, $s < f(y)$. In particular, $t + \varepsilon/2 \le \inf_G f \le f_{\text{lsc}}(x)$. So (x,t) is outside the epigraph of f_{lsc}. □

Lower semicontinuous regularization can also be expressed in terms of pointwise alterations of the values of f.

Exercise 2.10. Assume $\mathcal{X}$ is a metric space. Show that if $x_n \to x$, then $f_{\text{lsc}}(x) \le \underline{\lim}\, f(x_n)$. Prove that for each $x \in \mathcal{X}$ there is a sequence $x_n \to x$ such that $f(x_n) \to f_{\text{lsc}}(x)$. (The constant sequence $x_n = x$ is allowed here.) This gives the alternate definition $f_{\text{lsc}}(x) = \min(f(x), \underline{\lim}_{y \to x} f(y))$.

Now we apply this to large deviation rate functions. The next lemma shows that rate functions can be assumed to be lower semicontinuous.

Lemma 2.11. *Suppose I is a function such that* (2.1) *holds for all measurable sets A. Then* (2.1) *continues to hold if I is replaced by I_{lsc}.*

Proof. $I_{\text{lsc}} \le I$ and the upper bound is immediate. For the lower bound observe that $\inf_G I_{\text{lsc}} = \inf_G I$ when G is open. □

Due to Lemma 2.11 we will call a $[0,\infty]$-valued function I a *rate function* only when it is lower semicontinuous. Here is the precise definition of a *large deviation principle (LDP)* for the remainder of the text.

Definition 2.12. Let $I : \mathcal{X} \to [0,\infty]$ be a lower semicontinuous function and $r_n \nearrow \infty$ a sequence of positive real constants. A sequence of probability measures $\{\mu_n\} \subset \mathcal{M}_1(\mathcal{X})$ is said to satisfy a *large deviation principle with rate function I and normalization r_n* if the following inequalities hold:

$$\overline{\lim_{n\to\infty}} \frac{1}{r_n} \log \mu_n(F) \le - \inf_{x \in F} I(x) \qquad \forall \text{ closed } F \subset \mathcal{X}, \tag{2.3}$$

$$\underline{\lim_{n\to\infty}} \frac{1}{r_n} \log \mu_n(G) \ge - \inf_{x \in G} I(x) \qquad \forall \text{ open } G \subset \mathcal{X}. \tag{2.4}$$

We will abbreviate LDP(μ_n, r_n, I) if all of the above hold. When the sets $\{I \le c\}$ are compact for all $c \in \mathbb{R}$, we say I is a *tight* rate function.

Lower semicontinuity makes a rate function unique. For this we assume of $\mathcal{X}$ a little bit more than Hausdorff. A topological space is *regular* if points and closed sets can be separated by disjoint open neighborhoods. In particular, metric spaces are regular topological spaces.

Theorem 2.13. *If $\mathcal{X}$ is a regular topological space, then there is at most one (lower semicontinuous) rate function satisfying the large deviation bounds* (2.3) *and* (2.4).

Proof. We show that I satisfies

$$I(x) = \sup\Big\{ -\underline{\lim}\, \frac{1}{r_n} \log \mu_n(B) : B \ni x \text{ and } B \text{ is open}\Big\}.$$

One direction is easy: for all open $B \ni x$

$$-\underline{\lim}\, \frac{1}{r_n} \log \mu_n(B) \le \inf_B I \le I(x).$$

For the other direction, fix x and let $c < I(x)$. One can separate x from $\{I \le c\}$ by disjoint neighborhoods. Thus, there exists an open set G containing x and such that $\overline{G} \subset \{I > c\}$. (Note that this is true also for $c < 0$, which is relevant in case $I(x) = 0$.) Then

$$\begin{aligned}
&\sup\Big\{ -\underline{\lim}\, \frac{1}{r_n} \log \mu_n(B) : B \ni x \text{ and } B \text{ is open}\Big\} \\
&\quad \ge -\underline{\lim}\, \frac{1}{r_n} \log \mu_n(G) \ge -\overline{\lim}\, \frac{1}{r_n} \log \mu_n(\overline{G}) \ge \inf_{\overline{G}} I \ge c.
\end{aligned}$$

Increasing c to $I(x)$ concludes the proof. □

Remark 2.14. Tightness of a rate function is a very useful property, as illustrated by the two exercises below. In a large part of the large deviation literature a rate function I is called *good* when the sets $\{I \le c\}$ are compact for $c \in \mathbb{R}$. We prefer the term tight as more descriptive and because of the connection with exponential tightness: see Theorem 2.19 below.

***Exercise 2.15.** Suppose $\mathcal{X}$ is a Hausdorff topological space and let $\mathcal{E} \subset \mathcal{X}$ be a closed set. Assume that the relative topology on $\mathcal{E}$ is metrized by the metric d. Let $I : \mathcal{E} \to [0, \infty]$ be a tight rate function and fix an arbitrary closed set $F \subset \mathcal{E}$. Prove that

$$\lim_{\varepsilon \searrow 0} \inf_{F^\varepsilon} I = \inf_F I,$$

where $F^\varepsilon = \{x \in \mathcal{E} : \exists y \in F \text{ such that } d(x, y) < \varepsilon\}$.

***Exercise 2.16.** $\mathcal{X}$ and $\mathcal{E}$ as in the exercise above. Suppose ξ_n and η_n are $\mathcal{E}$-valued random variables defined on $(\Omega, \mathscr{F}, P)$, and for any $\delta > 0$ there exists an $n_0 < \infty$ such that $d(\xi_n(\omega), \eta_n(\omega)) < \delta$ for all $n \ge n_0$ and $\omega \in \Omega$.

(a) Show that if the distributions of ξ_n satisfy the lower large deviation bound (2.4) with some rate function $I : \mathcal{E} \to [0, \infty]$, then so do the distributions of η_n.

(b) Show that if the distributions of ξ_n satisfy the upper large deviation bound (2.3) with some tight rate function $I : \mathcal{E} \to [0, \infty]$, then so do the distributions of η_n.

2.3. Weak large deviation principle

It turns out that it is sometimes difficult to satisfy the upper bound (2.3) for all closed sets. A useful weakening of the LDP requires the upper bound only for compact sets.

Definition 2.17. A sequence of probability measures $\{\mu_n\} \subset \mathcal{M}_1(\mathcal{X})$ satisfies a *weak large deviation principle* with lower semicontinuous rate function $I : \mathcal{X} \to [0, \infty]$ and normalization $\{r_n\}$ if the lower large deviation bound (2.4) holds for all open sets $G \subset \mathcal{X}$ and the upper large deviation bound (2.3) holds for all compact sets $F \subset \mathcal{X}$.

With enough control on the tails of the measures μ_n, a weak LDP is sufficient for the full LDP.

Definition 2.18. We say $\{\mu_n\} \subset \mathcal{M}_1(\mathcal{X})$ is *exponentially tight* with normalization r_n if for each $0 < b < \infty$ there exists a compact set K_b such that

$$\varlimsup_{n\to\infty} \frac{1}{r_n} \log \mu_n(K_b^c) \le -b. \tag{2.5}$$

Theorem 2.19. *Assume the upper bound* (2.3) *holds for compact sets and* $\{\mu_n\}$ *is exponentially tight with normalization* r_n*. Then the upper bound* (2.3) *holds for all closed sets with the same rate function* I*.*

If the weak LDP(μ_n, r_n, I) *holds and* $\{\mu_n\}$ *is exponentially tight with normalization* r_n*, then the full* LDP(μ_n, r_n, I) *holds and* I *is a tight rate function.*

Proof. Let F be a closed set.

$$\begin{aligned}
\varlimsup_{n\to\infty} \frac{1}{r_n} \log \mu_n(F) &\le \varlimsup_{n\to\infty} \frac{1}{r_n} \log\bigl(\mu_n(F \cap K_b) + \mu_n(K_b^c)\bigr) \\
&\le \max\Bigl(-b, \varlimsup_{n\to\infty} \frac{1}{r_n} \log \mu_n(F \cap K_b)\Bigr) \\
&\le \max\bigl(-b, -\inf_{F\cap K_b} I\bigr) \\
&\le \max\bigl(-b, -\inf_F I\bigr).
\end{aligned}$$

Letting $b \nearrow \infty$ proves the upper large deviation bound (2.3).

The weak LDP already contains the lower large deviation bound (2.4) and so we have both bounds. From the lower bound and exponential tightness it follows that

$$\inf_{K_{b+1}^c} I \geq - \varlimsup_{n\to\infty} \frac{1}{r_n} \log \mu_n(K_{b+1}^c) \geq b+1.$$

This implies that $\{I \leq b\} \subset K_{b+1}$. As a closed subset of a compact set $\{I \leq b\}$ is compact. □

The connection between a tight rate function and exponential tightness is an equivalence if we assume a little more of the space. To prove the other implication in Theorem 2.21 below we give an equivalent reformulation of exponential tightness in terms of open balls. In a metric space $(\mathcal{X}, d)$, $B(x,r) = \{y \in \mathcal{X} : d(x,y) < r\}$ is the open r-ball centered at x.

Lemma 2.20. *Let $\{\mu_n\}$ be a sequence of probability measures on a Polish space $\mathcal{X}$. (A Polish space is a complete and separable metric space.) Then $\{\mu_n\}$ is exponentially tight if and only if for every $b < \infty$ and $\delta > 0$ there exist finitely many δ-balls $B_1, \ldots, B_m$ such that*

$$\mu_n\Big(\Big[\bigcup_{i=1}^m B_i\Big]^c\Big) \leq e^{-r_n b} \quad \forall n \in \mathbb{N}.$$

Proof. Ulam's theorem (page 282) says that on a Polish space an individual probability measure ν is tight, which means that $\forall \varepsilon > 0$ there exists a compact set A such that $\nu(A^c) < \varepsilon$. Consequently on such a space exponential tightness is equivalent to the stronger statement that for all $b < \infty$ there exists a compact set K_b such that $\mu_n(K_b^c) \leq e^{-r_n b}$ for all $n \in \mathbb{N}$.

Since a compact set can be covered by finitely many δ-balls, the ball condition is a consequence of this stronger form of exponential tightness.

Conversely, assume the ball condition and let $1 \leq b < \infty$. We need to produce the compact set K_b. For each $k \in \mathbb{N}$, find m_k balls $B_{k,1}, \ldots, B_{k,m_k}$ of radius k^{-1} such that

$$\mu_n\Big(\Big[\bigcup_{i=1}^{m_k} B_{k,i}\Big]^c\Big) \leq e^{-2kr_n b} \quad \forall n \in \mathbb{N}.$$

Let $K = \bigcap_{k=1}^\infty \bigcup_{i=1}^{m_k} \overline{B}_{k,i}$. As a closed subset of $\mathcal{X}$, K is complete. By its construction K is totally bounded. This means that for any $\varepsilon > 0$ it can be covered by finitely many ε-balls. Completeness and total boundedness are equivalent to compactness in a metric space [**26**, Theorem 2.3.1]. By explicitly evaluating the geometric series and some elementary estimation,

$$\mu_n(K^c) \leq \sum_{k=1}^{\infty} e^{-2kr_nb} \leq e^{-r_nb}$$

as long as $r_n \geq 1$. Exponential tightness has been verified. □

Theorem 2.21. *Suppose $\mathcal{X}$ is a Polish space. Assume probability measures $\{\mu_n\}$ satisfy the upper large deviation bound* (2.3) *with a tight rate function I. Then $\{\mu_n\}$ is exponentially tight.*

Proof. Let $\{x_i\}_{i\in\mathbb{N}}$ be a countable dense set in $\mathcal{X}$. Suppose we can show that for every $b < \infty$ and $\varepsilon > 0$ there exists $m \in \mathbb{N}$ such that

$$\overline{\lim_{n\to\infty}}\, r_n^{-1} \log \mu_n\Big(\Big[\bigcup_{i=1}^{m} B(x_i, \varepsilon)\Big]^c\Big) \leq -b. \tag{2.6}$$

This is sufficient for exponential tightness by Lemma 2.20. (See Exercise 2.22 below.)

To show (2.6), take m large enough so that the compact set $\{I \leq b\}$ is covered by $G = B(x_1, \varepsilon) \cup \cdots \cup B(x_m, \varepsilon)$. (Since $\{x_i\}$ is dense, the entire space is covered by $\bigcup_{i\geq 1} B(x_i, \varepsilon)$, and by compactness $\{I \leq b\}$ has a finite subcover.) By the upper large deviation bound,

$$\overline{\lim_{n\to\infty}}\, r_n^{-1} \log \mu_n(G^c) \leq -\inf_{x\in G^c} I(x) \leq -b. \qquad \square$$

Here is the missing detail from the proof.

***Exercise 2.22.** Show that the condition of Lemma 2.20 follows from the condition established in the proof above. The fact that the balls $B(x_i, \varepsilon)$ cover the entire space is again crucial.

The results of this section offer a strategy for proving an LDP. First prove a weak LDP and then verify exponential tightness. A weak LDP may be easier to prove because it reduces entirely to analyzing asymptotics of $r_n^{-1} \log \mu_n(B(x, \varepsilon))$ for small neighborhoods. This idea already appeared in the proof of Example 2.1 where we reduced the proof to asymptotics of point probabilities. Here is an example where this method applies.

Exercise 2.23. Prove the large deviation principle for the distribution of the sample mean S_n/n of an i.i.d. sequence of $\mathbb{R}^d$-valued normal random variables with mean m and nonsingular covariance matrix A.

Hint: The density of S_n/n suggests $I(x) = \frac{1}{2}(x - m) \cdot A^{-1}(x - m)$. Note that this is different from the one-dimensional case in Exercise 2.4 because one cannot use monotonicity of I and split closed sets F into a part below m and a part above m.

We end the section with an important theoretical exercise.

***Exercise 2.24.** For $x \in \mathcal{X}$, define upper and lower local rate functions by

$$\overline{\kappa}(x) = - \inf_{G \subset \mathcal{X}:\ G \text{ open},\ x \in G} \ \overline{\lim_{n\to\infty}} \frac{1}{r_n} \log \mu_n(G) \tag{2.7}$$

and

$$\underline{\kappa}(x) = - \inf_{G \subset \mathcal{X}:\ G \text{ open},\ x \in G} \ \underline{\lim_{n\to\infty}} \frac{1}{r_n} \log \mu_n(G). \tag{2.8}$$

Show that if $\overline{\kappa} = \underline{\kappa} = \kappa$, then the weak LDP holds with rate function κ. Note that, by monotonicity, the same infimum in (2.7) and (2.8) can be taken over any base of open neighborhoods at x.

2.4. Aspects of Cramér's theorem

Cramér's theorem is the LDP for the sample mean

$$S_n/n = (X_1 + \cdots + X_n)/n$$

of i.i.d. random variables $\{X_n\}$ with values in $\mathbb{R}$ or $\mathbb{R}^d$. Discussion around this theorem raises several basic themes of large deviation theory: moment generating functions, compactness, convexity, minimax theorems, and the change of measure argument. We prove partial results here and formulate many statements as exercises with hints for hands-on practice. The important themes appear again later, so this section can be skipped, though we would recommend that the reader at least skim the main points. A complete proof of Cramér's theorem in $\mathbb{R}^d$ is given in Section 4.3.

We start by stating the one-dimensional theorem. Let $\{X_n\}$ be i.i.d. real-valued random variables, and let X be another random variable with the same distribution. The moment generating function is $M(\theta) = E[e^{\theta X}]$ for $\theta \in \mathbb{R}$. $M(\theta) > 0$ always and $M(\theta) = \infty$ is possible. Define

$$I(x) = \sup_{\theta \in \mathbb{R}} \{\theta x - \log M(\theta)\}. \tag{2.9}$$

Since $M(0) = 1$, $I : \mathbb{R} \to [0, \infty]$ is a well-defined function.

Cramér's theorem on $\mathbb{R}$. *Let $\{X_n\}$ be a sequence of i.i.d. real-valued random variables. Let μ_n be the distribution of the sample mean S_n/n. Then the large deviation principle* LDP(μ_n, n, I) *is satisfied with I defined in* (2.9).

A proof of this general one-dimensional Cramér theorem that applies to all i.i.d. sequences can be found in [**15**]. The case where M is finite in a neighborhood of 0 is covered by our multidimensional Cramér theorem in Section 4.3. Here we develop the upper bound and some related facts as a series of exercises. Then we turn to discuss parts of the multidimensional Cramér theorem under stronger assumptions.

Using Chebyshev's inequality

$$P\{S_n \ge nb\} \le e^{-n\theta b} E[e^{\theta S_n}] = e^{-n\theta b} M(\theta)^n \quad \text{for } \theta \ge 0 \tag{2.10}$$

and

$$P\{S_n \le na\} \le e^{-n\theta a} E[e^{\theta S_n}] = e^{-n\theta a} M(\theta)^n \quad \text{for } \theta \le 0. \tag{2.11}$$

From the above we get immediately the upper bounds

$$\overline{\lim_{n\to\infty}} \frac{1}{n} \log P\{S_n \ge nb\} \le -\sup_{\theta \ge 0}\{\theta b - \log M(\theta)\}$$

and

$$\overline{\lim_{n\to\infty}} \frac{1}{n} \log P\{S_n \le na\} \le -\sup_{\theta \le 0}\{\theta a - \log M(\theta)\}.$$

***Exercise 2.25.** Suppose X has a finite mean $\bar{x} = E[X]$. Prove that if $a \le \bar{x} \le b$, then

$$\sup_{\theta \ge 0}\{\theta b - \log M(\theta)\} = \sup_{\theta \in \mathbb{R}}\{\theta b - \log M(\theta)\}$$

and

$$\sup_{\theta \le 0}\{\theta a - \log M(\theta)\} = \sup_{\theta \in \mathbb{R}}\{\theta a - \log M(\theta)\}.$$

Hint: Use Jensen's inequality to show that $\theta b - \log M(\theta) \le 0$ for $\theta < 0$ and $\theta a - \log M(\theta) \le 0$ for $\theta > 0$.

Definition 2.26. A subset A of a vector space $\mathcal{X}$ is *convex* if for all $x, y \in A$ and $t \in [0,1]$, $tx + (1-t)y \in A$. A function $f : \mathcal{X} \to [-\infty, \infty]$ is *convex* if $f(tx+(1-t)y) \le tf(x)+(1-t)f(y)$ for all $x, y \in \mathcal{X}$ and $t \in [0,1]$ such that the right-hand side of the inequality is well-defined (that is, not $\infty - \infty$).

***Exercise 2.27.** Prove that I is lower semicontinuous and convex and that if $\bar{x} = E[X]$ is finite, then I achieves its minimum at $\bar{x}$ with $I(\bar{x}) = 0$.

Hint: I is a supremum of lower semicontinuous convex functions. $I(x) \ge 0$ for all x, but by Jensen's inequality $I(\bar{x}) \le 0$.

***Exercise 2.28.** Suppose $M(\theta) < \infty$ in some open neighborhood around the origin. Show that then $\bar{x}$ is the *unique* zero of I: that is, $x \ne \bar{x}$ implies $I(x) > 0$.

Hint: For any $x > \bar{x}$, $(\log M(\theta))' < x$ for θ in some interval $(0, \delta)$.

Exercise 2.29. Check that the rate functions found in Example 1.1 and Exercises 2.4 and 2.5 match (2.9).

Exercise 2.27 together with the earlier observations shows that when $\bar{x}$ is finite $I(x)$ is nonincreasing for $x < \bar{x}$ and nondecreasing for $x > \bar{x}$. In particular, if $a \le \bar{x} \le b$, then $I(a) = \inf_{x \le a} I(x)$ and $I(b) = \inf_{x \ge b} I(x)$. This proves the upper bound for the sets $F = (-\infty, a]$ and $F = [b, \infty)$ in the case where the mean is finite.

Exercise 2.30. Prove that the sample mean S_n/n of i.i.d. real-valued random variables satisfies the upper large deviation bound (2.3) with normalization n and rate I defined in (2.9), with no further assumptions on the distribution.

Hint: The case of finite mean is almost done above. Then consider separately the cases where the mean is infinite and where the mean does not exist.

While Cramér's theorem is valid in general, it does not give much information unless the variables have exponentially decaying tails. This point is explored in the next exercise.

Exercise 2.31. Let $\{X_i\}$ be an i.i.d. real-valued sequence. Assume $E[X_1^2] < \infty$ but, for any $\varepsilon > 0$, $P\{X_1 > b\} > e^{-\varepsilon b}$ for all large enough b. Show that

(a) $\lim_{n\to\infty} \frac{1}{n} \log P\{S_n/n > E[X_1] + \delta\} = 0$ for any $\delta > 0$,

(b) the rate function is identically 0 on $[E(X_1), \infty)$.

Hint: For (a), deduce

$$P\{S_n/n \ge E[X_1] + \delta\} \ge P\{S_{n-1} \ge (n-1)E[X_1]\} P\{X_1 \ge n\delta + E[X_1]\}$$

and apply the central limit theorem. For (b), first find $M(\theta)$ for $\theta > 0$. Then observe that for $\theta \le 0$ and $x \ge E[X_1]$,

$$\theta x - \log M(\theta) \le \theta(x - E[X_1]) \le 0.$$

Exercise 2.32. Let $\{X_i\}$ be an i.i.d. real-valued sequence. Prove that the closure of the set $\{I < \infty\}$ is the same as the closure of the convex hull of the support of the distribution of X. (The *convex hull* of a set is the intersection of all convex sets containing it.)

Hint: Let K be the latter set and let $y \notin K$. To show that $I(y) = \infty$, find $\theta \in \mathbb{R}$ such that $\theta y - \varepsilon > \sup_{x \in K} x\theta$. For the other direction, take y in the interior of $\{I = \infty\}$. To get $y \notin K$, show first that there exists a sequence θ_n converging to either ∞ or $-\infty$ such that $\phi_y(\theta_n) = \theta_n y - \log M(\theta_n)$ converges to infinity. Assume $\theta_n \to \infty$. Show that for some ε, $|x - y| \le \varepsilon$ implies $\phi_x(\theta) \to \infty$ as $\theta \to \infty$. Then, for $\theta > 0$, $\theta(y - \varepsilon) - \log M(\theta) \le -\log \mu\{x : |x - y| \le \varepsilon\}$ where μ is the distribution of X. Let $\theta \to \infty$.

Cramér's theorem is quite crude because only the exponentially decaying terms of a full expansion affect the result. In some cases one can derive much more precise asymptotics.

Exercise 2.33. Prove that if the $\{X_k\}$ are i.i.d. standard normal, then for any $k \in \mathbb{N}$ and $a > 0$

$$\log P\{S_n \geq an\} \sim -\frac{a^2 n}{2} - \frac{1}{2}\log(2\pi n a^2) + \log\Big(1 - \frac{1}{a^2 n} + \frac{1\cdot 3}{a^4 n^2} - \cdots + (-1)^k \frac{1\cdot 3\cdot 5\cdots(2k-1)}{a^{2k} n^k}\Big).$$

Hint: Observe that

$$\frac{d}{dx}\Big(e^{-x^2/2}\sum_{k=0}^{n}(-1)^k(1\cdot 3\cdots(2k-1))x^{-2k-1}\Big)\begin{cases} < -e^{-x^2/2} & \text{if } n \text{ is even,} \\ > -e^{-x^2/2} & \text{if } n \text{ is odd.}\end{cases}$$

Exercise 2.34. Continuing Exercise 2.29, derive Cramér rate functions for further basic distributions.

(a) For real $\alpha > 0$, the rate α exponential distribution has density $f(x) = \alpha e^{-\alpha x}$ on $\mathbb{R}_+$. Derive the Cramér rate function

$$I(x) = \alpha x - 1 - \log \alpha x \qquad \text{for } x > 0.$$

(b) For real $\lambda > 0$, the mean λ Poisson distribution has probability mass function $p(k) = e^{-\lambda}\lambda^k/k!$ for $k \in \mathbb{Z}_+$. Derive the Cramér rate function

$$I(x) = x\log(x/\lambda) - x + \lambda \qquad \text{for } x \geq 0.$$

We turn to Cramér's theorem in multiple dimensions. When the $\{X_n\}$ are $\mathbb{R}^d$-valued, the moment generating function is given by $M(\theta) = E[e^{\theta\cdot X}]$ for $\theta \in \mathbb{R}^d$. Again, $M(\theta) \in (0, \infty]$. Define

$$I(x) = \sup_{\theta\in\mathbb{R}^d}\{\theta\cdot x - \log M(\theta)\}. \tag{2.12}$$

Exercise 2.35. Check that Exercises 2.27 and 2.28 apply to the multi-dimensional case as well.

Hölder's inequality implies that $\log M(\theta)$ is a convex function: with $t \in (0,1)$, $p = 1/t$, and $q = 1/(1-t)$,

$$\begin{aligned} M(t\theta_1 + (1-t)\theta_2) &= E[e^{t\theta_1\cdot X}e^{(1-t)\theta_2\cdot X}] \\ &\leq E[e^{\theta_1\cdot X}]^t E[e^{\theta_2\cdot X}]^{1-t} = M(\theta_1)^t M(\theta_2)^{1-t}. \end{aligned} \tag{2.13}$$

The full LDP of the one-dimensional Cramér theorem does not generalize to multiple dimensions without an additional assumption. Counterexamples appear in [**20**].

Cramér's theorem on $\mathbb{R}^d$. *Let $\{X_n\}$ be a sequence of i.i.d. $\mathbb{R}^d$-valued random variables and let μ_n be the distribution of the sample mean S_n/n. Then without further assumptions weak* $\mathrm{LDP}(\mu_n, n, I)$ *holds with I defined in* (2.12). *If, moreover, $M(\theta) < \infty$ in a neighborhood of 0, then* $\mathrm{LDP}(\mu_n, n, I)$ *holds and I is a tight rate function.*

At this point we prove the upper bound for compact sets without assumptions on M and then exponential tightness assuming that M is finite near the origin. Then we give a proof of the lower bound under the restrictive assumption

$$M(\theta) < \infty \text{ for all } \theta \in \mathbb{R}^d \text{ and } |\theta|^{-1} \log M(\theta) \to \infty \text{ as } |\theta| \to \infty. \tag{2.14}$$

Both proofs introduce important techniques. Assumption (2.14) ensures that the supremum in (2.12) is achieved. This is precisely the issue that needs to be overcome when no assumptions on M are present. In Section 4.3 we revisit the theorem and prove its final version.

Proof of the upper bound for compact sets and exponential tightness. For any Borel set C and $\theta \in \mathbb{R}^d$,

$$\begin{aligned} P\{S_n/n \in C\} &= E[\mathbb{1}\{S_n/n \in C\}] \le e^{-\inf_{y\in C} n\theta\cdot y} E[e^{\theta\cdot S_n}] \\ &= e^{-n\inf_{y\in C} \theta\cdot y} M(\theta)^n. \end{aligned}$$

This shows that

$$\frac{1}{n} \log P\{S_n/n \in C\} \le -\sup_{\theta} \inf_{y\in C} \{\theta \cdot y - \log M(\theta)\}. \tag{2.15}$$

We would like to interchange the sup and the inf to find $I(y)$ on the right-hand side. This can be done if C is a compact convex set.

Minimax theorem on $\mathbb{R}^d$. *Let $C \subset \mathbb{R}^d$ be compact and convex. Let $D \subset \mathbb{R}^d$ be convex. Let $f : C \times D \to \mathbb{R}$ be such that for each $\theta \in D$, $f(y, \theta)$ is convex and continuous in $y \in C$, and for each $y \in C$, $f(y, \theta)$ is concave in $\theta \in D$. Then*

$$\sup_{\theta\in D} \inf_{y\in C} f(y, \theta) = \inf_{y\in C} \sup_{\theta\in D} f(y, \theta).$$

This theorem is a special case of the more general minimax theorem proved in Section A.5 of Appendix A. To have a feeling for the theorem above think of a horse saddle in $\mathbb{R}^3$. We have a smooth function that is convex in one direction and concave in the other. Taking sup in the concave direction and inf in the convex direction will result in the saddle point regardless of the order.

The set $D = \{\theta : M(\theta) < \infty\}$ is convex by (2.13), C is a compact convex set by assumption, and $f(y, \theta) = \theta \cdot y - \log M(\theta)$ satisfies the assumptions of the minimax theorem. Thus the sup and the inf can be switched in (2.15) to give

$$\frac{1}{n} \log P\{S_n/n \in C\} \le -\inf_{y\in C} \sup_{\theta} \{\theta \cdot y - \log M(\theta)\} = -\inf_{y\in C} I(y). \tag{2.16}$$

We have the upper bound with rate function I of (2.12) for compact convex sets, even without taking the $n \to \infty$ limit.

We extend the upper bound to an arbitrary compact set K. Let $\alpha < \inf_K I$. Since I is lower semicontinuous, $\{I > \alpha\}$ is open. For each $x \in K \subset \{I > \alpha\}$ pick a compact ball C_x centered at x with nonempty interior and such that $C_x \subset \{I > \alpha\}$. Cover K with a finite collection $C_{x_1}, \dots, C_{x_N}$ of such balls. The upper bound for compact convex sets gives

$$P\{S_n/n \in K\} \le \sum_{i=1}^{N} P\{S_n/n \in C_{x_i}\} \le \sum_{i=1}^{N} e^{-n \inf_{C_{x_i}} I} \le N e^{-n\alpha}.$$

Taking $n \nearrow \infty$ and then $\alpha \nearrow \inf_K I$ gives the upper bound (2.3) in weak LDP(μ_n, n, I).

Last, we verify exponential tightness under the assumption that M is finite near the origin. Theorem 2.19 then implies the upper bound for closed sets. To this end, from (2.10) and (2.11) it follows that for any $b > 0$ we can find a large enough $a = a(b) > 0$ such that

$$P\{|S_n^{(i)}| \ge na\} \le e^{-bn} \quad \text{for } i = 1, 2, \dots, d \text{ and all } n \in \mathbb{N}.$$

Here $y^{(i)}$ denotes the ith coordinate of a vector $y \in \mathbb{R}^d$. Definition 2.18 of exponential tightness is satisfied with $r_n = n$ and $K_b = \{y : |y^{(i)}| \le a(b) \text{ for all } i = 1, \dots, d\}$. $\square$

Exercise 2.36. The minimax theorem was used above to turn (2.15) into the nonasymptotic upper bound (2.16) for compact convex sets. This was done to illustrate the minimax trick and because bounds that are valid for finite n are useful. However, we can proceed directly from (2.15) to the upper large deviation bound for a general compact set K. Fill in the details in the following outline. With notation as above, for each $x \in K$ find θ_x such that $\theta_x \cdot x - \log M(\theta_x) > \alpha$. Pick a compact convex ball U_x centered at x and with nonempty interior such that $\theta_x \cdot y - \log M(\theta_x) > \alpha - \varepsilon$ for $y \in U_x$. Proceed as in the proof above.

Proof of Cramér's lower bound under (2.14). We introduce the classical *change of measure* argument for the lower bound. Let our random variables $\{X_k\}$ be defined on a probability space $(\Omega, \mathscr{F}, P)$.

On any open set where $M(\theta)$ is finite it is differentiable and $\nabla M(\theta) = E[X e^{\theta \cdot X}]$. This is by dominated convergence. Thus $\theta \cdot x - \log M(\theta)$ is a concave differentiable function of θ that, by (2.14), achieves its maximum $I(x)$ at some θ_x. Then $\nabla M(\theta_x) = x M(\theta_x)$.

Define the probability measure ν_x on $\mathbb{R}^d$ by

$$\nu_x(B) = \frac{1}{M(\theta_x)} E[e^{\theta_x \cdot X} \mathbb{1}\{X \in B\}], \quad B \in \mathscr{B}_{\mathbb{R}^d}.$$

The mean of ν_x is

$$\int_{\mathbb{R}^d} y\,\nu_x(dy) = \frac{E[Xe^{\theta_x\cdot X}]}{M(\theta_x)} = \frac{\nabla M(\theta_x)}{M(\theta_x)} = x.$$

Let $Q_{x,n}$ be the probability measure on Ω defined by

$$Q_{x,n}(A) = \frac{E[\mathbb{1}_A \cdot e^{\theta_x\cdot S_n}]}{E[e^{\theta_x\cdot S_n}]} \quad \text{for } A \in \mathscr{F}.$$

Now for the lower bound. Take an open set $G \subset \mathbb{R}^d$, $x \in G$, and $\varepsilon > 0$ such that $\{y : |y - x| < \varepsilon\} \subset G$. Then

$$\begin{aligned} P\{S_n/n \in G\} &\geq P\{|S_n - nx| < \varepsilon n\} \\ &\geq e^{-n\theta_x\cdot x - n\varepsilon|\theta_x|} E[e^{\theta_x\cdot S_n}\mathbb{1}\{|S_n - nx| < \varepsilon n\}] \\ &= e^{-n\theta_x\cdot x - n\varepsilon|\theta_x|} M(\theta_x)^n Q_{x,n}\{|S_n - nx| < \varepsilon n\}. \end{aligned}$$

The key observation is that under $Q_{x,n}$ the variables $X_1, X_2, \dots, X_n$ are i.i.d. ν_x-distributed: for $B_1, \dots, B_n \in \mathscr{B}_{\mathbb{R}^d}$,

$$Q_{x,n}\Bigl(\bigcap_{k=1}^{n}\{X_k \in B_k\}\Bigr) = \prod_{k=1}^{n} \frac{E[\mathbb{1}_{B_k}(X)e^{\theta_x\cdot X}]}{E[e^{\theta_x\cdot X}]} = \prod_{k=1}^{n} \nu_x(B_k).$$

By the law of large numbers $Q_{x,n}\{|S_n - nx| < \varepsilon n\} \to 1$, and we get the bound

$$\varliminf_{n\to\infty} \frac{1}{n} \log P\{S_n/n \in G\} \geq -I(x) - \varepsilon|\theta_x|.$$

Taking $\varepsilon \to 0$ and sup over $x \in G$ on the right proves the lower bound (2.4). □

The measure ν_x is called the *tilted* measure. The dependence on n in $Q_{x,n}$ is an artifact we can eliminate by using a single infinite product measure on a sequence space. This is what we do in Section 5.2 on Sanov's theorem.

The change of measure argument replaced the original measure P by a new measure $Q_{x,n}$ under which outcome x became typical rather than rare. In the proof this appears to be merely a trick, but we shall see later that there is more to it. Namely, to produce the deviation $S_n \approx nx$ the process $\{X_k\}$ actually behaves like an i.i.d. ν_x-sequence. This is an interesting conclusion. A priori one could also imagine that the system prefers to deviate a small number of variables while letting most X_k's behave in a typical fashion. (See Exercises 2.38 and 6.19 and the related maximum entropy principle in Section 5.3.) A lesson of large deviation theory is that a deviation is not produced in an arbitrary manner, but rather in the most probable way, and this can be captured by the rate function.

Exercise 2.37. Let $\{X_n\}$ be i.i.d. Bernoulli random variables with success probability $p \in [0,1]$. Show that for $s \in [0,1]$ the measure ν_s in the proof above is the Bernoulli measure with success probability s. Investigate ν_x for your other favorite distributions.

Exercise 2.38. Let $S_n = X_1 + \cdots + X_n$ be a simple symmetric random walk on $\mathbb{Z}$. That is, the $\{X_k\}$ are i.i.d. with distribution $P(X_k = \pm 1) = 1/2$. Let $a \in [0,1]$. With elementary calculation find the limit of the process $\{X_k\}$ conditioned on $|S_n - \lfloor na \rfloor| \le 1$, as $n \to \infty$.

Hint: Fix $x_1, \dots, x_m \in \{\pm 1\}$, write the probability $P(X_1 = x_1, \dots, X_m = x_m \mid |S_n - \lfloor na \rfloor| \le 1)$ in terms of factorials, and observe the asymptotics. Note that the conditioning event cannot always be written $S_n = \lfloor na \rfloor$ because S_n must have the parity of n.

2.5. Limits, deviations, and fluctuations

Let $\{Y_n\}$ be a sequence of random variables with values in a metric space $(\mathcal{X}, d)$ and let μ_n be the distribution of Y_n, that is, $\mu_n(B) = P\{Y_n \in B\}$ for $B \in \mathscr{B}_{\mathcal{X}}$. Naturally an LDP for the sequence $\{\mu_n\}$ is related to the asymptotic behavior of Y_n. Suppose LDP(μ_n, r_n, I) holds and $Y_n \to \bar{y}$ in probability. Then the limit $\bar{y}$ does not represent a deviation. The rate function I recognizes this with the value $I(\bar{y}) = 0$ that follows from the upper bound. For any open neighborhood G of $\bar{y}$ we have $\mu_n(G) \to 1$. Consequently for the closure

$$0 \le \inf_{\overline{G}} I \le -\overline{\lim}\, r_n^{-1} \log \mu_n(\overline{G}) = 0.$$

Let G shrink down to $\bar{y}$. Lower semicontinuity forces $I(\bar{y}) = 0$.

Every LDP satisfies $\inf I = 0$, as can be seen by taking $F = \mathcal{X}$ in the upper bound (2.3). But the zero set of I does not necessarily represent limit values. It may simply be that the probability of a deviation decays slower than exponentially in r_n, which leads to $I = 0$.

Exercise 2.39. In case the reader prefers an off-the-shelf example rather than playing with his or her own examples, here is one. Fix a sequence $0 < a_n \nearrow \infty$, let m denote Lebesgue measure, and define $\{\mu_n\}$ on $\mathbb{R}$ by

$$\mu_n(A) = (1 - a_n^{-1})\mathbb{1}_A(0) + a_n^{-1} m(A \cap (0,1]).$$

Clearly $\mu_n \to \delta_0$ weakly or, equivalently, if Y_n has distribution μ_n, then $Y_n \to 0$ in probability. Given any $c \in [0, \infty]$, show that by an appropriate choice of r_n we can have the LDP with rate function

$$I(x) = \begin{cases} 0, & x = 0, \\ c, & x \in (0,1], \\ \infty, & x \notin [0,1]. \end{cases}$$

Returning to the general discussion, an LDP can imply convergence of the random variables if the rate function has good properties. Assume that I is a tight rate function and has a unique zero $I(\bar{y}) = 0$. Let $A = \{y : d(y, \bar{y}) \geq \varepsilon\}$. Compactness and lower semicontinuity ensure that the infimum $u = \inf_A I$ is achieved. Since $\bar{y} \notin A$, it must be that $u > 0$. Then, for n large enough, the upper large deviation bound (2.3) implies

$$P\{d(Y_n, \bar{y}) \geq \varepsilon\} \leq e^{-r_n(\inf_A I - u/2)} = e^{-r_n u/2}.$$

Thus, $Y_n \to \bar{y}$ in probability. If, moreover, r_n grows fast enough so that $\sum e^{-cr_n} < \infty \;\forall c > 0$, then the Borel-Cantelli lemma implies that $Y_n \to \bar{y}$ almost surely.

For i.i.d. variables Cramér's theorem should also be understood in relation to the central limit theorem (CLT). Consider the case where $M(\theta)$ is finite in a neighborhood of the origin so that X has finite mean $\bar{x} = E[X]$ and finite variance σ^2, and $I(x) > 0$ for $x \neq \bar{x}$ (Exercise 2.28). Then for each $\delta > 0$ we have the large deviation bound

$$P\{S_n/n - \bar{x} \geq \delta\} \leq e^{-nI(\bar{x}+\delta)}. \tag{2.17}$$

(Recall (2.10) and Exercise 2.25.)

By contrast, the CLT tells us that small deviations of order $n^{-1/2}$ converge to a limit distribution: for $r \in \mathbb{R}$,

$$P\{S_n/n - \bar{x} \geq rn^{-1/2}\} \underset{n\to\infty}{\longrightarrow} \int_r^\infty \frac{e^{-s^2/2\sigma^2}}{\sqrt{2\pi\sigma^2}}\, ds.$$

This distinction is sometimes expressed by saying that the CLT describes *fluctuations* as opposed to deviations. There is a significant qualitative difference between Cramér's theorem and the CLT. The CLT is an example of *universality*: the Gaussian limit is valid for all distributions with finite variance. The Cramér rate function I on the other hand depends on the entire distribution. (From convex analysis we will learn that I determines M.)

There are also results on *moderate deviations* that fall between large deviations and CLT fluctuations. For example, if $d = 1$ and M is finite in a neighborhood of 0, then for any $\alpha \in (0, 1/2)$

$$n^{-2\alpha} \log P\{|S_n/n - \bar{x}| \geq \delta n^{-1/2+\alpha}\} \underset{n\to\infty}{\longrightarrow} -\frac{\delta^2}{2\sigma^2}.$$

Note that this limit picks the leading exponential factor from the Gaussian. In Chapter 11 we discuss refinements to Cramér's theorem and moderate deviations.

Chapter 3

Large deviations and asymptotics of integrals

This chapter takes up two general issues: transferring LDPs from one space to another by a mapping and asymptotics of integrals. In the last section we discuss our first example from statistical mechanics.

3.1. Contraction principle

When $f : \mathcal{X} \to \mathcal{Y}$ is a measurable mapping, a measure μ on $\mathcal{X}$ can be "pushed forward" to a measure ν on $\mathcal{Y}$ by the definition $\nu(B) = \mu(f^{-1}(B))$ for measurable subsets $B \subset \mathcal{Y}$. This definition is abbreviated as $\nu = \mu \circ f^{-1}$. It preserves total mass, so it transforms probability measures into probability measures. The contraction, or push-forward, principle applies this same idea to transfer an LDP from $\mathcal{X}$ to $\mathcal{Y}$. In formula (3.1) below note that by convention the infimum of an empty set is infinite. Recall also the definition of the lower semicontinuous regularization u_{lsc} of a function u: $u_{\text{lsc}}(y) = \sup\{\inf_G u : \text{open } G \ni y\}$.

Contraction principle. *Let $\mathcal{X}$ and $\mathcal{Y}$ be Hausdorff spaces and $f : \mathcal{X} \to \mathcal{Y}$ a continuous mapping. Assume* LDP(μ_n, r_n, I) *on $\mathcal{X}$. Let $\nu_n = \mu_n \circ f^{-1}$. Set*

$$\widetilde{J}(y) = \inf_{x: f(x) = y} I(x) \quad \textit{for } y \in \mathcal{Y} \tag{3.1}$$

and $J = \widetilde{J}_{\rm lsc}$*. Then*

(a) $\mathrm{LDP}(\nu_n, r_n, J)$ *holds on* $\mathcal{Y}$,

(b) *if* I *is tight, then* $J = \widetilde{J}$ *and* J *is tight as well.*

Proof. By Lemma 2.11 it suffices to prove that $\widetilde{J}$ satisfies the large deviation bounds (2.3) and (2.4). Take a closed set $F \subset \mathcal{Y}$. Then

$$\overline{\lim_{n\to\infty}} \frac{1}{r_n} \log \mu_n(f^{-1}(F)) \le -\inf_{x\in f^{-1}(F)} I(x) = -\inf_{y\in F} \inf_{f(x)=y} I(x) = -\inf_{y\in F} \widetilde{J}(y).$$

The lower bound is proved similarly and (a) follows.

Assume now that I is tight. Observe that if $\widetilde{J}(y) < \infty$, then $f^{-1}(y)$ is nonempty and closed, and the nested nonempty compact sets $\{I \le \widetilde{J}(y) + 1/n\} \cap f^{-1}(y)$ have a nonempty intersection. Hence $\widetilde{J}(y) = I(x)$ for some $x \in f^{-1}(y)$. Consequently, $\{\widetilde{J} \le c\} = f(\{I \le c\})$ is a compact subset of $\mathcal{Y}$. In particular, $\{\widetilde{J} \le c\}$ is closed and $\widetilde{J}$ is lower semicontinuous and is hence identical to J. ☐

If the rate function I is not tight, then $\widetilde{J}$ may fail to be lower semicontinuous (and hence $J \ne \widetilde{J}$).

Exercise 3.1. Let $\mathcal{X} = [0, \infty)$ and $\mu_n(dx) = \phi_n(x)dx$ where

$$\phi_n(x) = nx^{-2}e^{1-n/x}\mathbb{1}_{(0,n)}(x).$$

Show that $\{\mu_n\}$ are not tight on $[0,\infty)$ but $\mathrm{LDP}(\mu_n, n, I)$ holds with $I(x) = x^{-1}$ for $x > 0$ and infinite otherwise. (Tightness is discussed in Section B.4 of Appendix B.) Note that I is not a tight rate function.

***Exercise 3.2.** Let $f : [0,\infty) \to S^1 = \{y \in \mathbb{C} : |y| = 1\}$ be $f(x) = e^{2\pi i \frac{x}{x+1}}$ and $\nu_n = \mu_n \circ f^{-1}$, with μ_n defined in the previous exercise. Prove that $\widetilde{J}(z) = \inf_{f(x)=z} I(x)$ is not lower semicontinuous and that the $\{\nu_n\}$ are tight and converge weakly to δ_1. Prove also that $\mathrm{LDP}(\nu_n, n, J)$ holds with $J(e^{2\pi i t}) = \frac{1-t}{t}$ for $t \in (0,1]$.

The simplest situation when the contraction principle is applied is when $\mathcal{X}$ is a subspace of $\mathcal{Y}$.

***Exercise 3.3.** Suppose $\mathrm{LDP}(\mu_n, r_n, I)$ holds on $\mathcal{X}$ and that $\mathcal{X}$ is a Hausdorff space contained in the larger Hausdorff space $\mathcal{Y}$. Find J so that $\mathrm{LDP}(\mu_n, r_n, J)$ holds on $\mathcal{Y}$. What happens when I is tight on $\mathcal{X}$?

Hint: A natural way to extend I is to simply set it to infinity outside $\mathcal{X}$. However, this may fail to be lower semicontinuous.

The next example is basically a tautology but it is an example of Sanov's theorem from Section 5.2.

Example 3.4. Fix $0 < p < 1$ and let $\{X_n\}$ be an i.i.d. Bernoulli sequence with success probability p. Take $\mathcal{X} = [0,1]$ and express the common distribution of $\{X_n\}$ as $p\,\delta_1 + (1-p)\delta_0$. Here δ_x is the probability measure that puts all mass at the single point x, equivalently

$$\delta_x(A) = \mathbb{1}_A(x) = \begin{cases} 1, & x \in A, \\ 0, & x \notin A. \end{cases}$$

Example 2.1 gave LDP(μ_n, n, I_p) for the distribution μ_n of $S_n/n = (X_1 + \cdots + X_n)/n$ with rate I_p from (1.2). Consider now the *empirical measures*

$$L_n = \frac{1}{n}\sum_{k=1}^{n} \delta_{X_k}\,.$$

L_n is a random variable with values in $\mathcal{Y} = \mathcal{M}_1(\{0,1\})$. Let ν_n be its distribution. The empirical measure usually contains more information than the sample mean S_n/n, but in the Bernoulli case

$$L_n = \frac{S_n}{n}\delta_1 + \Bigl(1 - \frac{S_n}{n}\Bigr)\delta_0.$$

Hence $\nu_n = \mu_n \circ f^{-1}$ for $f : \mathcal{X} \to \mathcal{Y}$ defined by $f(s) = s\delta_1 + (1-s)\delta_0$. The contraction principle gives LDP(ν_n, n, H) with rate function defined for $\alpha \in \mathcal{M}_1(\{0,1\})$ by

$$H(\alpha) = I_p(s) \text{ for } \alpha = s\delta_1 + (1-s)\delta_0 \text{ with } s \in [0,1].$$

In Chapter 5 we will see that $H(\alpha)$ is a relative entropy and that the LDP for the empirical measure holds in general for i.i.d. processes.

3.2. Varadhan's theorem

For a measurable function $f : \mathcal{X} \to \mathbb{R}$ bounded above, a probability measure μ, and a sequence $r_n \to \infty$, the moment generating function obeys the following asymptotics:

$$\lim_{n\to\infty} \tfrac{1}{r_n} \log \int e^{r_n f}\, d\mu = \mu\text{-ess sup}\, f.$$

With $c = \mu$-ess sup f the argument is

$$c \ge \tfrac{1}{r_n} \log \int e^{r_n f}\, d\mu \ge \tfrac{1}{r_n} \log \int_{f > c-\varepsilon} e^{r_n f}\, d\mu \ge \tfrac{1}{r_n} \log \mu\{f > c - \varepsilon\} + c - \varepsilon.$$

Let us replace μ with a sequence μ_n. If $\{\mu_n\}$ satisfies a large deviation principle with normalization r_n, then the rate function I comes into the picture. The result is known as Varadhan's theorem. It is a probabilistic analogue of the well-known Laplace method for asymptotics of integrals illustrated by the next simple exercise.

Exercise 3.5 (Stirling's formula). Use induction to show that

$$n! = \int_0^\infty e^{-x} x^n \, dx.$$

Observe that $e^{-x}x^n$ has a unique maximum at $x = n$. Prove that

$$\lim_{n\to\infty} \frac{n!}{\sqrt{2\pi n}\, e^{-n} n^n} = 1.$$

Hint: Change variables $y = x/n$ to reduce the problem to one of estimating an integral of the form $\int_0^\infty e^{nf(y)}\, dy$. Show that the main contribution to this integral comes from $y \in [1-\varepsilon, 1+\varepsilon]$ and use Taylor expansion of f near $y = 1$.

Varadhan's theorem. *Suppose* LDP(μ_n, r_n, I) *holds,* $f : \mathcal{X} \to [-\infty, \infty]$ *is a continuous function, and*

$$\lim_{b\to\infty} \overline{\lim_{n\to\infty}} \tfrac{1}{r_n} \log \int_{f \geq b} e^{r_n f}\, d\mu_n = -\infty. \tag{3.2}$$

Then

$$\lim_{n\to\infty} \tfrac{1}{r_n} \log \int e^{r_n f}\, d\mu_n = \sup_{x: f(x)\wedge I(x) < \infty} \{f(x) - I(x)\}.$$

Exercise 3.6. A function f bounded above is the trivial case that satisfies (3.2). More generally, (3.2) follows if there exists $\alpha > 1$ such that

$$\sup_n \Big(\int e^{\alpha r_n f}\, d\mu_n \Big)^{1/r_n} < \infty. \tag{3.3}$$

Note that even though $f(x) = \infty$ is allowed, condition (3.2) forces $\mu_n\{f = \infty\} = 0$ for large enough n.

Informally, here is the idea behind Varadhan's theorem. Suppose that we can partition the space into small sets U_i with points $x_i \in U_i$ such that $f \approx f(x_i)$ on U_i and $\mu_n(U_i) \approx e^{-r_n I(x_i)}$. Then for large n the following approximations are valid:

$$\begin{aligned} \tfrac{1}{r_n} \log \int e^{r_n f}\, d\mu_n &\approx \tfrac{1}{r_n} \log \sum_i e^{r_n f(x_i)} \mu_n(U_i) \\ &\approx \tfrac{1}{r_n} \log \sum_i e^{r_n (f(x_i) - I(x_i))} \approx \max_i [f(x_i) - I(x_i)]. \end{aligned}$$

We can get stronger statements for separate upper and lower bounds. Hence we split the proof into two parts.

Lemma 3.7. *Let $f : \mathcal{X} \to [-\infty, \infty]$ be a lower semicontinuous function. Assume $\{\mu_n\}$ satisfies the lower large deviation bound (2.4) with rate function I and normalization r_n. Then,*

$$\varliminf_{n\to\infty} \frac{1}{r_n} \log \int e^{r_n f}\, d\mu_n \geq \sup_{f \wedge I < \infty} (f - I).$$

Proof. Suppose $f(x) \wedge I(x) < \infty$. We may assume $f(x) > -\infty$. Let $-\infty < c < f(x)$. Then $G = \{f > c\} \ni x$ is open, and $\inf_G f \geq c$. So

$$\begin{aligned}
\varliminf_{n\to\infty} \frac{1}{r_n} \log \int e^{r_n f}\, d\mu_n &\geq \varliminf_{n\to\infty} \frac{1}{r_n} \log \int_G e^{r_n f}\, d\mu_n \\
&\geq c + \varliminf_{n\to\infty} \frac{1}{r_n} \log \mu_n(G) \\
&\geq c - \inf_G I \geq c - I(x).
\end{aligned}$$

Now let $c \nearrow f(x)$ and take the supremum over x. □

Lemma 3.8. *Let $f : \mathcal{X} \to [-\infty, \infty]$ be an upper semicontinuous function (that is, $-f$ is lower semicontinuous). Assume $\{\mu_n\}$ satisfies the upper large deviation bound (2.3) with rate function I and normalization r_n. Assume also that*

$$\lim_{b\to\infty} \varlimsup_{n\to\infty} \frac{1}{r_n} \log \int_{f \geq b} e^{r_n f}\, d\mu_n = -\infty.$$

Then,

$$\varlimsup_{n\to\infty} \frac{1}{r_n} \log \int e^{r_n f}\, d\mu_n \leq \sup_{f \wedge I < \infty} (f - I).$$

Proof. First, assume $f(x) < b$ for all x. Let m and M be positive integers. Set $B_{k,m} = \{x : b - k/m \leq f(x) < b - (k-1)/m\}$ and write

$$\int e^{r_n f}\, d\mu_n = \sum_{k=1}^{M} \int_{B_{k,m}} e^{r_n f}\, d\mu_n + \int_{f < b - M/m} e^{r_n f}\, d\mu_n .$$

Upper semicontinuity of f implies that $\{f \geq b - k/m\}$ is a closed set and allows us to use the upper large deviation bound (2.3). Then

$$\begin{aligned}
&\varlimsup_{n\to\infty} \frac{1}{r_n} \log \int e^{r_n f}\, d\mu_n \\
&\quad\leq \max_{1\leq k\leq M} \Big\{ \varlimsup_{n\to\infty} \frac{1}{r_n} \log \int_{B_{k,m}} e^{r_n f}\, d\mu_n \Big\} \vee \Big\{ \varlimsup_{n\to\infty} \frac{1}{r_n} \log \int_{f < b - M/m} e^{r_n f}\, d\mu_n \Big\} \\
&\quad\leq \max_{1\leq k\leq M} \Big\{ b - \tfrac{k-1}{m} - \inf_{f \geq b - k/m} I \Big\} \vee (b - M/m) \\
&\quad\leq \max_{1\leq k\leq M} \Big\{ \sup_{f \geq b - k/m} (f - I) + 1/m \Big\} \vee (b - M/m).
\end{aligned}$$

After $M \nearrow \infty$ we have

$$\overline{\lim_{n\to\infty}} \frac{1}{r_n} \int e^{r_n f} d\mu_n \le \sup_{k\ge 1} \Big\{ \sup_{f\ge b-k/m} (f-I) + 1/m \Big\} = \sup(f-I) + 1/m.$$

Let $m \nearrow \infty$ and we have proved the statement for f bounded above.

Let $b < \infty$ and apply the step above to $f \wedge b$. Then

$$\begin{aligned}
&\overline{\lim_{n\to\infty}} \frac{1}{r_n} \log \int e^{r_n f} d\mu_n \\
&\quad \le \overline{\lim_{n\to\infty}} \frac{1}{r_n} \log \Big(\int e^{r_n (f\wedge b)} d\mu_n + \int_{f\ge b} e^{r_n f} d\mu_n \Big) \\
&\quad \le \sup(f\wedge b - I) \vee \overline{\lim_{n\to\infty}} \frac{1}{r_n} \log \int_{f\ge b} e^{r_n f} d\mu_n \\
&\quad \le \sup_{f\wedge I<\infty} (f-I) \vee \overline{\lim_{n\to\infty}} \frac{1}{r_n} \log \int_{f\ge b} e^{r_n f} d\mu_n \,.
\end{aligned}$$

On the second to last line the supremum can be restricted to $I < \infty$. On that set $f\wedge b - I \le f - I$. Then the supremum can be further enlarged to the set $f \wedge I < \infty$. Take b to infinity to conclude the proof. □

***Exercise 3.9.** Suppose $\mathrm{LDP}(\mu_n, r_n, I)$ holds and f is a bounded continuous function on $\mathcal{X}$. Define the probability measures ν_n by

$$\nu_n(A) = \frac{E^{\mu_n}[e^{r_n f} \mathbb{1}_A]}{E^{\mu_n}[e^{r_n f}]}\,.$$

Prove that $\mathrm{LDP}(\nu_n, r_n, J)$ holds with $J(x) = I(x) - f(x) - \inf(I - f)$.

Hint: For a closed set F, replacing f by $-\infty$ outside F gives an upper semicontinuous function.

In the exercise above the minimizers of the new rate function J are the main contributors to the integrals $\int e^{r_n f} d\mu_n$. These are precisely the maximizers of $f - I$.

Varadhan's theorem provides an explanation for the following curious phenomenon.

Exercise 3.10. Let Z_k be i.i.d. with $P\{Z_k = 1\} = P\{Z_k = 2\} = 1/2$. Set $W_n = Z_1 Z_2 \cdots Z_n$. Then, $E[W_n] = (3/2)^n$, but show that W_n grows like $(3/2)^n$ with exponentially small probability, or more precisely, that for small enough $\varepsilon > 0$, $P\{W_n > (3/2-\varepsilon)^n\}$ converges to 0 exponentially fast. Show that the typical growth rate is $\sqrt{2}$, in the sense that $P\{(\sqrt{2}-\varepsilon)^n < W_n < (\sqrt{2}+\varepsilon)^n\}$ converges to 1 exponentially fast.

Remark 3.11. Another interesting fact about W_n is that $(2/3)^n W_n \to 0$ almost surely (by the strong law of large numbers) while $(2/3)^{np} E[W_n^p] \to \infty$

for any $p > 1$. This is related to an important phenomenon called *intermittency*. For more about this see Section 8.1 of [**82**].

Example 3.12. Let $X_k = \log Z_k$, with Z_k as above. Then, $W_n = e^{S_n}$. What the exercise above shows is that even though paths $\{X_k\}$ such that $(\sqrt{2}-\varepsilon)^n < W_n < (\sqrt{2}+\varepsilon)^n$ have overwhelming probability, they nevertheless do not contribute the most to the expectation $E[W_n]$. Taking products seems to have suppressed the role of these paths while amplifying the importance of others. This is precisely what Varadhan's theorem explains:

$$\log \tfrac{3}{2} = \lim_{n\to\infty} \tfrac{1}{n} \log E[W_n] = \lim_{n\to\infty} \tfrac{1}{n} \log E[e^{S_n}] = \sup_{0\le x\le \log 2} (x - I(x)),$$

where

$$I(x) = \frac{x}{\log 2} \log \frac{x}{\log 2} + \frac{\log 2 - x}{\log 2} \log \frac{\log 2 - x}{\log 2} + \log 2$$

is the Cramér rate for the distribution of S_n/n. (You can obtain this from (1.2) by the contraction principle.) The supremum above is attained at $x = \frac{2}{3}\log 2$ instead of $E[X_1] = \log\sqrt{2}$. Thus, the paths that contribute the most are the ones for which $S_n/n \sim \frac{2}{3}\log 2$, that is, $W_n \sim (2^{2/3})^n$. These paths occur with an exponentially small probability, but when they do occur, they make an exponentially large contribution. The two balance out to produce the mean value $(3/2)^n$.

Integrals of the form $\int e^{nf}\, d\mu_n$ are encountered in statistical mechanics. The distributions ν_n of Exercise 3.9 are examples of Gibbs measures. We will see more of this in the second part of the book. For an appetizer, Section 3.4 applies large deviation theory to a mean-field model from statistical mechanics.

3.3. Bryc's theorem

Varadhan's theorem gives asymptotics of integrals as a consequence of an LDP. The converse is Bryc's theorem: knowing the asymptotics of sufficiently many integrals is equivalent to an LDP. We state and prove this theorem for the sake of completeness. It is not applied later in the text. Let $\mathscr{C}_b(\mathcal{X})$ denote the set of bounded and continuous real-valued functions on $\mathcal{X}$.

Bryc's theorem. *Let $\{\mu_n\}$ be a sequence of probability measures on a metric space $\mathcal{X}$. Assume $\{\mu_n\}$ is exponentially tight with normalization r_n. Suppose the limit*

$$\Gamma(f) = \lim_{n\to\infty} \frac{1}{r_n} \log \int e^{r_n f}\, d\mu_n$$

exists for all $f \in \mathscr{C}_b(\mathcal{X})$*. Then, LDP*$(\mu_n, r_n, I)$ *holds with the tight rate function*

$$I(x) = \sup_{f \in \mathscr{C}_b(\mathcal{X})} \{f(x) - \Gamma(f)\}. \tag{3.4}$$

Given the LDP, Varadhan's theorem implies that $\Gamma(f) = \sup(f - I)$. Note, however, that the relation between Γ and I is not the convex duality we study in Chapter 4, where the functions f are continuous linear functions.

Proof of Bryc's theorem. Function I is lower semicontinuous since it is a sup over continuous functions. Taking $f = 0$ shows $I(x) \geq 0$. Since the $\{\mu_n\}$ are exponentially tight, we only need to prove the weak LDP. We start with the lower bound.

Let G be an open set and fix $x \in G$. Let $f : \mathcal{X} \to [0, 1]$ be a continuous function such that $f(x) = 1$ and f vanishes outside G. Take $a > 0$ and define $f_a = a(f - 1)$. Then,

$$\int e^{r_n f_a} d\mu_n \leq e^{-a r_n} + \mu_n(G).$$

Thus

$$\begin{aligned}\max\Big\{\underline{\lim}_{n\to\infty} \tfrac{1}{r_n} \log \mu_n(G), -a\Big\} &\geq \underline{\lim}_{n\to\infty} \tfrac{1}{r_n} \log \int e^{r_n f_a} d\mu_n = \Gamma(f_a) \\ &= -\{f_a(x) - \Gamma(f_a)\} \geq -I(x).\end{aligned}$$

Take a to infinity and then the sup over x.

For the upper bound, let C be any measurable set and let f be a bounded continuous function. Then,

$$\overline{\lim_{n\to\infty}} \tfrac{1}{r_n} \log \mu_n(C) \leq \overline{\lim_{n\to\infty}} \tfrac{1}{r_n} \log \Big(e^{-r_n \inf_C f} \int e^{r_n f} d\mu_n\Big) \leq -\inf_C f + \Gamma(f).$$

Since f is not necessarily convex, a minimax theorem like the one on page 30 cannot be used. Let us proceed with the proof of the upper bound for an arbitrary compact set K. Fix

$$c < \inf_K I = \inf_{x \in K} \sup_{f \in \mathscr{C}_b(\mathcal{X})} \{f(x) - \Gamma(f)\}.$$

Then, for each $x \in K$ there exists a bounded continuous function f_x such that $f_x(x) - \Gamma(f_x) > c$. Since f_x is continuous, there exists an open neighborhood of x, say B_x, such that $f_x(y) - \Gamma(f_x) > c$ for all $y \in B_x$. One can then cover K with a finite number of these neighborhoods, say $B_{x_1}, \ldots, B_{x_N}$.

But then,

$$\varlimsup_{n\to\infty} \frac{1}{r_n} \log \mu_n(K) \le \max_{1\le k\le N} \varlimsup_{n\to\infty} \frac{1}{r_n} \log \mu_n(B_{x_k})$$
$$\le - \min_{1\le k\le N} \left\{ \inf_{B_{x_k}} f_{x_k} - \Gamma(f_{x_k}) \right\} \le -c.$$

Taking c up to $\inf_K I$ proves the upper bound for compact sets. The theorem is proved. □

Exercise 3.13. Let $\eta_k > 0$ be any sequence converging to 0, and let ρ_k be a probability measure supported on the interval $[-\eta_k, \eta_k]$. Let $\{X_k\}$ be an arbitrary sequence of random variables such that X_k has distribution ρ_k. Let S_n be the partial sum as before and μ_n the distribution of S_n/n. Show $\mathrm{LDP}(\mu_n, n, I)$ holds with rate function $I(0) = 0$, $I(x) = \infty$ for $x \neq 0$.

We generalize this in Chapter 15. For example, the result above will hold for independent $\{X_k\}$ under weak convergence $\rho_k \to \delta_0$, as long as these distributions have a common compact support.

But show that if X_k has distribution $P\{X_k = 0\} = 1 - \eta_k$ and $P\{X_k = k\} = \eta_k$, then the rate function above does not work if η_k converges to zero slowly enough.

3.4. Curie-Weiss model of ferromagnetism

This section is our first taste of statistical mechanics. Imagine that a piece of material is magnetized by subjecting it to a magnetic field. The field is turned off. Does the magnetization persist? We introduce a toy model for this phenomenon, the Curie-Weiss ferromagnet, and solve it with large deviation theory. We find that at low enough temperature magnetization persists. This is called *spontaneous magnetization*. We will also see natural examples of nonconvex rate functions.

Statistical mechanics explains macroscopic properties of matter from microscopic models. Microscopic reality is taken as random, and so states of the system are probability measures called Gibbs measures. In a limit to an infinite system laws of large numbers take hold and yield deterministic values for macroscopic observables.

The Curie-Weiss model begins with n atoms that each have a ± 1 valued *spin* ω_i, $1 \le i \le n$. The space of n-spin configurations is $\Omega_n = \{-1, 1\}^n$. The energy of the system is given by the *Hamiltonian*

$$\mathcal{H}_n(\omega) = -\frac{J}{2n} \sum_{1\le i,j\le n} \omega_i \omega_j - h \sum_{j=1}^{n} \omega_j. \tag{3.5}$$

A ferromagnet has a positive coupling constant $J > 0$ and $h \in \mathbb{R}$ is the external magnetic field. Nature prefers low energy, so in a ferromagnet spins tend to align with each other and with the magnetic field h if $h \neq 0$. The *Gibbs measure* for n spins is

$$\gamma_n(\omega) = \frac{1}{Z_n} e^{-\beta \mathcal{H}_n(\omega)} P_n(\omega), \quad \omega \in \Omega_n.$$

$P_n(\omega) = 2^{-n}$ is the a priori measure under which the spins are oriented entirely randomly as i.i.d. fair coin flips. The parameter $\beta > 0$ is the *inverse temperature.* The normalization constant $Z_n = \int e^{-\beta \mathcal{H}_n} dP_n$ is called the *partition function.*

The Gibbs measure captures the competition between the ordering tendency of the energy term $\mathcal{H}(\omega)$ and the randomness represented by P_n. It is instructive to consider the extremes of β, with $h = 0$. The zero temperature limit $\lim_{\beta\to\infty} \gamma_n(\omega) = \frac{1}{2}(\delta_{\omega\equiv 1} + \delta_{\omega\equiv -1})$ is concentrated on the two *ground states.* Thus at low temperature energy dominates and complete order reigns at the limit. By contrast, $\lim_{\beta\to 0} \gamma_n(\omega) = P_n$, so at high temperature thermal noise dominates and complete disorder reigns at the limit. The key question is the existence of phase transition: namely, is there a *critical temperature* β_c^{-1} (*Curie point*) at which the infinite model undergoes a transition that reflects the order/disorder dichotomy of the finite model.

An excess of $+$ or $-$ spins indicates magnetization, so we define magnetization as the expectation $M_n(\beta, h) = E^{\gamma_n}[S_n]$ of the total spin $S_n = \sum_{i=1}^n \omega_i$ under the Gibbs measure γ_n. If $h = 0$, $\gamma_n(\omega) = \gamma_n(-\omega)$ and so $M_n(\beta, 0) = 0$. This is a probability model on a finite space so $M_n(\beta, h)$ is clearly a continuous function of (β, h). To see something interesting we take $n \to \infty$. We show that the limit

$$m(\beta, h) = \lim_{n\to\infty} \frac{1}{n} M_n(\beta, h) \tag{3.6}$$

exists and even that the probability distribution of S_n/n converges.

Large deviation analysis will show that the asymptotic magnetization values are found from the equation

$$\tfrac{1}{2} \log \frac{1+z}{1-z} = J\beta z + \beta h \qquad \text{for } z \in [-1, 1]. \tag{3.7}$$

A qualitative change happens at $\beta_c = J^{-1}$.

Theorem 3.14. *Let $0 < \beta, J < \infty$ and let $h \in \mathbb{R}$.*

(i) *For $h \neq 0$, the limit $m(\beta, h)$ of* (3.6) *is the unique solution of* (3.7) *that has the same sign as h.*

(ii) *Let $h = 0$ and $\beta \leq J^{-1}$. Then $z = 0 = m(\beta, 0)$ is the unique solution of* (3.7) *and $m(\beta, \tilde{h}) \to 0$ as $\tilde{h} \to 0$.*

(iii) *Let* $h = 0$ *and* $\beta > J^{-1}$. *Then* (3.7) *has two nonzero solutions* $m(\beta, +) > 0$ *and* $m(\beta, -) = -m(\beta, +)$. *Spontaneous magnetization happens: for* $\beta > J^{-1}$,

$$\lim_{0<\tilde{h}\searrow 0} m(\beta, \tilde{h}) = m(\beta, +) \quad \textit{and} \quad \lim_{0>\tilde{h}\nearrow 0} m(\beta, \tilde{h}) = m(\beta, -). \tag{3.8}$$

Exercise 3.15. Prove the claims made in the theorem above, except point (i) about the limit (3.6) equaling the solution of (3.7). The rest is just calculus. It is helpful to sketch the left- and right-hand sides of equation (3.7) and observe how the locations of their intersections vary with h and the slope $J\beta$.

Remark 3.16. Before turning to large deviations and the proof of Theorem 3.14(i), we give a brief discussion of Curie-Weiss as the mean-field approximation of the Ising model. The feature that makes the Curie-Weiss model easy is that in (3.5) each pair of spins ω_i, ω_j interacts with the same strength, regardless of the distance $|i - j|$. However, physical experiments show that spatial structure does matter. The more realistic *Ising model* in a finite-volume $\Lambda \subset \mathbb{Z}^d$ has Hamiltonian

$$\mathcal{H}_\Lambda^{\text{Ising}}(\omega) = -J \sum_{\{x,y\}\subset\Lambda:\, |x-y|_1=1} \omega_x \omega_y - h \sum_{x\in\Lambda} \omega_x$$

where the first summation is over nearest-neighbor pairs of spins in Λ. Curie-Weiss is called the *mean-field approximation* of the Ising model because its Hamiltonian can be expressed as

$$\mathcal{H}_n(\omega) = -\frac{J}{2} \sum_{i=1}^n \omega_i \Big(\frac{1}{n} \sum_{j=1}^n \omega_j \Big) - h \sum_{i=1}^n \omega_i$$

where each spin interacts with the "mean field" S_n/n created by all the spins.

Returning to the Curie-Weiss model, the key to large deviations is that the Hamiltonian is a function of S_n/n:

$$-\mathcal{H}_n(\omega) = n\tfrac{1}{2}J(S_n/n)^2 + nhS_n/n.$$

Let λ_n be the distribution of S_n/n under P_n and let μ_n be the distribution of S_n/n under γ_n. By Cramér's theorem (page 26) LDP$(\lambda_n, n, I_{\text{BER}})$ holds with rate function

$$I_{\text{BER}}(z) = \begin{cases} \frac{1}{2}(1-z)\log(1-z) + \frac{1}{2}(1+z)\log(1+z) & \text{if } -1 \le z \le 1, \\ \infty & \text{otherwise.} \end{cases}$$

To take advantage of this, relate μ_n to λ_n: for Borel $B \subset [0,1]$,

$$\begin{aligned}\mu_n(B) = \gamma_n\{S_n/n \in B\} &= \frac{1}{Z_n}\int \mathbb{1}_B(S_n/n)e^{-\beta\mathcal{H}_n}\,dP_n \\ &= \frac{1}{Z_n}\int \mathbb{1}_B(S_n/n)e^{n\left(\frac{1}{2}J\beta(S_n/n)^2+\beta h(S_n/n)\right)}\,dP_n \\ &= \frac{1}{Z_n}\int \mathbb{1}_B(z)e^{n\left(\frac{1}{2}J\beta z^2+\beta hz\right)}\,\lambda_n(dz).\end{aligned}$$

By Exercise 3.9, LDP$(\mu_n, n, I_{\mathrm{CW}})$ holds with rate function

$$I_{\mathrm{CW}}(z) = I_{\mathrm{BER}}(z) - \tfrac{1}{2}J\beta z^2 - \beta hz - c \quad \text{where } c = \inf_z\{I_{\mathrm{BER}}(z) - \tfrac{1}{2}J\beta z^2 - \beta hz\}. \tag{3.9}$$

In rate functions I_{CW} and I_{BER}, CW stands for Curie-Weiss model and BER stands for Bernoulli distribution.

To understand limits of S_n/n under γ_n we find the minimizers of I_{CW}. Critical points satisfy $I'_{\mathrm{CW}}(z) = 0 \iff$ (3.7). We collect the calculus in the next exercise. See Figure 3.1 for plots of I_{CW}. Recall the definitions of $m(\beta, h)$ and $m(\beta, \pm)$ in Theorem 3.14.

Exercise 3.17. Prove the following.

(a) If $h \neq 0$, then I_{CW} is uniquely minimized at $z = m(\beta, h)$.

(b) If $h = 0$ and $J\beta \le 1$, then I_{CW} is uniquely minimized at $z = m(\beta, 0) = 0$.

(c) If $h = 0$ and $J\beta > 1$, then I_{CW} is minimized by $m(\beta, \pm)$.

Here are the limits of the distribution of S_n/n. Existence of limit (3.6) and claim (i) of Theorem 3.14 follow from this because S_n/n is bounded, so a weak limit of its distribution implies the limit of its expectation.

Theorem 3.18. *The following hold.*

(a) *Suppose that either $h \neq 0$, or $h = 0$ and $\beta \le 1/J$. Then $\mu_n \to \delta_{m(\beta,h)}$ weakly. For all $\varepsilon > 0$, $\gamma_n\{|S_n/n - m(\beta, h)| \ge \varepsilon\} \to 0$ exponentially fast.*

(b) *If $h = 0$ and $\beta > 1/J$, then $\mu_n \to \frac{1}{2}(\delta_{m(\beta,+)} + \delta_{m(\beta,-)})$ weakly. If A is a closed set such that $m(\beta, \pm) \notin A$ then $\gamma_n\{S_n/n \in A\} \to 0$ exponentially fast.*

Parts (a) and (b) clarify the phase transition. Even though the limiting magnetization $m(\beta, 0) = 0\ \forall \beta > 0$, in low temperature this comes from two equally likely nonzero limit points $m(\beta, -) < 0 < m(\beta, +)$ where either + or − spins dominate. This is the effect of the ordering tendency at low temperature. Figure 3.2 of the phase diagram summarizes the results. Discussion about phase transition continues on page 108.

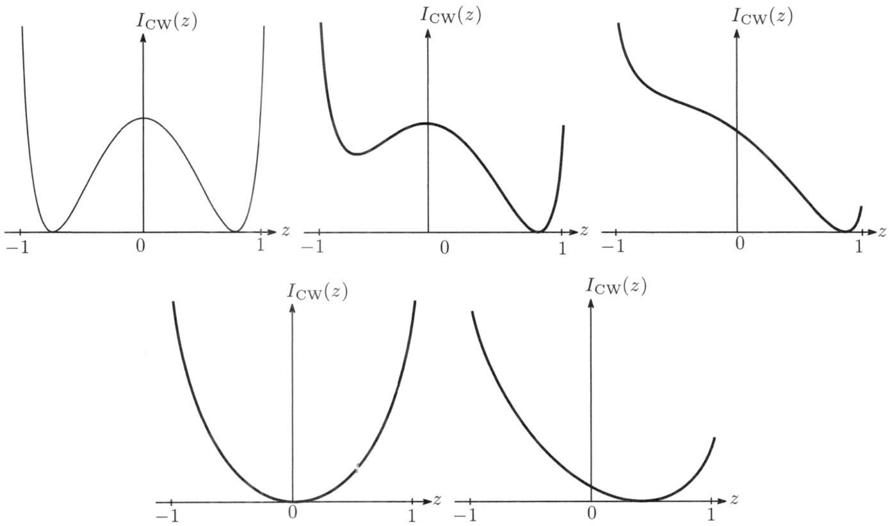

Figure 3.1. The Curie-Weiss rate function. Top plots have $\beta > J^{-1}$ while bottom plots have $\beta \le J^{-1}$. Top left to right: $h = 0$, $0 < h < h_0(J, \beta)$, and $h > h_0(J, \beta)$. Bottom left to right, $h = 0$ and $h > 0$. The case $h < 0$ is symmetric to that of $h > 0$.

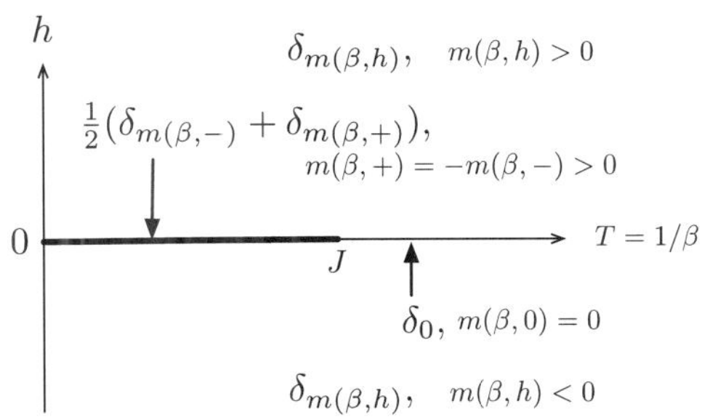

Figure 3.2. The Curie-Weiss phase diagram.

Exercise 3.19. Prove the following further properties of the rate function $I_{\rm CW}$, referring to Figure 3.1.

(a) $I_{\rm CW}$ is convex if and only if $J\beta \le 1$.

(b) If $J\beta > 1$, then $I_{\rm CW}$ has two local minima (one of which is the global minimum) if

$$|h| < h_0(J, \beta) = J\sqrt{1 - 1/(J\beta)} - (2\beta)^{-1} \log \frac{1 + \sqrt{1 + (J\beta)^{-1}}}{1 - \sqrt{1 - (J\beta)^{-1}}}.$$

Otherwise, it has a unique local minimum (that is also the global minimum).

Proof of Theorem 3.18. Part (a) is immediate because I_{CW} has a unique minimizer:

$$\overline{\lim_{n\to\infty}} \frac{1}{n} \log \gamma_n\{|S_n/n - m(\beta,h)| \ge \varepsilon\} \le - \inf_{|z-m(\beta,h)|\ge\varepsilon} I_{\mathrm{CW}}(z) < 0.$$

For part (b) the large deviation upper bound gives similarly

$$\lim_{n\to\infty} \gamma_n\{|S_n/n - m(\beta,-)| < \varepsilon \text{ or } |S_n/n - m(\beta,+)| < \varepsilon\} = 1.$$

From $\gamma_n(\omega) = \gamma_n(-\omega)$ it follows that μ_n is symmetric and so

$$\lim_{n\to\infty} \gamma_n\{|S_n/n - m(\beta,-)| < \varepsilon\} = \lim_{n\to\infty} \gamma_n\{|S_n/n - m(\beta,+)| < \varepsilon\} = \tfrac{1}{2}.$$

This shows the weak convergence $\mu_n \to \frac{1}{2}(\delta_{m(\beta,-)} + \delta_{m(\beta,+)})$. □

Literature notes. The hypothesis of Bryc's theorem can be relaxed to require the limit $\Gamma(f)$ only for some rich enough class of functions. A general result of this type is Theorem 4.4.10 of [**15**], and a particular application of Bryc's theorem in the context of optimization problems appears in [**76**].

Our application of large deviations to study the Curie-Weiss model follows Section IV.4 of Ellis [**32**]. For further discussion of the mean field approximation, see Section V.9 of Ellis [**32**].

Chapter 4

Convex analysis in large deviation theory

We have seen that convex functions occur naturally in large deviations. This chapter presents the basics of the general theory of convex functions and two immediate applications to large deviation theory. In particular, we give a proof of Cramér's theorem in $\mathbb{R}^d$. Convexity remains central to much of the rest of the text. For a deeper study of convex analysis the reader is referred to [**30**] and [**68**], and to [**32**] in the context of large deviation theory.

4.1. Some elementary convex analysis

It is most useful to approach this material in an abstract setting, while keeping key examples in mind Let $\mathcal{X}$ and $\mathcal{Y}$ be two real vector spaces. This means that for $x, x' \in \mathcal{X}$ and scalars $a \in \mathbb{R}$, $x + x'$ and ax are defined in $\mathcal{X}$, and familiar rules of algebra are in force. Assume that $\mathcal{X}$ and $\mathcal{Y}$ are in duality, that is, there is a bilinear function $\langle \cdot\,, \cdot \rangle : \mathcal{X} \times \mathcal{Y} \to \mathbb{R}$. The weak topology $\sigma(\mathcal{X}, \mathcal{Y})$ on $\mathcal{X}$ is the minimal topology under which the functions $\{x \mapsto \langle x, y\rangle : y \in \mathcal{Y}\}$ are continuous, and similarly for $\sigma(\mathcal{Y}, \mathcal{X})$ on $\mathcal{Y}$.

Assumption 4.1. *Assume that for each nonzero $x \in \mathcal{X}$ there exists $y \in \mathcal{Y}$ such that $\langle x, y\rangle \neq 0$ and, conversely, for each nonzero $y \in \mathcal{Y}$ there exists $x \in \mathcal{X}$ such that $\langle x, y\rangle \neq 0$.*

Under the assumption above the topologies $\sigma(\mathcal{X}, \mathcal{Y})$ and $\sigma(\mathcal{Y}, \mathcal{X})$ are Hausdorff. Here are the two most important examples.

Example 4.2. $\mathcal{X} = \mathcal{Y} = \mathbb{R}^d$, $\langle x, y\rangle = \sum_{i=1}^d x_i y_i$. Then $\sigma(\mathcal{X}, \mathcal{Y}) = \sigma(\mathcal{Y}, \mathcal{X})$ is the usual Euclidean topology on $\mathbb{R}^d$.

Example 4.3. Let $(\mathcal{S}, \mathscr{B})$ be a measurable space, for example a metric space with its Borel σ-algebra. Let $\mathcal{X} = \mathcal{M}(\mathcal{S})$, the space of real-valued measures on $(\mathcal{S}, \mathscr{B})$, and let $\mathcal{Y}$ be some vector space of bounded measurable functions on $\mathcal{S}$. The duality is integration: $\langle \mu, f \rangle = \int f \, d\mu$. Natural choices for $\mathcal{Y}$ are $\mathcal{Y} = b\mathscr{B}$, the space of all bounded measurable $f : \mathcal{X} \to \mathbb{R}$, and if $\mathcal{S}$ is metric, $\mathcal{Y} = \mathscr{C}_b(\mathcal{S})$, the space of bounded continuous $f : \mathcal{X} \to \mathbb{R}$.

Remark 4.4. To cover the example above we are forced to consider weak topologies $\sigma(\mathcal{X}, \mathcal{Y})$ that do not have a metric. To see why, let $\mathcal{S}$ be a metric space. We claim that the weak topology $\sigma(\mathcal{M}(\mathcal{S}), \mathscr{C}_b(\mathcal{S}))$ is not metrizable unless $\mathcal{S}$ is a finite set. This follows from two facts.

First, in general $\sigma(\mathcal{X}, \mathcal{Y})$ cannot be metrizable unless $\mathcal{Y}$, as a vector space, has a finite or countable Hamel basis, that is, a set of basis vectors in the algebraic sense (Exercise B.7).

Second, this cannot be true for $\mathcal{Y} = \mathscr{C}_b(\mathcal{S})$ if $\mathcal{S}$ is infinite. Briefly, here is the reason. If $\mathscr{C}_b(\mathcal{S})$ had a countable basis, it would be a union of countably many finite-dimensional subspaces. If $\mathcal{S}$ is infinite, finite-dimensional subspaces of $\mathscr{C}_b(\mathcal{S})$ are closed sets with empty interior in the standard sup-norm topology of $\mathscr{C}_b(\mathcal{S})$. Under the sup-norm $\mathscr{C}_b(\mathcal{S})$ is a complete metric space. Hence by the Baire category theorem [**26**, Theorem 2.5.2] $\mathscr{C}_b(\mathcal{S})$ cannot be a countable union of closed sets with empty interior.

Despite the remark above, the space $\mathcal{M}_1(\mathcal{S})$ of probability measures is metrizable when $\mathcal{S}$ is a separable metric space (Section B.4).

Example 4.5. Generally, if $\mathcal{X}$ is a Banach space and $\mathcal{Y} = \mathcal{X}^*$ (the dual of $\mathcal{X}$, the space of bounded linear functionals on $\mathcal{X}$), then $\sigma(\mathcal{X}, \mathcal{Y})$ and $\sigma(\mathcal{Y}, \mathcal{X})$ are, respectively, the weak and the weak* topology studied in functional analysis.

The next proposition lists some consequences of the definitions. We leave the proof to the reader, either as an exercise or to be looked up from a functional analysis text, such as Chapter 3 in Rudin [**69**].

Proposition 4.6. *Assume the setting as above.*

(a) *A base for $\sigma(\mathcal{X}, \mathcal{Y})$ is given by the collection of sets of the type*

$$\{x \in \mathcal{X} : |\langle x, y_i \rangle - \langle x_0, y_i \rangle| < \varepsilon, \ i = 1, \dots, m\}$$

for $m \in \mathbb{N}$, $x_0 \in \mathcal{X}$, $y_1, \dots, y_m \in \mathcal{Y}$, and $\varepsilon > 0$. In particular, $\sigma(\mathcal{X}, \mathcal{Y})$ is a locally convex topology, which means that every open neighborhood of $x \in \mathcal{X}$ contains a convex neighborhood of x.

(b) *$x_n \to x$ in $\sigma(\mathcal{X}, \mathcal{Y})$ if and only if $\langle x_n, y \rangle \to \langle x, y \rangle$ for all $y \in \mathcal{Y}$.*

(c) *The maps $(a, x) \mapsto ax$ from $\mathbb{R} \times \mathcal{X}$ into $\mathcal{X}$ and $(x, x') \mapsto x + x'$ from $\mathcal{X} \times \mathcal{X}$ into $\mathcal{X}$ are continuous in the product topologies. In* *other words, the vector space operations are continuous. This is expressed by saying that $\mathcal{X}$ is a topological vector space.*

(d) *Suppose $f : \mathcal{X} \to \mathbb{R}$ is a continuous linear functional. Then there exists a unique $y \in \mathcal{Y}$ such that $f(x) = \langle x, y \rangle$ for all $x \in \mathcal{X}$. This gives an isomorphism between $\mathcal{Y}$ and the dual space $\mathcal{X}^*$ (the vector space of continuous linear functionals on $\mathcal{X}$).*

(e) *All of the above also hold if the roles of $\mathcal{X}$ and $\mathcal{Y}$ are reversed.*

Recall that a function $f : \mathcal{X} \to [-\infty, \infty]$ is *convex* if

$$f(\alpha x_1 + (1-\alpha)x_2) \le \alpha f(x_1) + (1-\alpha) f(x_2)$$

for all $x_1, x_2 \in \mathcal{X}$ and $\alpha \in [0,1]$ such that the right-hand side is well-defined. A *proper* convex function maps into $(-\infty, \infty]$ and is not identically ∞. A function $f : \mathcal{X} \to \mathbb{R}$ is *affine* if $f(x) = a + \langle x, y \rangle$ for some $a \in \mathbb{R}$ and $y \in \mathcal{Y}$.

***Exercise 4.7.** Suppose $\mathcal{E}$ is a convex subset of $\mathcal{X}$ and let $f : \mathcal{E} \to (-\infty, \infty]$ be lower semicontinuous. Prove that f is convex on $\mathcal{E}$ if and only if for all $x, y \in \mathcal{E}$,

$$f\Big(\frac{x+y}{2}\Big) \le \frac{f(x) + f(y)}{2}.$$

Hint: First take care of dyadic rationals $\alpha = k2^{-n}$ by induction on n.

***Exercise 4.8.** If you have never done so, check that a finite convex function on an open interval is continuous.

Hint: Fix $x_0 < x < x_1$ and let $y \to x$. Use convexity to bound $f(y)$ from above and below to control $\overline{\lim} f(y)$ and $\underline{\lim} f(y)$ as $y \to x$.

Continuity of convex functions on convex open sets is true also in higher dimensions. See Theorem 10.1 in Rockafellar [**68**].

Definition 4.9. For any function $f : \mathcal{X} \to [-\infty, \infty]$, define the *convex conjugate* $f^* : \mathcal{Y} \to [-\infty, \infty]$ by

$$f^*(y) = \sup_{x \in \mathcal{X}} \{\langle x, y \rangle - f(x)\}.$$

Similarly, if $g : \mathcal{Y} \to [-\infty, \infty]$, define $g^* : \mathcal{X} \to [-\infty, \infty]$ by

$$g^*(x) = \sup_{y \in \mathcal{Y}} \{\langle x, y \rangle - g(y)\}.$$

Inductively, $f^{**} = (f^*)^*$ is the *convex biconjugate* and $f^{*n} = (f^{*(n-1)})^*$. Similarly for g. Alternative terms for the convex conjugate are convex dual and Legendre(-Fenchel) conjugate.

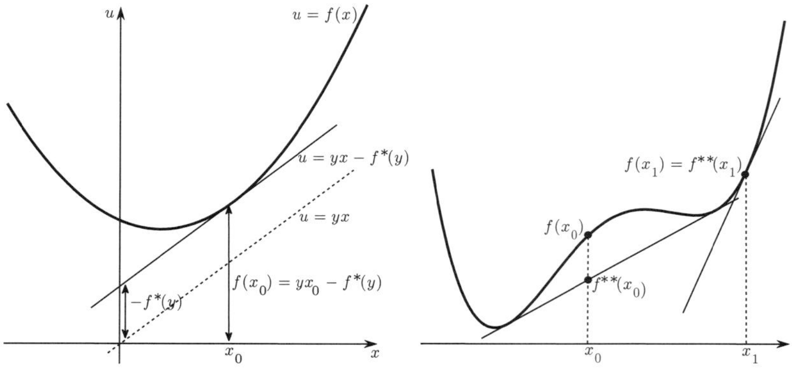

Figure 4.1. f^* and f^{**} in one dimension. On the left: draw a line with slope y under the graph of f and move it as high as possible. Its vertical intercept is $-f^*(y)$. Point x_0 is a maximizer of $f^*(y) = \sup\{yx - f(x)\}$. On the right: f^{**} is the supremum over all lines below the graph of f.

Proposition 4.10. *For arbitrary $f : \mathcal{X} \to [-\infty, \infty]$ the following are true.*

(a) *f^* is convex and lower semicontinuous, where the constants $f^* = \infty$ and $f^* = -\infty$ also qualify as convex lower semicontinuous functions.*

(b) *$f^{**}(x) \le f(x)$ for all $x \in \mathcal{X}$.*

(c) *$f^{*n} = f^*$ if $n \ge 1$ is odd and $f^{*n} = f^{**}$ if $n \ge 2$ is even.*

Proof. If f takes the value $-\infty$, then $f^* = \infty$. If $f = \infty$, then $f^* = -\infty$. In both cases part (a) is true. If, on the other hand, $f(x) > -\infty$ for all x and f is not identically infinite, then f^* is the pointwise supremum of a nonempty family of affine continuous functions $y \mapsto \langle x, y\rangle - f(x)$, $x \in \mathcal{X}$ with $f(x) < \infty$. Hence the claim of part (a) is true in this case too.

Part (b) follows from Exercise 4.11 below or more immediately from

$$\langle x, y\rangle - f^*(y) \le \langle x, y\rangle - \big(\langle x, y\rangle - f(x)\big) = f(x).$$

If $f_1 \le f_2$, then $f_1^* \ge f_2^*$. Thus part (b) implies both $f^{*3} \le f^*$ and $f^{*3} \ge f^*$. So $f^{*3} = f^*$ and part (c) follows by induction. $\square$

Figure 4.1 illustrates f^* and f^{**} in $\mathcal{X} = \mathbb{R}$. The connection between the slope and f^* comes in Theorem 4.21 below. The next exercise verifies the picture for f^{**}.

***Exercise 4.11.** Prove that f^{**} is the supremum of affine minorants of f. Precisely,

$$f^{**}(x) = \sup\{a + \langle x, y\rangle : a \in \mathbb{R},\, y \in \mathcal{Y},\, a + \langle z, y\rangle \le f(z)\ \forall z \in \mathcal{X}\}.$$

An immediate consequence of the definitions is the *Fenchel-Young inequality*: for any $x \in \mathcal{X}$ and $y \in \mathcal{Y}$

$$\langle x, y\rangle \leq f(x) + f^*(y), \tag{4.1}$$

whenever the right-hand side makes sense.

One of the main points we reach below is that f^{**} is the maximal convex lower semicontinuous minorant of f. Here is the main step.

Fenchel-Moreau theorem. *Suppose $f : \mathcal{X} \to (-\infty, \infty]$ is not identically ∞. Then, $f = f^{**}$ if and only if f is convex and lower semicontinuous.*

Proof. Given Proposition 4.10, we only need to show that $f^{**} \geq f$ for a proper lower semicontinuous convex function f. We start by showing that f^* is proper.

Pick $x_0 \in \mathcal{X}$ and $t_0 > -\infty$ such that $t_0 < f(x_0) < \infty$. From this

$$f^*(y) \geq \langle x_0, y\rangle - f(x_0) > -\infty$$

for all $y \in \mathcal{Y}$. Thus, we only need to show the existence of $y_0 \in \mathcal{Y}$ such that $f^*(y_0) < \infty$. Since f is convex lower semicontinuous, its *epigraph* $\operatorname{epi} f = \{(x,t) \in \mathcal{X} \times \mathbb{R} : f(x) \leq t < \infty\}$ is a closed convex set. Since $\mathcal{X} \times \mathbb{R}$ is a locally convex topological vector space and (x_0, t_0) is outside the epigraph, the Hahn-Banach separation theorem (Section A.4) implies that there exist a $\phi \in (\mathcal{X} \times \mathbb{R})^*$ and $\gamma \in \mathbb{R}$ such that

$$\phi(x_0, t_0) > \gamma > \phi(x, t)$$

for all $(x, t) \in \operatorname{epi} f$.

***Exercise 4.12.** Prove that $(\mathcal{X} \times \mathbb{R})^*$ is isomorphic to $\mathcal{X}^* \times \mathbb{R}$ (and hence to $\mathcal{Y} \times \mathbb{R}$).

By the exercise above, there exists a pair $(y, s) \in \mathcal{Y} \times \mathbb{R}$ such that $\phi(x, t) = \langle x, y\rangle + st$. Hence

$$\langle x_0, y\rangle + st_0 > \gamma > \langle x, y\rangle + st$$

for all (x, t) in the epigraph. In particular, since $(x_0, f(x_0))$ is in the epigraph,

$$\langle x_0, y\rangle + st_0 > \langle x_0, y\rangle + sf(x_0)$$

and thus $st_0 > sf(x_0)$ and $s < 0$. But then $\langle x, y/|s|\rangle - f(x) < \gamma/|s|$ whenever $f(x) < \infty$. It follows that

$$f^*(y/|s|) = \sup_{x: f(x)<\infty} \{\langle x, y/|s|\rangle - f(x)\} \leq \gamma/|s| < \infty.$$

This proves that f^* is proper. Consequently, f^{**} is also proper. Indeed,

$$f^{**}(x) \geq \langle x, y_0 \rangle - f^*(y_0) > -\infty$$

and $f^{**} \leq f$ is not identically ∞. We are now ready to prove that $f^{**} \geq f$.

Suppose there exists a point $x_0 \in \mathcal{X}$ such that $f^{**}(x_0) < f(x_0)$. Then $|f^{**}(x_0)| < \infty$ and we can use the Hahn-Banach separation theorem to separate $(x_0, f^{**}(x_0))$ from epi f. In other words, there exist $\gamma \in \mathbb{R}$ and a pair $(y, s) \in \mathcal{Y} \times \mathbb{R}$ such that

$$\langle x_0, y \rangle + s f^{**}(x_0) > \gamma > \langle x, y \rangle + st$$

for all $(x, t) \in \operatorname{epi} f$. Now pick x such that $f(x) < \infty$ and let t grow to infinity. This proves that $s \leq 0$. Suppose $s < 0$. Then,

$$\begin{aligned} f^{**}(x_0) + f^*(-y/s) &= f^{**}(x_0) + \sup_{x: f(x) < \infty} \{\langle x, y/|s| \rangle - f(x)\} \\ &= f^{**}(x_0) + |s|^{-1} \sup_{x: f(x) < \infty} \{\langle x, y \rangle + s f(x)\} \\ &< f^{**}(x_0) + |s|^{-1}(\langle x_0, y \rangle + s f^{**}(x_0)) \\ &= \langle x_0, -y/s \rangle, \end{aligned}$$

contradicting the Fenchel-Young inequality (4.1). Hence, $s = 0$ and

$$\langle x_0, y \rangle > \gamma > \langle x, y \rangle$$

for all x such that $f(x) < \infty$. Observe that this implies that $f(x_0)$ cannot be finite. Now let $\alpha > 0$. Since f^* is proper we can choose y_0 such that $f^*(y_0) < \infty$. Then

$$\begin{aligned} f^*(y_0 + \alpha y) &= \sup_x \{\langle x, y_0 + \alpha y \rangle - f(x)\} \\ &\leq \sup_x \{\langle x, y_0 \rangle - f(x)\} + \alpha \sup_{x: f(x) < \infty} \langle x, y \rangle \\ &\leq f^*(y_0) + \alpha \gamma. \end{aligned}$$

From this

$$f^{**}(x_0) \geq \langle x_0, y_0 + \alpha y \rangle - f^*(y_0 + \alpha y) \geq \langle x_0, y_0 \rangle + \alpha(\langle x_0, y \rangle - \gamma) - f^*(y_0).$$

Recalling that $f^*(y_0) < \infty$ and taking α to infinity shows that $f^{**}(x_0) = \infty$, which contradicts the choice of x_0 and concludes the proof. ☐

The first part of the proof of the theorem above contains the following.

Lemma 4.13. *If $f : \mathcal{X} \to (-\infty, \infty]$ is a proper lower semicontinuous convex function, then so is f^*.*

Corollary 4.14. *Consider $f : \mathcal{X} \to (-\infty, \infty]$ that is not identically infinite. Let g be a lower semicontinuous convex function such that $-\infty < g(x) \leq f(x)$ for all $x \in \mathcal{X}$. Then $g \leq f^{**}$.*

Proof. $g \le f$ implies $g^* \ge f^*$ which implies $g^{**} \le f^{**}$. But $g^{**} = g$. □

Remark 4.15. It is not always possible to put a proper convex function below f. Take, for example, $f(x) = -x^2$.

Since part (b) of Proposition 4.10 says that $f^{**} \le f$, the corollary above tells us that the biconjugate f^{**} is the maximal lower semicontinuous convex minorant of f. This can be rephrased in terms of epigraphs. The *convex hull* of a set is the intersection of all the convex sets containing it.

Lemma 4.16. *Consider $f : \mathcal{X} \to (-\infty, \infty]$ that is not identically infinite. The epigraph of f^{**} is the closure of the convex hull of the epigraph of f.*

Proof. Since $f^{**} \le f$ the epigraph of f is included in that of f^{**}. The latter is a closed convex set and hence contains the closure of the convex hull of the former. For the other direction, let $f^{**}(x_0) \le t_0 < \infty$ with (x_0, t_0) outside the closure of the convex hull. There are a y, an s, and a γ such that

$$\langle x, y \rangle + st < \gamma < \langle x_0, y \rangle + st_0$$

for all $t \ge f(x)$. Pick x with $f(x)$ finite and let $t \to \infty$. This shows that $s \le 0$. If $s < 0$, then

$$f(x) > \gamma/s - \langle x, y/s \rangle$$

for all x such that $f(x) < \infty$ and thus, in fact, for all x. Since the right-hand side is affine in x,

$$t_0 \ge f^{**}(x_0) \ge \gamma/s - \langle x_0, y/s \rangle,$$

which contradicts the separation. So $s = 0$ and

$$\langle x, y \rangle < \gamma < \langle x_0, y \rangle$$

for all x such that $f(x) < \infty$. Thus $f(x)$ is infinite for all x with $\langle x, y \rangle > \gamma$. But resetting f^{**} to infinity on this region creates a lower semicontinuous convex minorant of f. Thus f^{**} is also infinite on this region and in particular at x_0. This contradicts the choice of x_0. □

If f is already convex, then f^{**} recovers the lower semicontinuous regularization f_{lsc} of definition (2.2).

Theorem 4.17. *Suppose $f : \mathcal{X} \to (-\infty, \infty]$ is convex and not identically infinite. Then $f_{\text{lsc}} = f^{**}$.*

Proof. Note that if f^{**} takes the value $-\infty$, then $f^{***} = \infty$ and thus $f^* = \infty$ and $f^{**} = -\infty$. By Lemma 4.16, this implies that the closure of the epigraph of f is all of $\mathcal{X} \times \mathbb{R}$ and thus, by Lemma 2.9, that the epigraph of f_{lsc} is all of $\mathcal{X} \times \mathbb{R}$ and $f_{\text{lsc}} = -\infty = f^{**}$. We can, therefore, assume that $f^{**} > -\infty$.

Since $f^{**} \le f$ and f^{**} is lower semicontinuous, Lemma 2.8 implies then that $f^{**} \le f_{\rm lsc}$. In particular, $f_{\rm lsc} > -\infty$.

If we now show that $f_{\rm lsc}$ is convex, then Corollary 4.14 implies that $f_{\rm lsc} \le f^{**}$. Let G be an open neighborhood of $\alpha x_1 + (1-\alpha)x_2$ such that $\inf_G f > -\infty$. By the continuity of the map $(x_1, x_2) \mapsto \alpha x_1 + (1-\alpha)x_2$ there exists a neighborhood G_1 of x_1 and a neighborhood G_2 of x_2 such that $\alpha G_1 + (1-\alpha)G_2 \subset G$. Then,

$$\begin{aligned}\inf_G f &\le \inf_{z_1 \in G_1} \inf_{z_2 \in G_2} f(\alpha z_1 + (1-\alpha)z_2) \le \alpha \inf_{G_1} f + (1-\alpha) \inf_{G_2} f \\ &\le \alpha f_{\rm lsc}(x_1) + (1-\alpha) f_{\rm lsc}(x_2).\end{aligned}$$

Taking the supremum over G proves the convexity of $f_{\rm lsc}$ and concludes the proof of the theorem. □

The last item in the section is differentiation. On an interval a convex function $f : (a, b) \to \mathbb{R}$ has left and right derivatives at each point because its slope is nondecreasing. There can be a nontrivial interval of tangent slopes at a point. For example, all numbers $t \in [-1, 1]$ are tangent slopes for $f(x) = |x|$ at $x = 0$, in the sense that

$$f(x) \ge f(0) + t(x - 0) \quad \forall x \in \mathbb{R}.$$

This possible multiplicity of tangent lines and tangent hyperplanes is captured by the notion of subdifferential.

Definition 4.18. Let $f : \mathcal{X} \to [-\infty, \infty]$. The *subdifferential* of f at x is the *multivalued mapping* $\partial f : \mathcal{X} \to \mathcal{Y}$ defined by

$$\partial f(x) = \{y \in \mathcal{Y} : \forall u \in \mathcal{X}, f(u) \ge f(x) + \langle u - x, y \rangle\}.$$

In general $\partial f(x)$ is a closed convex set, possibly empty.

Exercise 4.19. Prove that if f is not identically ∞ and $f(x) = \infty$, then $\partial f(x) = \varnothing$. Then prove that if f is convex and proper and bounded above on an open neighborhood of x, then $\partial f(x) \ne \varnothing$. Find a counterexample where $f(x) < \infty$ but $\partial f(x) = \varnothing$.

Hint: Since epi f has nonempty interior, it is contained in the closure of its interior. Separate $(x, f(x))$ from the interior of epi f.

Exercise 4.20. Let $\mathcal{X} = \mathbb{R}^d$ and assume f is proper, convex, and differentiable at x. Prove that $\partial f(x) = \{\nabla f(x)\}$.

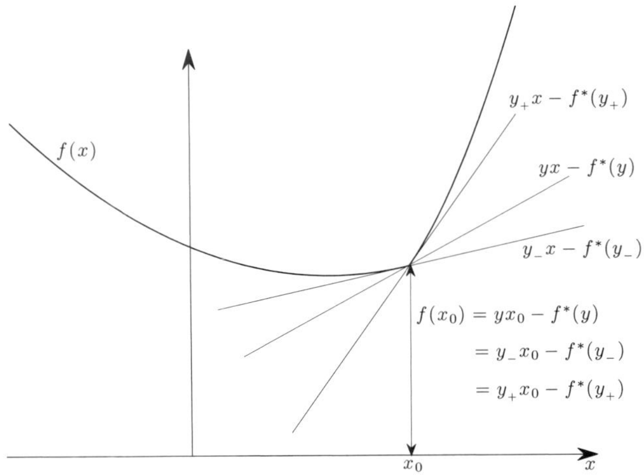

Figure 4.2. If $y_- = f'(x_0-) < f'(x_0+) = y_+$, then f has a corner at x_0, $\partial f(x_0) = [y_-, y_+]$, and $f^*(y)$ is linear over $y \in [y_-, y_+]$.

Hint: Think of directional derivatives. (You can find the proof in Lemma 12.11.)

Theorem 4.21. *Let $f : \mathcal{X} \to (-\infty, \infty]$ be not identically infinite. Then the following are equivalent.*

(a) $y \in \partial f(x)$.

(b) $f(x) + f^*(y) \le \langle x, y \rangle$.

(c) $f(x) + f^*(y) = \langle x, y \rangle$

If f is also lower semicontinuous and convex, then (a)–(c) *are equivalent to*

(d) $x \in \partial f^*(y)$.

Proof. The hypotheses on f guarantee that $f^* > -\infty$ so the left-hand sides of (b) and (c) are well-defined. (b) implies (c) by the Fenchel-Young inequality (4.1). By the definition of f^*, (c) implies (a) and (a) implies (b). Applying the equivalence of (a) and (c) to f^* one sees that (d) is equivalent to $f^*(y)+f^{**}(x) = \langle x, y \rangle$. But $f = f^{**}$ if f is a proper lower semicontinuous convex function. Hence (d) is equivalent to (c). □

Exercise 4.22. Let f be a function on an interval. Show that if $[a, b] \subset \partial f(x)$, then f^* is linear on $[a, b]$ with slope x.

***Exercise 4.23.** Let f be a finite convex function on (a, b). The following properties follow quickly from the definition.

(a) $\partial f(x)$ is a nonempty closed interval. $f'(x)$ exists if and only if $\partial f(x)$ is a singleton, and then $\partial f(x) = \{f'(x)\}$.

(b) Let $x \neq y$ and $\alpha \in \partial f(x)$. Then $f(y) = f(x) + \alpha(y - x)$ implies that $\alpha \in \partial f(y)$.

(c) If $x < y$, $\alpha \in \partial f(x)$, and $\beta \in \partial f(y)$, then

$$\alpha \leq \frac{f(y) - f(x)}{y - x} \leq \beta.$$

$\partial f(x)$ and $\partial f(y)$ have at most one point in common. If that happens, then this point is an endpoint for both sets and over $[x, y]$ the graph of f is a line segment.

(d) Suppose $x_j > x$, $\alpha_j \in \partial f(x_j)$, and $x_j \to x$. Then $\alpha_j \to \sup \partial f(x)$.

(e) If f is differentiable on (a, b), then f' is continuous on (a, b).

4.2. Rate function as a convex conjugate

We prove a general upper large deviation bound (2.3) for compact sets and investigate when the rate function of an LDP can be obtained by convex conjugation. The setting for this section is the same as before: $\mathcal{X}$ and $\mathcal{Y}$ are real vector spaces in duality and $\mathcal{X}$ is topologized by $\sigma(\mathcal{X}, \mathcal{Y})$. Let $\mathcal{E}$ be a closed convex subset of $\mathcal{X}$. We are given a sequence of probability measures $\{\mu_n\} \subset \mathcal{M}_1(\mathcal{E})$. Define

$$\bar{p}(y) = \overline{\lim_{n\to\infty}} \tfrac{1}{r_n} \log \int_{\mathcal{E}} e^{r_n \langle x, y\rangle} \mu_n(dx) \in [-\infty, \infty] \text{ for } y \in \mathcal{Y}$$

and

$$\bar{p}^*(x) = \sup_{y \in \mathcal{Y}} \{\langle x, y\rangle - \bar{p}(y)\} \in [0, \infty] \text{ for } x \in \mathcal{X}.$$

Theorem 4.24. *$\bar{p}$ is convex and $\bar{p}^* : \mathcal{X} \to [0, \infty]$ is lower semicontinuous and convex. For any compact subset $F \subset \mathcal{E}$,*

$$\overline{\lim_{n\to\infty}} \frac{1}{r_n} \log \mu_n(F) \leq -\inf_F \bar{p}^*.$$

Proof. $\bar{p}$ is convex by Hölder's inequality. As a convex conjugate $\bar{p}^*$ is lower semicontinuous and convex, and $\bar{p}(0) = 0$ implies $\bar{p}^* \geq 0$.

The upper bound proof is familiar by now. Let $\delta > 0$ and $c < \inf_F \bar{p}^*$. For each $x \in F$, there is $y \in \mathcal{Y}$ such that $\langle x, y\rangle - \bar{p}(y) > c$. Let $B_x = \{u \in \mathcal{X} : |\langle u, y\rangle - \langle x, y\rangle| < \delta\}$, an open neighborhood of x. Since F is compact, it can be covered with $B_{x_1}, \dots, B_{x_m}$, with corresponding $y_1, \dots, y_m$. Write

$$\mu_n(B_{x_i}) = \int_{B_{x_i}} e^{-r_n\langle u, y_i\rangle + r_n\langle u, y_i\rangle} \mu_n(du) \leq e^{r_n(-\langle x_i, y_i\rangle + \delta)} \int_{B_{x_i}} e^{r_n\langle u, y_i\rangle} \mu_n(du)$$

and thus

$$\overline{\lim_{n\to\infty}} \tfrac{1}{r_n} \log \mu_n(B_{x_i}) \leq -\langle x_i, y_i\rangle + \delta + \bar{p}(y_i) \leq -c + \delta.$$

Consequently,

$$\overline{\lim_{n\to\infty}} \frac{1}{r_n} \log \mu_n(F) \le \max_{1\le i\le m} \Big\{ \overline{\lim_{n\to\infty}} \frac{1}{r_n} \mu_n(B_{x_i}) \Big\} \le -c + \delta.$$

Take $\delta \searrow 0$ and $c \nearrow \inf_F \bar{p}^*$. □

If an LDP holds, Varadhan's theorem (page 38) gives a sufficient condition for the existence of the limit

$$p(y) = \lim_{n\to\infty} \frac{1}{r_n} \log \int e^{r_n \langle x,y\rangle} \mu_n(dx). \tag{4.2}$$

This function is sometimes called the *pressure*. By the previous theorem, p^* is a candidate for the rate function. When is the rate $I = p^*$? Under a bit of regularity, the rate $I = p^*$ basically always when I is convex (but see Exercise 4.29).

Theorem 4.25. *Assume* LDP(μ_n, r_n, I) *holds on the closed convex set* $\mathcal{E}$ *and assume* I *is convex. Extend* I *to* $\mathcal{X}$ *by setting* $I \equiv \infty$ *outside* $\mathcal{E}$.

(a) LDP(μ_n, r_n, I) *holds on* $\mathcal{X}$ *with lower semicontinuous convex rate function* I.

(b) *Assume additionally that*

$$\sup_n \Big(\int_{\mathcal{E}} e^{r_n\langle x,y\rangle} \mu_n(dx) \Big)^{1/r_n} < \infty \text{ for all } y \in \mathcal{Y}. \tag{4.3}$$

Then $\forall y \in \mathcal{Y}$ *limit* (4.2) *exists and*

$$p(y) = \sup_{x\in\mathcal{E}} \{\langle x, y\rangle - I(x)\}. \tag{4.4}$$

For the extended I, *we have* $p = I^*$ *and* $I = p^*$.

Proof. No new ideas are needed and we leave some details to the reader.

(a) Closedness of $\mathcal{E}$ renders I lower semicontinuous and convexity of $\mathcal{E}$ makes I convex.

(b) Assumption (4.3) applied to αy implies condition (3.3). Varadhan's theorem gives (4.4), which is the same as $p = I^*$. Double duality gives $I = I^{**} = p^*$. □

When $I = p^*$ the set of zeroes of the rate I is the subdifferential of the pressure at 0:

$$I(x) = 0 \iff p^*(x) = 0 \iff p^*(x) + p(0) = 0 = \langle x, 0\rangle \iff x \in \partial p(0).$$

Exercise 4.26. Theorem 4.25 gives a unique rate function. Show that $\sigma(\mathcal{X}, \mathcal{Y})$ is a regular topology: that is, given a closed set A and a point $x \notin A$, there are disjoint open sets U and V such that $x \in U$ and $A \subset V$. In other words, the uniqueness of I here does not give us anything beyond Theorem 2.13.

The Curie-Weiss model gave us examples of nonconvex rate functions (Exercise 3.17(c)). Here is a simple way to construct such rate functions.

Exercise 4.27. Let μ_n and ν_n be probability measures on $\mathcal{X}$ such that $\mathrm{LDP}(\mu_n, r_n, I)$ and $\mathrm{LDP}(\nu_n, r_n, J)$ hold. Prove that for any $\alpha \in (0,1)$, $\mathrm{LDP}(\alpha\mu_n + (1-\alpha)\nu_n, r_n, I \wedge J)$ holds. Observe that $I \wedge J$ may fail to be convex even if I and J are convex.

The next exercise connects superexponential convergence in the LDP with a linear pressure functional.

Exercise 4.28. Let $\{\eta_n\}$ be $\mathbb{R}^d$-valued random variables and let $m \in \mathbb{R}^d$. Assume the $\{\eta_n\}$ are uniformly bounded. Prove that the following are equivalent.

(a) $\forall \varepsilon > 0$, $\lim_{n\to\infty} r_n^{-1} \log P\{|\eta_n - m| \geq \varepsilon\} = -\infty$.

(b) $\forall t \in \mathbb{R}^d$, $p(t) = \lim_{n\to\infty} r_n^{-1} \log E[e^{r_n t \cdot \eta_n}] = t \cdot m$.

(c) $\mathrm{LDP}(P\{\eta_n \in \cdot\}, r_n, I)$ holds with $I(m) = 0$ and $I(x) = \infty$ for $x \neq m$.

Note that boundedness is only needed for (c)$\Rightarrow$(b).

Hint: To prove that (b) implies (a) find finitely many $t_1, \dots, t_k \in \mathbb{R}^d$ and $\delta > 0$ such that $\{x : |x - m| \geq \varepsilon\} \subset \bigcup_{i=1}^k \{x : x \cdot t_i \geq c_i\}$ and $m \cdot t_i + \delta \leq c_i$ for all i.

If (4.3) is violated, then even when all limits exist, Varadhan's theorem and the convex conjugate representation of the rate may fail. Consequently, the upper bound of Theorem 4.24 may not be the optimal one.

Exercise 4.29 ([**15**, page 140])**.** Let $0 < p_n < 1/2$, $b_n \in \mathbb{R}$, and $\mu_n \in \mathcal{M}_1(\mathbb{R})$ such that $\mu_n\{0\} = 1 - 2p_n$ and $\mu_n\{b_n\} = \mu_n\{-b_n\} = p_n$. Prove the following.

(a) If $n^{-1} \log p_n \to -\infty$, then $\mathrm{LDP}(\mu_n, n, I)$ holds with the lower semicontinuous convex rate I such that $I(0) = 0$ and $I(x) = \infty$ for $x \neq 0$.

(b) If furthermore $b_n = n^{-1} \log p_n$, then (4.3) does not hold and

$$p(t) = \lim_{n\to\infty} n^{-1} \log \int e^{ntx} \mu_n(dx)$$

exists for $t \in \mathbb{R}$, but $p^*(x) < I(x)$ for all $x \neq 0$, and consequently $p^* \neq I$ and $p \neq I^*$.

4.3. Multidimensional Cramér theorem

In this section we establish the final form of Cramér's theorem in $\mathbb{R}^d$. Convexity figures prominently in the proof. Let us recall the setting. X and $\{X_n\}$ are i.i.d. $\mathbb{R}^d$-valued random variables, $S_n = X_1 + \cdots + X_n$, and $\mu_n(B) = P\{S_n/n \in B\}$ is the distribution of the sample mean. The moment generating function is $M(\theta) = E[e^{\theta \cdot X}]$, and

$$p(\theta) = \log M(\theta)$$

is a $(-\infty, \infty]$-valued convex, lower semicontinuous function on $\mathbb{R}^d$. Lower semicontinuity comes from Fatou's lemma.

Cramér's theorem on $\mathbb{R}^d$. *Let $\{X_n\}$ be a sequence of i.i.d. $\mathbb{R}^d$-valued random variables and let μ_n be the distribution of the sample mean S_n/n. Then without further assumptions weak* $\mathrm{LDP}(\mu_n, n, I)$ *holds with $I = p^*$. If, moreover, $M(\theta) < \infty$ in a neighborhood of 0, then* $\mathrm{LDP}(\mu_n, n, I)$ *holds and I is a tight rate function.*

Proof. This proof gives us the opportunity to display another method for obtaining LDPs, namely subadditivity.

Step 1. We show that there exists a convex rate function I such that the weak $\mathrm{LDP}(\mu_n, n, I)$ holds.

This is where subadditivity comes in. We claim that for each open ball $B \subset \mathbb{R}^d$, the limit

$$j(B) = -\lim_{n\to\infty} n^{-1} \log P\{S_n/n \in B\} \in [0, \infty] \tag{4.5}$$

exists. Any convex open set would in fact work but we need only balls. Let

$$a_n = -\log P\{S_n/n \in B\}.$$

We establish two properties for this sequence, namely

$$\text{either} \quad a_n = \infty \ \forall n \quad \text{or} \quad \exists N < \infty \text{ such that } a_n < \infty \ \forall n \geq N \tag{4.6}$$

and subadditivity

$$a_{m+n} \leq a_m + a_n. \tag{4.7}$$

To check (4.6), suppose there exists k such that $P\{S_k/k \in B\} > 0$. This k is kept fixed while we verify (4.6). Since B is the union of countably many closed balls, we can find a closed ball $K \subset B$ such that $P\{S_k/k \in K\} > 0$. Fix $\varepsilon > 0$ smaller than the distance from K to B^c. Then $x \in K$ and $|y| < \varepsilon$ imply $x + y \in B$.

Abbreviate $S_{m,n} = X_{m+1} + X_{m+2} + \cdots + X_n$. Note that $S_{m,n}$ has the same distribution as S_{n-m}.

Decompose any n as $n = mk + \ell$ with $0 \le \ell < k$, and then write

$$S_{mk+\ell} = S_{0,k} + S_{k,2k} + \cdots + S_{(m-1)k,mk} + S_{mk,mk+\ell}.$$

The terms above are independent and the first m of them also identically distributed. Note the inequality

$$\left| \frac{S_n}{n} - \frac{S_{mk}}{mk} \right| = \left| \frac{\ell S_{mk}}{nmk} - \frac{S_{mk,mk+\ell}}{n} \right| \le \frac{|S_{mk}|}{mn} + \frac{|S_{mk,mk+\ell}|}{n}.$$

Since $S_1, \dots, S_{k-1}$ are finite random variables, there exists N_0 such that

$$\max_{0 \le \ell < k} P\{|S_\ell| \ge \tfrac{1}{2}\varepsilon n\} \le \tfrac{1}{2} \qquad \text{for } n \ge N_0.$$

Pick a radius r such that $K \subset \{x : |x| < r\}$. Then for $n > 2kr/\varepsilon$, $(mk)^{-1} S_{mk} \in K$ implies $(mk)^{-1} S_{mk} < n\varepsilon/(2k)$. Fix any $N > N_0 \vee (2kr/\varepsilon)$ and let $n \ge N$. Below use the choice of ε, independence, convexity of K, and the i.i.d. assumption:

$$\begin{aligned}
P\{S_n/n \in B\} &\ge P\left\{ \frac{S_{mk}}{mk} \in K, \left| \frac{S_n}{n} - \frac{S_{mk}}{mk} \right| < \varepsilon \right\} \\
&\ge P\left\{ \frac{S_{mk}}{mk} \in K, \frac{|S_{mk}|}{mn} < \frac{\varepsilon}{2}, \frac{|S_{mk,mk+\ell}|}{n} < \frac{\varepsilon}{2} \right\} \\
&= P\left\{ \frac{S_{mk}}{mk} \in K, \frac{|S_{mk}|}{mk} < \frac{n\varepsilon}{2k} \right\} P\left\{ \frac{|S_\ell|}{n} < \frac{\varepsilon}{2} \right\} \\
&\ge \tfrac{1}{2} \cdot P\left\{ \frac{S_{mk}}{mk} \in K \right\} \\
&\ge \tfrac{1}{2} \cdot P\left\{ \frac{S_{0,k}}{k} \in K, \frac{S_{k,2k}}{k} \in K, \dots, \frac{S_{(m-1)k,mk}}{k} \in K \right\} \\
&= \tfrac{1}{2} \cdot \big(P\{k^{-1} S_k \in K\}\big)^m > 0.
\end{aligned}$$

Property (4.6) has been verified.

Property (4.7) follows also from convexity of B and the i.i.d. assumption:

$$\begin{aligned}
a_{m+n} &= -\log P\{(m+n)^{-1} S_{m+n} \in B\} \\
&\le -\log P\{m^{-1} S_m \in B\,,\; n^{-1} S_{m,m+n} \in B\} \\
&= -\log P\{m^{-1} S_m \in B\} - \log P\{n^{-1} S_n \in B\} \\
&= a_m + a_n.
\end{aligned}$$

Note that the inequalities above work even if some probabilities vanish.

With these properties checked we can apply the next classic fact. In Section 6.2 we generalize it to a multidimensional index set.

Fekete's lemma. *Let* $(a_n)_{n\geq 1}$ *be a sequence in* $(-\infty, \infty]$ *with properties* (4.6) *and* (4.7). *Then*

$$\lim_{n\to\infty} \frac{a_n}{n} = \inf_n \frac{a_n}{n} \quad \textit{with values in } [-\infty, \infty].$$

Proof. For the identically infinite case the claimed property is trivially true. So only the case where the sequence is eventually finite needs proof. The inequality

$$\varliminf_{n\to\infty} \frac{a_n}{n} \geq \inf_n \frac{a_n}{n}$$

needs no proof.

Fix any $k \in \mathbb{N}$ such that $a_k < \infty$. Fix m_0 such that $m_0 k \geq N$ for N from assumption (4.6). Consider $n > m_0 k$ and write $n = mk + m_0 k + \ell$ for some $0 \leq \ell < k$. Subadditivity gives

$$a_n \leq m a_k + a_{m_0 k+\ell} \leq m a_k + \max_{j: m_0 k \leq j < (m_0+1)k} a_j.$$

The last term above is a finite, fixed constant independent of n. Divide by n and let $n \to \infty$ to get

$$\varlimsup_{n\to\infty} \frac{a_n}{n} \leq \frac{a_k}{k}.$$

An infimum over k on the right finishes the proof of the lemma. □

The limits $j(B)$ in (4.5) have been shown to exist. Define

$$I(x) = \sup_{B: B\ni x} j(B) \tag{4.8}$$

where the supremum is taken over open balls that contain x. The weak LDP(μ_n, n, I) now follows from Exercise 2.24. In particular, I is lower semicontinuous.

To complete Step 1 we check convexity of I. By Exercise 4.7 it suffices to consider the case $x = (y+z)/2$. Given a ball B that contains x we can find small balls $B_1 \ni y$ and $B_2 \ni z$ such that $\frac{1}{2}B_1 + \frac{1}{2}B_2 \subset B$. Utilizing convexity and independence once again

$$\begin{aligned} P\{(2n)^{-1}S_{2n} \in B\} &\geq P\{n^{-1}S_n \in B_1\,,\ n^{-1}S_{n,2n} \in B_2\} \\ &= P\{n^{-1}S_n \in B_1\}\, P\{n^{-1}S_n \in B_2\} \end{aligned}$$

from which

$$\begin{aligned} &-\frac{1}{2n}\log P\{(2n)^{-1}S_{2n} \in B\} \\ &\quad\leq -\frac{1}{2n}\log P\{n^{-1}S_n \in B_1\} - \frac{1}{2n}\log P\{n^{-1}S_n \in B_2\}. \end{aligned}$$

Taking $n \to \infty$ leads to

$$j(B) \leq \tfrac{1}{2}j(B_1) + \tfrac{1}{2}j(B_2) \leq \tfrac{1}{2}I(y) + \tfrac{1}{2}I(z).$$

A supremum over $B \ni x$ gives the inequality $I(x) \le (I(y) + I(z))/2$. This completes the verification of Step 1.

Step 2. We show that the rate function I from Step 1 equals p^*. We cannot use Theorem 4.25 to identify the rate because assumption (4.3) can fail without further hypotheses on M (see the end of the present proof).

From Lemma 3.7 (lower bound for Varadhan's theorem) we get one inequality for free:

$$p(\theta) \ge I^*(\theta). \tag{4.9}$$

To prove the other inequality $p(\theta) \le I^*(\theta)$ start by observing that for an open ball B and any finite n,

$$n^{-1} \log P\{S_n/n \in B\} \le - \inf_{x \in \overline{B}} I(x). \tag{4.10}$$

This is a consequence of subadditivity and the upper large deviation bound.

Let us first restrict the calculation of $p(\theta)$ to a δ-ball around a point $z \in \mathbb{R}^d$ and develop an estimate for it:

$$\begin{aligned} &n^{-1} \log E\big[e^{\theta \cdot S_n},\, n^{-1} S_n \in B(z,\delta)\big] \\ &\quad \le \theta \cdot z + |\theta|\,\delta + n^{-1} \log P\{n^{-1} S_n \in B(z,\delta)\} \\ &\quad \le \theta \cdot z + |\theta|\,\delta - \inf_{x \in \overline{B(z,\delta)}} I(x) \;\le\; \sup_{x \in \overline{B(z,\delta)}} \{\theta \cdot x - I(x)\} + 2\,|\theta|\,\delta \\ &\quad \le I^*(\theta) + 2\,|\theta|\,\delta. \end{aligned}$$

From small balls we go to a large ball of radius R centered at the origin. Cover the R-ball with δ-balls $B(z_i, \delta)$, $1 \le i \le m$. Then

$$\begin{aligned} &n^{-1} \log E\big[e^{\theta \cdot S_n},\, \big|n^{-1} S_n\big| \le R\big] \\ &\quad \le n^{-1} \log m + \max_{1 \le i \le m} n^{-1} \log E\big[e^{\theta \cdot S_n},\, n^{-1} S_n \in B(z_i,\delta)\big] \\ &\quad \le n^{-1} \log m + I^*(\theta) + 2\,|\theta|\,\delta. \end{aligned}$$

Define

$$\widetilde{p}(\theta, R) = \varlimsup_{n \to \infty} n^{-1} \log E\big[e^{\theta \cdot S_n},\, \big|n^{-1} S_n\big| \le R\big].$$

From above we have the bound

$$\widetilde{p}(\theta, R) \le I^*(\theta) + 2\,|\theta|\,\delta.$$

It remains to argue that $\widetilde{p}(\theta, R)$ is a good approximation to the true $p(\theta)$. Independence makes this easy:

$$\begin{aligned} &n^{-1}\log E\big[e^{\theta\cdot S_n},\,\big|n^{-1}S_n\big|\le R\big] \\ &\ge n^{-1}\log E\big[e^{\theta\cdot S_n},\,|X_i|\le R\;\forall i\in\{1,\dots,n\}\big] \\ &= n^{-1}\log E\Big[\prod_{i=1}^{n}\big(e^{\theta\cdot X_i}\mathbb{1}\{|X_i|\le R\}\big)\Big] = \log E\big[e^{\theta\cdot X},\,|X|\le R\big]. \end{aligned}$$

Combining the last two displays gives

$$\log E\big[e^{\theta\cdot X},\,|X|\le R\big]\le I^*(\theta)+2\,|\theta|\,\delta.$$

Take $\delta \searrow 0$ and $R \nearrow \infty$ to get $p(\theta) \le I^*(\theta)$. Together with (4.9) we have $p = I^*$. Since I is convex, lower semicontinuous, and not identically infinite (the upper bound guarantees that a rate function has $\inf I = 0$), $I = I^{**} = p^*$. This completes the proof of Step 2.

Step 3. The step from weak LDP to full LDP comes from exponential tightness. Under the assumption that M is finite in a neighborhood of the origin, exponential tightness was shown in the partial proof of Cramér's theorem given on page 31 of Section 2.4. By Theorem 2.19 we have the full LDP and the tight rate function. Cramér's theorem is proved.

Step 2 is not needed if $M(\theta) < \infty$ for all $\theta \in \mathbb{R}^d$. In this case $I = p^*$ follows from Theorem 4.25. We leave the details as an exercise. □

Exercise 4.30. Let G be a convex open subset of $\mathbb{R}^d$. Show that under the assumptions of Cramér's theorem,

$$P\{S_n/n \in G\} \le e^{-n\cdot\inf_{x\in\overline{G}} I(x)} \qquad \text{for all } n \ge 1.$$

Exercise 4.31. A process $\{X_n\}$ is *exchangeable* if its distribution is invariant under finite permutations of the random variables. According to de Finetti's theorem (page 287), an infinite exchangeable sequence is a mixture of i.i.d. sequences. Use this, Cramér's theorem, and either Bryc's theorem or Exercise 4.27 to derive an LDP for the sample mean S_n/n of a real-valued exchangeable process under some suitable hypotheses.

Chapter 5

Relative entropy and large deviations for empirical measures

This chapter introduces relative entropy and shows that it is the rate function in one of the basic large deviation principles, namely Sanov's theorem. The results here are fundamental to several later parts of the book. Sanov's theorem is extended to Markov chains in Chapter 13 and to a nonstationary process in Chapter 15. This chapter concludes with some exercises of a statistical mechanical flavor that reveal how entropy minimization leads naturally to Gibbs measures.

5.1. Relative entropy

Let $\mathcal{M}(\mathcal{S})$ be the space of finite signed measures on a measurable space $(\mathcal{S}, \mathscr{B})$ and let $\mathcal{M}_1(\mathcal{S})$ be the subspace of probability measures. Let $b\mathscr{B}$ be the space of bounded measurable functions $g : \mathcal{S} \to \mathbb{R}$. As discussed in Example 4.3, there is a natural duality between $\mathcal{M}(\mathcal{S})$ and $b\mathscr{B}$:

$$\langle \nu, g \rangle = \int g \, d\nu \text{ for } g \in b\mathscr{B} \text{ and } \nu \in \mathcal{M}(\mathcal{S}).$$

The resulting weak topologies separate points.

***Exercise 5.1.** Show that for each $g \in b\mathscr{B}$ such that $g \neq 0$ there exists a measure $\nu \in \mathcal{M}_1(\mathcal{S})$ such that $\langle \nu, g \rangle \neq 0$. Then show that for each $\nu \in \mathcal{M}(\mathcal{S})$ such that $\nu \neq 0$ (ν is not the zero measure) there exists a $g \in b\mathscr{B}$ such that $\langle \nu, g \rangle \neq 0$. Assume that $\mathcal{S}$ is a metric space and $\mathscr{B} = \mathscr{B}_{\mathcal{S}}$ its Borel σ-algebra.

Recall that $\mathscr{C}_b(\mathcal{S})$ is the space of bounded continuous functions $g : \mathcal{S} \to \mathbb{R}$. Show that for each $\nu \in \mathcal{M}(\mathcal{S})$ such that $\nu \neq 0$ there exists $g \in \mathscr{C}_b(\mathcal{S})$ such that $\langle \nu, g \rangle \neq 0$.

Definition 5.2. For $\nu, \lambda \in \mathcal{M}_1(\mathcal{S})$, the *entropy of ν relative to λ* is defined by

$$H(\nu \,|\, \lambda) = \begin{cases} \int \phi \log \phi \, d\lambda & \text{if } \nu \ll \lambda \text{ and } \phi = \frac{d\nu}{d\lambda}, \\ \infty & \text{otherwise.} \end{cases}$$

Note that $x \log x \geq -1/e$, so the integral above is well-defined.

Lemma 5.3. *$H(\nu \,|\, \lambda) \geq 0$ and $H(\nu \,|\, \lambda) = 0$ if and only if $\nu = \lambda$.*

Proof. Jensen's inequality applied to the convex function $x \log x$ implies that H is nonnegative. The strict convexity of $x \log x$ implies that equality holds if and only if ϕ is constant λ-a.s. □

Let us fix the reference probability measure $\lambda \in \mathcal{M}_1(\mathcal{S})$ and then write $H(\nu) = H(\nu \,|\, \lambda)$ when it suits us. Let $p : b\mathscr{B} \to \mathbb{R}$ be defined by

$$p(g) = \log \int e^g \, d\lambda.$$

Extend H to all of $\mathcal{M}(\mathcal{S})$ by setting $H(\nu) = \infty$ for $\nu \in \mathcal{M}(\mathcal{S}) \smallsetminus \mathcal{M}_1(\mathcal{S})$.

Theorem 5.4. *p and H are convex conjugates of one another. In particular, H is convex and has the variational representation*

$$H(\nu \,|\, \lambda) = \sup_{g \in b\mathscr{B}} \{ E^\nu[g] - \log E^\lambda[e^g] \}. \tag{5.1}$$

Proof. Step 1. We prove $p(g) = H^*(g)$. First put $d\nu = \frac{e^g}{E^\lambda[e^g]} d\lambda$, and derive

$$\begin{aligned} H^*(g) \geq \langle \nu, g \rangle - H(\nu \,|\, \lambda) &= \frac{E^\lambda[g e^g]}{E^\lambda[e^g]} - E^\lambda\Big[\frac{e^g}{E^\lambda[e^g]} (g - \log E^\lambda[e^g]) \Big] \\ &= \log E^\lambda[e^g] = p(g). \end{aligned}$$

On the other hand, let $\nu \in \mathcal{M}_1(\mathcal{S})$ satisfy $H(\nu \,|\, \lambda) < \infty$. Then $\nu \ll \lambda$ and the Radon-Nikodym derivative $\phi = \frac{d\nu}{d\lambda}$ satisfies $E^\nu[|\log \phi|] < \infty$. By an application of Jensen's inequality to log,

$$\begin{aligned} \langle \nu, g \rangle - H(\nu \,|\, \lambda) = E^\nu[g - \log \phi] &= E^\nu[\log \tfrac{e^g}{\phi}] \leq \log E^\nu[\tfrac{e^g}{\phi}] \\ &= \log E^\lambda[e^g \mathbb{1}\{\phi > 0\}] \leq \log E^\lambda[e^g] = p(g). \end{aligned}$$

This shows $H^*(g) \leq p(g)$.

Step 2. We show that $H \leq p^*$. First there are the simple cases of $H(\nu) = \infty$ to deal with.

Let $\nu \in \mathcal{M}(\mathcal{S})$. Suppose there exists a nonnegative $g \in b\mathscr{B}$ such that $E^\nu[g] < 0$. Then for $0 < c < \infty$,

$$p^*(\nu) \geq E^\nu[-cg] - \log E^\lambda[e^{-cg}] \geq -cE^\nu[g].$$

Taking $c \nearrow \infty$ shows that $p^*(\nu) = \infty$. In the rest, we assume ν is a nonnegative measure.

Suppose that $\nu(\mathcal{S}) > 1$. Then

$$p^*(\nu) \geq E^\nu[c] - \log E^\lambda[e^c] = c(\nu(\mathcal{S}) - 1)$$

and once again taking c to infinity shows that $p^*(\nu) = \infty$. Similarly, if $\nu(\mathcal{S}) < 1$, then $p^*(\nu) = \infty$. Thus we can assume $\nu \in \mathcal{M}_1(\mathcal{S})$.

Suppose there exists a measurable set A such that $\lambda(A) = 0 < \nu(A)$. Then $\forall\, 0 < c < \infty$

$$p^*(\nu) \geq E^\nu[c\mathbb{1}_A] - \log E^\lambda[e^{c\mathbb{1}_A}] = c\nu(A)$$

and again $p^*(\nu) = \infty$. We thus assume $\nu \ll \lambda$.

Suppose $\phi = \frac{d\nu}{d\lambda}$ and let $\phi_c^t = a \vee (\phi \wedge b)$, for $0 < a < 1 < b < \infty$. Then

$$p^*(\nu) \geq E^\nu[\log \phi_a^b] - \log E^\lambda[\phi_a^b] \geq E^\nu[\log \phi_a^b] - \log E^\lambda[\phi \vee a].$$

Since $\log \phi_a^b \geq \log a$, monotone convergence implies that $E^\nu[\log \phi_a^b]$ converges, as $b \to \infty$, to $E^\nu[\log(\phi \vee a)] \geq E^\nu[\log \phi] = H(\nu \,|\, \lambda)$. Thus,

$$p^*(\nu) \geq H(\nu \,|\, \lambda) - \log E^\lambda[\phi \vee a].$$

Since $0 \leq \phi \vee a \leq \phi + 1$, dominated convergence implies that $\log E^\lambda[\phi \vee a]$ converges to $\log E^\lambda[\phi] = 0$, as $a \to 0$. This completes the proof of $p^* \geq H$.

Step 3. $H \leq p^* = H^{**} \leq H$, where the last inequality follows from part (b) of Proposition 4.10. □

It is very valuable for the existence of unique minimizers that relative entropy turns out to be strictly convex.

***Exercise 5.5.** Show that if $\mu, \nu \in \mathcal{M}_1(\mathcal{S})$ and $\theta \in (0,1)$, then

$$H(\theta\mu + (1-\theta)\nu \,|\, \lambda) = \theta H(\mu \,|\, \lambda) + (1-\theta) H(\nu \,|\, \lambda) < \infty$$

is equivalent to $\mu = \nu$.

Hint: Use strict convexity of $x \log x$.

When $\mathcal{S}$ is a metric space, $\mathscr{C}_b(\mathcal{S})$ is also in duality with $\mathcal{M}(\mathcal{S})$ and, by Exercise 5.1, the topologies separate points. Restricted to $\mathcal{M}_1(\mathcal{S})$ the topology $\sigma(\mathcal{M}(\mathcal{S}), \mathscr{C}_b(\mathcal{S}))$ is the standard weak topology of probability measures and is very convenient to work with. (See Section B.4.) The duality of p and H works also with $\mathscr{C}_b(\mathcal{S})$.

Theorem 5.6. *Let $\mathcal{S}$ be a metric space. Then*

$$H(\nu\,|\,\lambda) = \sup_{f\in\mathscr{C}_b(\mathcal{S})}\{E^\nu[f] - \log E^\lambda[e^f]\}. \tag{5.2}$$

In particular, H and p are convex conjugates in the duality of the spaces $\mathcal{M}(\mathcal{S})$ and $\mathscr{C}_b(\mathcal{S})$. On the space $\mathcal{M}_1(\mathcal{S})$ of probability measures, H is lower semicontinuous in the weak topology generated by $\mathscr{C}_b(\mathcal{S})$.

The proof requires the following technical lemma which we prove in Appendix A.

Lemma A.11. *Let $\mathcal{S}$ be a metric space and let $\mathcal{H}$ be a class of bounded functions that contains the space $\mathscr{U}_b(\mathcal{S})$ of bounded uniformly continuous functions and is closed under uniformly bounded pointwise convergence (that is, $f_n \in \mathcal{H}$ for all n, $\max_n \sup_x |f_n(x)| < \infty$, and $f_n(x) \to f(x)$ for all $x \in \mathcal{S}$ together imply $f \in \mathcal{H}$). Then $b\mathscr{B}_{\mathcal{S}} \subset \mathcal{H}$.*

Proof of Theorem 5.6. Let

$$\alpha = \sup_{f\in\mathscr{C}_b(\mathcal{S})}\{E^\nu[f] - \log E^\lambda[e^f]\}$$

and let $\mathcal{H} = \{f \in b\mathscr{B}_{\mathcal{S}} : E^\nu[f] - p(f) \le \alpha\}$. $\mathcal{H}$ is closed under uniformly bounded pointwise convergence and contains $\mathscr{C}_b(\mathcal{S})$, and hence $b\mathscr{B}_{\mathcal{S}} \subset \mathcal{H}$. Theorem 5.4 implies $H(\nu\,|\,\lambda) = \alpha$. We conclude that $H = p^*$ in the $\sigma(\mathcal{M}(\mathcal{S}), \mathscr{C}_b(\mathcal{S}))$-topology. That $p(f) = H^*(f)$ for $f \in \mathscr{C}_b(\mathcal{S})$ is a special case of Theorem 5.4. □

At this point we know that H qualifies as a convex rate function on $\mathcal{M}_1(\mathcal{S})$ when $\mathcal{S}$ is a metric space. By upgrading $\mathcal{S}$ to a complete, separable metric space we establish that H is even a tight rate function. Separable means that the space has a countable dense set. Complete, separable metric spaces are also called *Polish* spaces.

Proposition 5.7. *Assume $\mathcal{S}$ is Polish. Then for $c \in \mathbb{R}$, the sublevel set*

$$A = \{\nu \in \mathcal{M}_1(\mathcal{S}) : H(\nu\,|\,\lambda) \le c\}$$

is compact in the weak topology generated by $\mathscr{C}_b(\mathcal{S})$.

The proof makes use of uniform integrability of a set of functions and tightness of a set of probability measures.

Definition 5.8. A collection $\mathcal{D}$ of measurable functions is *uniformly integrable* if

$$\lim_{M\to\infty}\sup_{\varphi\in\mathcal{D}} E[\,|\varphi|\,\mathbb{1}\{|\varphi| \ge M\}] = 0.$$

***Exercise 5.9.** Assume there exists a nonnegative function $G : [0,\infty) \to [0,\infty)$ such that $G(x)/x \to \infty$ as $x \to \infty$ and $\sup_{\varphi\in\mathcal{D}} E[G(|\varphi|)] < \infty$. Show that $\mathcal{D}$ is uniformly integrable.

***Exercise 5.10.** Prove that if $\mathcal{D}$ is uniformly integrable, then for all $\varepsilon > 0$ there exists a $\delta > 0$ such that $P(B) < \delta$ implies $\sup_{\varphi\in\mathcal{D}} E[|\varphi|\mathbb{1}_B] < \varepsilon$.

Proof of Proposition 5.7. A is closed by the lower semicontinuity of H. By Prohorov's theorem (page 282), it suffices to show that A is tight. Let $\varepsilon > 0$.

Take $G(x) = x\log x$. By Exercise 5.9, $\mathcal{D} = \{\frac{d\nu}{d\lambda} : \nu \in A\}$ is uniformly integrable. By Exercise 5.10, $\exists\delta > 0$ such that $\lambda(B) < \delta$ implies $\sup_{\nu\in A}\nu(B) < \varepsilon$. By the regularity of probability measures on Polish spaces (Ulam's theorem, page 282), there exists a compact set $K \subset \mathcal{S}$ such that $\lambda(K^c) < \delta$. Then $\sup_{\nu\in A}\nu(K^c) < \varepsilon$, which says A is tight. □

We conclude this section with some exercises. But first we revisit Bernoulli variables. Example 3.4 showed that if the $\{X_k\}$ are i.i.d. Bernoulli random variables with success probability p, then the distributions of the empirical measures

$$L_n = \frac{1}{n}\sum_{k=1}^{n}\delta_{X_k}$$

satisfy a large deviation principle with rate function $H : \mathcal{M}_1(\{0,1\}) \to [0,\infty]$ given by

$$H(\alpha) = I_p(s) \text{ for } \alpha = s\delta_1 + (1-s)\delta_0 \text{ with } s \in [0,1], \tag{5.3}$$

where $I_p(s) = s\log\frac{s}{p} + (1-s)\log\frac{1-s}{1-p}$. In Sections 1.1 and 1.2 we saw that I_p is related to both thermodynamic and information-theoretic entropy. Now we can identify I_p as a relative entropy. Indeed, entropy of a measure $\alpha \in \mathcal{M}_1(\{0,1\})$ relative to the Bernoulli distribution with success probability p is exactly (5.3).

Exercise 5.11. Let ρ_s denote the rate s exponential distribution with density $f(x) = se^{-sx}$ on $\mathbb{R}_+$. Compute the entropy $H(\rho_s \,|\, \rho_t)$ and compare your answer with the Cramér rate function computed in Exercise 2.34(a). Show that $\nu = \rho_s$ uniquely minimizes $H(\nu \,|\, \rho_t)$ over probability measures ν on $\mathbb{R}_+$ with expectation $\int x\,\nu(dx) = s^{-1}$.

The general fact that relative entropy contracts to the Cramér rate function is explored in Exercise 5.18.

Exercise 5.12. For $0 \le f \in L^1(\mathbb{R})$ such that $\int f(x)\,dx = 1$, let

$$H_{Sh}(f) = -\int f \log f \, dx.$$

Sh is for Shannon.

(a) Show that the Gaussian $\varphi(x) = (2\pi\sigma^2)^{-1/2} e^{-x^2/2\sigma^2}$ is the unique maximizer of $H_{Sh}(f)$ among f that satisfy $\int x^2 f(x)\,dx = \sigma^2$.

(b) Show that the exponential $g(x) = \lambda e^{-\lambda x}$ is the unique maximizer among f supported on $[0, \infty)$ and satisfying $\int x f(x)\,dx = \lambda^{-1}$.

Hint: Compute entropies $H(f(x)dx \,|\, \varphi(x)dx)$, $H(f(x)dx \,|\, g(x)dx)$, $H_{Sh}(\varphi)$, and $H_{Sh}(g)$. The first two quantities are nonnegative.

***Exercise 5.13** (Conditional entropy formula)**.** Let μ and λ be probability measures on the measurable space $(\mathcal{S}, \mathscr{B})$. Let $\mathscr{D}$ be a sub-σ-algebra of $\mathscr{B}$, and let $\mu_{\mathscr{D}}$ and $\lambda_{\mathscr{D}}$ be the restrictions of μ and λ to $\mathscr{D}$. Suppose there exist conditional probability measures $\mu^x(\cdot) = \mu(\cdot\,|\mathscr{D})(x)$ and $\lambda^x(\cdot) = \lambda(\cdot\,|\mathscr{D})(x)$ of μ and λ, given $\mathscr{D}$. (See Theorem B.5 and the paragraph following it.) Prove that

$$H(\mu \,|\, \lambda) = H(\mu_{\mathscr{D}} \,|\, \lambda_{\mathscr{D}}) + \int_{\mathcal{S}} H(\mu^x \,|\, \lambda^x)\, \mu(dx).$$

Hint: Show that, if $\mu \ll \lambda$, then $\frac{d\mu_{\mathscr{D}}}{d\lambda_{\mathscr{D}}}(x) = E^\lambda[\frac{d\mu}{d\lambda}(x)|\mathscr{D}](x)$ and, for μ-almost every x, $\frac{d\mu^x}{d\lambda^x}(y) = \frac{d\mu}{d\lambda}(y)\big(\frac{d\mu_{\mathscr{D}}}{d\lambda_{\mathscr{D}}}(y)\big)^{-1}\mathbb{1}\{\frac{d\mu_{\mathscr{D}}}{d\lambda_{\mathscr{D}}}(y) > 0\}$. Or as a warmup, take $\mathcal{S}$ finite and $\mathscr{D}$ generated by a partition $\{A_1, \dots, A_m\}$ of $\mathcal{S}$. Then $\mu^x(z) = \mu(z \,|\, A_i)$ for $x \in A_i$, where $\mu(z \,|\, A_i) = \mu(\{z\} \cap A_i)/\mu(A_i)$ is the elementary conditional probability.

***Exercise 5.14.** Suppose $\mu(A) = 1$ and $\lambda(A) > 0$. Let $\pi = \lambda(\,\cdot \,|\, A)$. Show that

$$H(\mu \,|\, \lambda) = H(\mu \,|\, \pi) - \log \lambda(A).$$

Exercise 5.15. Let $\mathcal{S}$ be a finite space. Show that there is a maximizing function g in (5.1) if and only if ν and λ have the same support.

Exercise 5.16 (Convergence theorem for finite Markov chains with an entropy argument)**.** Let P be an irreducible aperiodic Markov transition matrix on a finite state space $\mathcal{S}$ with unique invariant distribution π. Let μ be an arbitrary initial distribution. Show that $\mu P^n \to \pi$ by completing the steps below. You need the existence of an integer $m \in \mathbb{N}$ such that $P^m > 0$

elementwise. Make this point part of the exercise or look it up and take it for granted (for example, Lemma 2.13.3 in [**67**] or Lemma 6.6.3 in [**27**]).

(a) $H(\cdot \,|\, \pi)$ is a continuous function on $\mathcal{M}_1(\mathcal{S})$.

(b) For all $\nu \in \mathcal{M}_1(\mathcal{S})$, $H(\nu P \,|\, \pi) \le H(\nu \,|\, \pi)$, and if $\nu \neq \pi$, then $H(\nu P^m \,|\, \pi) < H(\nu \,|\, \pi)$.
Hint: Use the variational formula (5.1) for entropy and Jensen's inequality. Since $\nu P^m > 0$ the variational formula for $H(\nu P^m \,|\, \pi)$ has maximizer $g(x) = \log \nu P^m(x)/\pi(x)$.

(c) Every limit point of μP^n must equal π.

5.2. Sanov's theorem

Let $\mathcal{S}$ be a Polish space and let $\lambda \in \mathcal{M}_1(\mathcal{S})$ be a probability measure on $\mathcal{S}$. Let $\{X_n\}$ be a sequence of i.i.d. $\mathcal{S}$-valued random variables with common distribution λ. The *empirical measure* L_n is a random probability measure on $\mathcal{S}$ defined by

$$L_n = \frac{1}{n} \sum_{k=1}^{n} \delta_{X_k}.$$

In other words, L_n is a random element of the space $\mathcal{M}_1(\mathcal{S})$, itself a Polish space in the weak topology generated by $\mathscr{C}_b(\mathcal{S})$. For the expectation of a function f under the empirical measure we use the equivalent notation $L_n(f) = E^{L_n}[f] = n^{-1} \sum_{k=1}^{n} f(X_k)$. The distribution of L_n is ρ_n, defined by $\rho_n(B) = P\{L_n \in B\}$ for Borel subsets $B \subset \mathcal{M}_1(\mathcal{S})$. See Exercise 5.17 for the measurability issue here.

Our aim is to prove an LDP for the distributions $\{\rho_n\}$. But prior to an LDP, we should develop some sense of a law of large numbers. For any particular $f \in \mathscr{C}_b(\mathcal{S})$, by the strong law of large numbers,

$$L_n(f) \to E^{\lambda}[f] \quad P\text{-a.s. as } n \to \infty. \tag{5.4}$$

Since $\mathcal{S}$ is separable, there is a countable set of functions $\{f_j\} \subset \mathscr{C}_b(\mathcal{S})$ that determines weak convergence (see page 280). Limit (5.4) holds with probability one, simultaneously for all f_j, and thereby we can conclude that

$$L_n \to \lambda \quad P\text{-a.s. as } n \to \infty. \tag{5.5}$$

We are ready to state the large deviation principle.

Sanov's theorem. *Let $\mathcal{S}$ be a Polish space and ρ_n the distribution of the empirical measure L_n. Then* LDP(ρ_n, n, H) *holds on the space $\mathcal{M}_1(\mathcal{S})$ with the tight convex rate function $H(\nu) = H(\nu \,|\, \lambda)$.*

To give the theorem some concrete feeling, let $\mathcal{S} = \mathbb{R}$ and suppose λ has a probability density function f, which means that $\lambda(B) = \int_B f(x)dx$ for

$B \in \mathscr{B}_{\mathbb{R}}$. A normalized histogram of the data points $(X_k)_{k=1}^n$ is supposed to resemble f. Sanov's theorem tells us that, given the null hypothesis that our data is i.i.d. with distribution λ, the probability that the histogram looks like the probability density of a different measure ν decays like e^{-cn}, where c is the entropy of ν relative to λ. We saw this for a Bernoulli measure λ in Example 3.4.

Proof. Since our only interest lies in statements concerning the distributions of the random variables $\{X_k\}$, we are free to use any particular probability space $(\Omega, \mathscr{F}, P)$ we please, as long as we are faithful to the assumption that the $\{X_k\}$ are i.i.d. λ-distributed. Consequently we choose to work in the *canonical setting*. We take $\Omega = \mathcal{S}^{\mathbb{N}} = \{\omega = (x_k)_{k\in\mathbb{N}} : x_k \in \mathcal{S}\ \forall k\}$, the space of $\mathcal{S}$-valued sequences. The $X_k(\omega) = x_k$ are the coordinate random variables. $\mathscr{F}$ is the product σ-algebra. $P = \lambda^{\otimes\mathbb{N}}$ is the product measure on Ω that, by definition, makes the coordinates $\{X_k\}$ i.i.d. λ-distributed. If the variables $\{X_k\}$ come to us defined on some other probability space $(\widetilde{\Omega}, \widetilde{\mathscr{F}}, \widetilde{P})$, we simply define $P(B) = \widetilde{P}\{(X_k)_{k\in\mathbb{N}} \in B\}$ for $B \in \mathscr{F}$, so that P is the distribution of the Ω-valued random sequence $(X_k)_{k\in\mathbb{N}}$.

Throughout the proof we have to consider expectations under more than one measure. We write $E^P[f] = \int f\,dP$ for expectation under P.

Step 1. Upper bound for compact sets. We follow the strategy given in Theorem 4.24. For $f \in \mathscr{C}_b(\mathcal{S})$,

$$\begin{aligned}\bar{p}(f) &= \varlimsup_{n\to\infty} \tfrac{1}{n} \log \int e^{n\langle \nu, f\rangle} \rho_n(d\nu) = \varlimsup_{n\to\infty} \tfrac{1}{n} \log E^P\Big[e^{\sum_{k=1}^n f(X_k)}\Big] \\ &= \lim_{n\to\infty} \log \int e^f\,d\lambda = p(f).\end{aligned}$$

Since $p^* = H$, the general upper bound in Theorem 4.24 implies the upper bound for compact sets.

Step 2. Exponential tightness. By Theorem 2.19, to get an upper bound for general closed sets it is enough to establish the exponential tightness of $\{\rho_n\}$. Use Ulam's theorem (page 282) to pick a compact set Γ_ℓ such that $\lambda(\Gamma_\ell^c) < e^{-2\ell^2}$. The set $A_\ell = \{\nu : \nu(\Gamma_\ell) \ge 1 - 1/\ell\}$ is closed in the weak topology. To justify this observe first that, by part (c) of the portmanteau theorem (page 279),

$$\nu(\Gamma_\ell) \ge \varlimsup_{j\to\infty} \nu_j(\Gamma_\ell) \quad \text{whenever } \nu_j \to \nu,$$

and then note that for a Polish space $\mathcal{S}$ the weak topology of $\mathcal{M}_1(\mathcal{S})$ is metrizable so closedness is determined by sequences. The set $K_L = \bigcap_{\ell\ge L} A_\ell$ is then also closed. Since K_L is also tight it is compact by Prohorov's theorem (page 282). Now we estimate.

$$\begin{aligned}
\rho_n(A_\ell^c) &= P\{L_n(\Gamma_\ell^c) > 1/\ell\} = P\Big\{\sum_{i=1}^n \mathbb{1}_{\Gamma_\ell^c}(X_i) > n/\ell\Big\} \\
&\le e^{-2n\ell} E^P\Big[e^{2\ell^2 \sum_{i=1}^n \mathbb{1}_{\Gamma_\ell^c}(X_i)}\Big] = e^{-2n\ell} E^\lambda\Big[e^{2\ell^2 \mathbb{1}_{\Gamma_\ell^c}(X_1)}\Big]^n \\
&\le e^{-2n\ell}(e^{2\ell^2} e^{-2\ell^2} + 1)^n \le e^{-n\ell}.
\end{aligned}$$

From this

$$\rho_n(K_L^c) \le \sum_{\ell \ge L} e^{-n\ell} \le 2e^{-nL},$$

which verifies exponential tightness of $\{\rho_n\}$.

Step 3. Lower bound for open sets. We use the change of measure argument. Basically, this amounts to computing the "cost" of forcing our process to behave according to a different marginal distribution μ. Replacement of λ by μ brings in Radon-Nikodym derivatives. This way relative entropy enters the calculation.

Let G be an open subset of $\mathcal{M}_1(\mathcal{S})$ and let $\mu \in G$. We can assume $H(\mu \,|\, \lambda) < \infty$ for otherwise μ is irrelevant for $\inf_G H$. Then $\mu \ll \lambda$ with Radon-Nikodym derivative $\phi = \frac{d\mu}{d\lambda}$. Let $Q = \mu^{\otimes\mathbb{N}}$ be the distribution of the i.i.d. sequence with marginal μ. As infinite product measures, Q and P are mutually singular. But if we consider only events that involve the first n coordinates, then we have a derivative again. Precisely speaking, the restrictions of the measures to the σ-algebra $\mathscr{F}_n = \sigma(X_1, \dots, X_n)$ have a derivative Φ_n given by

$$\Phi_n(\omega) = \frac{dQ}{dP}\Big|_{\mathscr{F}_n}(\omega) = \prod_{i=1}^n \phi(x_i)$$

Now the change of measure. Note that $Q\{\Phi_n > 0\} = 1$.

$$\begin{aligned}
\frac{1}{n}\log P\{L_n \in G\} &\ge \frac{1}{n}\log E^P\big[\mathbb{1}_G(L_n)\mathbb{1}\{\Phi_n > 0\}\big] \\
&= \frac{1}{n}\log E^Q\big[\mathbb{1}_G(L_n)\Phi_n^{-1}\big] \\
&= \frac{1}{n}\log\Big[\frac{1}{Q\{L_n \in G\}} E^Q[\mathbb{1}_G(L_n)\Phi_n^{-1}]\Big] + \frac{1}{n}\log Q\{L_n \in G\} \\
&\ge \frac{-1}{nQ\{L_n \in G\}} E^Q[\mathbb{1}_G(L_n)\log\Phi_n] + \frac{1}{n}\log Q\{L_n \in G\}.
\end{aligned}$$

The law of large numbers (5.5) applied to Q gives $Q\{L_n \in G\} \to 1$. In particular, we are not dividing by zero above. The last inequality above came from Jensen's inequality applied to the convex function $-\log x$. Since

$x \log x \geq -1/e$,

$$\begin{aligned} E^Q[\mathbb{1}_G(L_n)\log\Phi_n] &= E^Q[\log\Phi_n] - E^Q[\mathbb{1}_{G^c}(L_n)\log\Phi_n] \\ &= nE^\mu[\log\phi] - E^P[\mathbb{1}_{G^c}(L_n)\Phi_n\log\Phi_n] \\ &\leq nH(\mu\,|\,\lambda) + 1/e. \end{aligned}$$

Combining,

$$\frac{1}{n}\log P\{L_n \in G\} \geq \frac{1}{Q\{L_n \in G\}}\Big\{-H(\mu\,|\,\lambda) - \frac{1}{ne}\Big\} + \frac{1}{n}\log Q\{L_n \in G\}.$$

Letting $n \to \infty$ gives the bound

$$\varliminf_{n\to\infty} \frac{1}{n}\log \rho_n(G) \geq -H(\mu\,|\,\lambda).$$

Taking the sup over $\mu \in G$ finishes the proof of the lower bound. The LDP has been verified. □

***Exercise 5.17.** Let $\mathcal{S}$ be a Polish space. Put on $\mathcal{S}^n$ the product topology and on $\mathcal{M}_1(\mathcal{S})$ the weak topology generated by $\mathscr{C}_b(\mathcal{S})$. Show that the mapping $(x_1,\dots,x_n) \mapsto n^{-1}\sum_{k=1}^n \delta_{x_k}$ from $\mathcal{S}^n$ to $\mathcal{M}_1(\mathcal{S})$ is continuous. From this, conclude that if $X_1,\dots,X_n$ are $\mathcal{S}$-valued random variables on $(\Omega,\mathscr{F},P)$, then L_n is a measurable function from Ω to $\mathcal{M}_1(\mathcal{S})$.

Sanov's theorem and Cramér's theorem are closely related. Sanov's theorem can be derived from a more general version of Cramér's theorem in Polish spaces. (See Theorem 6.1.3, Corollary 6.2.3, and Lemma 6.2.6 all in [**15**].) In the next exercise we look at Cramér's theorem as a contraction from Sanov's theorem.

***Exercise 5.18.** Let $\mathcal{S}$ be a Polish space. Let $\{X_n\}$ be a sequence of i.i.d. $\mathcal{S}$-valued random variables with common distribution λ and let $\phi : \mathcal{S} \to \mathbb{R}^d$ be a continuous function. Assume that $E[e^{a|\phi(X_1)|}] < \infty$ for all $a > 0$. Let μ_n be the distribution of the sample mean $S_n/n = n^{-1}\sum_{k=1}^n \phi(X_k)$. Prove by a contraction from Sanov's theorem that $\mathrm{LDP}(\mu_n, n, I)$ holds with the tight convex rate function $I : \mathbb{R}^d \to [0,\infty)$ given by

$$I(z) = \inf\{H(\nu\,|\,\lambda) : E^\nu[\phi] = z\} = \sup_{\theta\in\mathbb{R}^d}\Big\{z\cdot\theta - \log\int_{\mathcal{S}} e^{\theta\cdot\phi(x)}\,\lambda(dx)\Big\}.$$

Hint: Start with ϕ bounded. The first formula for I is a result of the contraction principle. To prove the second equality compute I^*. Next, write $\phi(X) = \phi(X)\mathbb{1}\{|\phi(X)| \leq b\} + \phi(X)\mathbb{1}\{|\phi(X)| > b\}$. The unbounded part is controlled by the fact that $E[e^{a|\phi(X)|}, |\phi(X)| > b] \to 0$ as $b \to \infty$. The lower bound follows immediately. If $I^{(b)}$ is the rate function corresponding to the bounded part, then for closed $F \subset \mathbb{R}^d$ we have the upper bound

$$\varlimsup_{n\to\infty} n^{-1}\log P\{S_n/n \in F\} \leq -\varliminf_{\varepsilon\to 0}\varliminf_{b\to\infty}\inf_{z\in \overline{F^\varepsilon}} I^{(b)}(z),$$

where $F^\varepsilon = \{z + y : z \in F, |y| < \varepsilon\}$. Using the fact that $\mathbb{E}[e^{a|\phi(X)|}] < \infty$ conclude the upper bound in a similar fashion to Exercise 2.15.

The next exercise gives the relation $p = H^*$ a thermodynamical flavor.

***Exercise 5.19.** Let $\lambda \in \mathcal{M}_1(\mathcal{S})$ and fix a bounded continuous function $\mathcal{H} : \mathcal{S} \to \mathbb{R}$. (The letter $\mathcal{H}$ is to suggest *Hamiltonian* or energy, not to be confused with relative entropy.) For $\beta \in \mathbb{R}$ set $Z_\beta = E^\lambda[e^{-\beta\mathcal{H}}]$ and define the probability measure $\mu_\beta \in \mathcal{M}_1(\mathcal{S})$ by

$$\mu_\beta(dx) = \frac{e^{-\beta\mathcal{H}(x)}}{Z_\beta}\lambda(dx). \tag{5.6}$$

In statistical mechanics, Z_β is the *partition function* and μ_β is the *Gibbs measure* at inverse temperature β. Prove that

$$-\log Z_\beta = \inf\big\{\beta E^\nu[\mathcal{H}] + H(\nu\,|\,\lambda) : \nu \in \mathcal{M}_1(\mathcal{S})\big\}$$

is uniquely attained at μ_β. When $\beta > 0$, the above can be rewritten as

$$-\beta^{-1}\log Z_\beta = \inf\big\{E^\nu[\mathcal{H}] + \beta^{-1}H(\nu\,|\,\lambda) : \nu \in \mathcal{M}_1(\mathcal{S})\big\}. \tag{5.7}$$

Helmholtz free energy is the amount of thermodynamic energy in a system that can be converted into useful work at a constant temperature and volume. Thermodynamic entropy multiplied by temperature is the amount of unusable heat the system gives up when work is applied. Thus, free energy equals the system's internal energy less its thermodynamic entropy multiplied by its temperature, with the understanding that thermodynamic entropy is $-H(\nu\,|\,\lambda)$. (Think of a group of toddlers you want to convince of doing some project together. You will have to spend an unproductive amount of energy just to make them sit still, and then there is the amount of productive energy that goes into the actual project. The latter is the free energy and the former is the entropy times the temperature, while the total amount of energy you and the children spend is the internal or total energy.)

The exercise above characterizes the Gibbs measure by a variational principle. It is a mathematical statement of the thermodynamical principle that "left on its own, nature tends to minimize free energy (or work)". (On their own, children will minimize the amount of work they do.) Furthermore, (5.7) gives an expression of the Helmholtz free energy in terms of the partition function.

***Exercise 5.20.** Let the notation be as in the previous exercise. Let $\{X_k\}$ be i.i.d. with marginal λ. By Exercise 5.18 the large deviation principle holds for the distributions of

$$\overline{\mathcal{H}}_n = \frac{1}{n}\sum_{k=1}^n \mathcal{H}(X_k)$$

with rate function

$$p^*(z) = \inf\{H(\nu \,|\, \lambda) : \nu \in \mathcal{M}_1(\mathcal{S}),\ E^\nu[\mathcal{H}] = z\},$$

where p^* is the convex conjugate of $p(t) = \log E^\lambda[e^{t\mathcal{H}}] < \infty$, $t \in \mathbb{R}$.

(a) Prove that whenever $p^*(z) < \infty$, there is a unique $\nu_z \in \mathcal{M}_1(\mathcal{S})$ such that $E^{\nu_z}[\mathcal{H}] = z$ and $p^*(z) = H(\nu_z \,|\, \lambda)$.
Hint: Use lower semicontinuity and compact sublevel sets (Proposition 5.7) for existence of the minimizer, and use strict convexity (Exercise 5.5) for uniqueness.

(b) Prove that $p'(t) = E^{\mu_{-t}}[\mathcal{H}]$,

$$p''(t) = E^{\mu_{-t}}[\mathcal{H}^2] - E^{\mu_{-t}}[\mathcal{H}]^2$$

(the variance of $\mathcal{H}$ under μ_{-t}), and $\lim_{t\to 0} p'(t) = \int \mathcal{H}\, d\lambda$ (high temperature limit).

Define

$$A = \lambda\text{-ess inf}\, \mathcal{H} \quad \text{and} \quad B = \lambda\text{-ess sup}\, \mathcal{H},$$

that is, $A = \sup\{a : \lambda(\mathcal{H} < a) = 0\}$ and $B = \inf\{b : \lambda(\mathcal{H} > b) = 0\}$. Show that

$$\lim_{t\to-\infty} p'(t) = A \quad \text{and} \quad \lim_{t\to\infty} p'(t) = B.$$

(c) Assume $\mathcal{H}$ is not λ-a.s. constant so that $p'' > 0$. Prove that for $z \in (A, B)$, there is a unique $\beta = \beta_z$ such that $p'(-\beta) = z$ and $p^*(z) = -z\beta - p(-\beta)$. Furthermore, $\nu_z = \mu_\beta$ from (5.6). In other words, for each energy value $z \in (A, B)$, there is a unique inverse temperature β such that $E^{\mu_\beta}[\mathcal{H}] = z$, the Gibbs measure μ_β minimizes the entropy $H(\nu \,|\, \lambda)$ subject to the energy constraint $E^\nu[\mathcal{H}] = z$, and the minimum entropy is precisely the value $p^*(z)$ of the Cramér rate for $\overline{\mathcal{H}}_n$.

The exercise above illustrates the *Gibbs conditioning principle*, which says that "under an energy constraint, nature maximizes entropy". (Children will maximize disorder until they exhaust themselves.)

5.3. Maximum entropy principle

In the setting of Sanov's theorem $L_n \to \lambda$ a.s. as $n \to \infty$. Let C be a set of probability measures whose closure does not contain λ. Then $P\{L_n \in C\} \to 0$. But what happens to L_n asymptotically if we condition L_n to remain in C? Under some technical assumptions, L_n converges towards those element(s) of C that minimize $H(\,\cdot\,|\,\lambda)$. The principle is named "maximum entropy" because minimizing relative entropy corresponds to maximizing thermodynamic entropy.

Maximum entropy principle. *Suppose $C \subset \mathcal{M}_1(\mathcal{S})$ is closed, convex, and satisfies*

$$\inf_{\nu \in C} H(\nu \,|\, \lambda) = \inf_{\nu \in C^\circ} H(\nu \,|\, \lambda) < \infty. \tag{5.8}$$

Then, there is a unique $\tilde{\nu} \in C$ that minimizes $H(\,\cdot\,|\,\lambda)$ over C. The conditioned distributions of L_n converge weakly to a point mass at $\tilde{\nu}$; that is,

$$\lim_{n\to\infty} P\{L_n \in \cdot \,|\, L_n \in C\} = \delta_{\tilde{\nu}}(\cdot) \tag{5.9}$$

in the weak topology of $\mathcal{M}_1(\mathcal{M}_1(\mathcal{S}))$ generated by $\mathscr{C}_b(\mathcal{M}_1(\mathcal{S}))$. This convergence is exponentially fast: for any weak neighborhood U of $\tilde{\nu}$ $\exists b > 0$ such that

$$P\{L_n \in U^c \,|\, L_n \in C\} \le e^{-nb}$$

for large enough n. Furthermore, for any fixed k, the conditioned distribution of X_k converges weakly to $\tilde{\nu}$; that is,

$$\lim_{n\to\infty} E[f(X_k) \,|\, L_n \in C] = E^{\tilde{\nu}}[f]$$

for all $f \in \mathscr{C}_b(\mathcal{S})$.

As a heuristic principle maximum entropy is used in statistics to solve the following problem. Suppose our belief is that a random variable X obeys a distribution λ. Then we receive new information about X, perhaps by performing an experiment, and λ is no longer compatible with the new information. What should our new best guess be for the unknown distribution of X, among the compatible distributions (the set C)? The maximum entropy principle picks the distribution that is closest to λ in "entropy distance". (Quotes are in order because relative entropy is not a metric on probability measures.) The theorem above offers some theoretical justification for this principle.

Proof of the maximum entropy principle. The existence and uniqueness of $\tilde{\nu}$ follow from lower semicontinuity, compact sublevel sets, and strict convexity of H. Upper and lower large deviation bounds from Sanov's theorem together with assumption (5.8) imply that

$$n^{-1} \log P\{L_n \in C\} \to -H(\tilde{\nu} \,|\, \lambda) > -\infty.$$

Consequently $P\{L_n \in C\} > 0$ for large n and conditioning on this event is feasible.

Let U be any neighborhood of $\tilde{\nu}$. By the definition of conditional probability and since $\tilde{\nu} \notin U^c$ and U^c is closed,

$$\begin{aligned}
&\overline{\lim_{n\to\infty}} \frac{1}{n} \log P\{L_n \in U^c \,|\, L_n \in C\} \\
&\qquad = \overline{\lim_{n\to\infty}} \Big(\frac{1}{n} \log P\{L_n \in U^c \cap C\} - \frac{1}{n} \log P\{L_n \in C\} \Big) \\
&\qquad \le - \inf_{\nu \in U^c \cap C} H(\nu \,|\, \lambda) \;+\; H(\tilde{\nu} \,|\, \lambda) < 0.
\end{aligned}$$

We can verify the weak convergence (5.9) for example by observing that for closed sets $A \subset \mathcal{M}_1(\mathcal{S})$,

$$\overline{\lim_{n\to\infty}} P\{L_n \in A \,|\, L_n \in C\} \le \delta_{\tilde{\nu}}(A).$$

Let $f \in \mathscr{C}_b(\mathcal{S})$ and define $F(\nu) = \int f \, d\nu$, a bounded continuous function on $\mathcal{M}_1(\mathcal{S})$. Fix k and consider $n > k$. Permuting $X_1, \dots, X_n$ does not alter their joint distribution or L_n. Consequently

$$\begin{aligned}
E[f(X_k) \,|\, L_n \in C] &= \frac{E[f(X_k) \mathbb{1}\{L_n \in C\}]}{P\{L_n \in C\}} \\
&= E\Big[\frac{1}{n} \sum_{k=1}^{n} f(X_k) \,\Big|\, L_n \in C \Big] = E\big[L_n(f) \,|\, L_n \in C\big] \\
&= E[F(L_n) \,|\, L_n \in C\,] \longrightarrow \int_{\mathcal{M}_1(\mathcal{S})} F \, d\delta_{\tilde{\nu}} = F(\tilde{\nu}) = E^{\tilde{\nu}}[f].
\end{aligned}$$

The convergence on the last line is a consequence of (5.9). □

As an interesting application of the maximum entropy principle, we see how a Gibbs measure arises as a limit of conditional probabilities.

***Exercise 5.21** (Elementary equivalence of ensembles)**.** With notation as in Exercise 5.20, suppose $z \in (A, B)$ so there is a unique β such that $E^{\mu_\beta}[\mathcal{H}] = z$. Prove that for any fixed k,

$$\lim_{\delta \to 0} \lim_{n\to\infty} P\{X_k \in \cdot \;|\; |\overline{\mathcal{H}}_n - z| \le \delta\} = \mu_\beta.$$

Hint: Let $M_\delta = \{\nu : |E^\nu[\mathcal{H}] - z| \le \delta\}$. Show that there is a unique entropy minimizing measure ν_δ in M_δ. Let $m(\delta) = \inf_{\nu \in M_\delta} H(\nu \,|\, \lambda)$. Show that m is left-continuous and from this deduce that $C = M_\delta$ satisfies the assumptions of the maximum entropy principle. Show that $\nu_\delta \to \mu_\beta$ as $\delta \to 0$.

Exercise 5.21 says, roughly, that if a large number of particles governed by the free measure $P = \lambda^{\otimes \mathbb{N}}$ are constrained to have average energy $\overline{\mathcal{H}}_n = z$ (in a controlled experiment, say), then an individual particle obeys the Gibbs measure μ_β with the temperature $1/\beta$ chosen to produce the correct expected energy $E^{\mu_\beta}[\mathcal{H}] = z$.

In statistical mechanics, the measure $P\{\cdot\,|\,\overline{\mathcal{H}}_n = z\}$ is called the *micro-canonical ensemble*, and the Gibbs measure μ_β is the *canonical ensemble*. These measures were introduced by Josiah Willard Gibbs (1839–1903), who is credited with systematizing equilibrium statistical mechanics after the pioneering work of Maxwell and Boltzmann. The problem of equivalence of ensembles is whether these two ensembles give equivalent results in the infinite particle limit. The exercise gives an affirmative answer in a simple case where the particles do not interact and we look only at the marginal distribution of a single particle.

As the last item we present an elegant classic example of equivalence of ensembles.

Maxwell's principle. *For an integer $n \ge 1$ let $\mathbf{X}_n = (X_1^{(n)}, \dots, X_n^{(n)}) \in \mathbb{R}^n$ be uniformly distributed on the $(n-1)$-dimensional sphere of radius $\sigma\sqrt{n}$. For $k \le n$ let $P_{k,n}$ be the distribution of $(X_1^{(n)}, \dots, X_k^{(n)})$. Then, for each fixed k, as $n \to \infty$, $P_{k,n}$ converges weakly to the distribution of k i.i.d. normals with mean 0 and variance σ^2.*

Proof. Let $\{Z_k\}$ be an i.i.d. sequence of standard normal random variables. For $k \le n$ let

$$Y_k^{(n)} = \frac{\sigma\sqrt{n}Z_k}{\sqrt{Z_1^2 + \cdots + Z_n^2}}.$$

***Exercise 5.22.** Prove that $\mathbf{Y}_n = (Y_1^{(n)}, \dots, Y_n^{(n)})$ has the same distribution as $\mathbf{X}_n$.

By the strong law of large numbers $(Z_1^2 + \cdots + Z_n^2)/n$ converges almost surely to $E[Z_1^2] = 1$. The claim follows. □

Literature note. Further discussion and refinements of the Gibbs conditioning principle appear in Section 7.3 of [**15**].

Chapter 6

Process level large deviations for i.i.d. fields

In the previous chapter we studied Sanov's theorem, which quantifies the exponential decay of the probability that values of the random variables X_k occur at an atypical frequency. The next step is multivariate large deviations. For example, what is the exponential decay of the probability that the empirical distribution of pairs (X_k, X_{k+1}) approximates a particular two-dimensional distribution? In order to capture multivariate large deviations of any dimension, we take up the study of the full *empirical process*

$$R_n = \frac{1}{n}\sum_{k=1}^{n} \delta_{(X_j)_{j\geq k}} \in \mathcal{M}_1(\mathcal{S}^{\mathbb{N}}).$$

This is called *process level* large deviations, in contrast to the *position level* of Sanov's theorem. Another terminology separates large deviations into level 1 (Cramér's theorem), level 2 (Sanov's theorem), and level 3 (process level) large deviations.

To treat statistical mechanics models with spatial structure, we also generalize from a sequence $(X_j)_{j\in\mathbb{N}}$ to a *random field* $(X_i)_{i\in\mathbb{Z}^d}$ of i.i.d. random variables indexed by the d-dimensional integer lattice.

6.1. Setting

The state space of our process is a Polish space $\mathcal{S}$. Fix a dimension $d \in \mathbb{N}$. The *configuration space* is the Cartesian product space $\Omega = \mathcal{S}^{\mathbb{Z}^d}$. A generic element of Ω is a configuration $\omega = (\omega_i)_{i\in\mathbb{Z}^d}$ with coordinates $\omega_i \in \mathcal{S}$. Endowed with the product topology, Ω is Polish as well. Let $\mathscr{F}$ be the product σ-algebra on Ω, which is also the Borel σ-algebra on Ω. Theorem B.14 implies that $\mathcal{M}_1(\Omega)$, with the weak topology generated by $\mathscr{C}_b(\Omega)$, is also Polish.

Call a function $f : \Omega \to \mathbb{R}$ *local* if it is a function of only finitely many coordinates ω_i. Let $\mathscr{C}_{b,\text{loc}}(\Omega)$ be the space of bounded continuous local functions from Ω into $\mathbb{R}$.

***Exercise 6.1.** Show that $\mu_k \to \mu$ in $\mathcal{M}_1(\Omega)$ if and only if all finite-dimensional marginals converge.

Hint: Use the equivalence of (a) and (b) in the portmanteau theorem (page 279) along with a density argument.

***Exercise 6.2.** Prove that the topologies on $\mathcal{M}_1(\Omega)$ generated by $\mathscr{C}_b(\Omega)$ and $\mathscr{C}_{b,\text{loc}}(\Omega)$ coincide.

Hint: On each $\Omega_\Lambda = \mathcal{S}^\Lambda$, Λ finite, use the argument on page 280 to find countably many functions that determine weak convergence. Put all these functions together to form a metric of the type (B.3) that has only local functions. Use Exercise B.9.

Fix a probability measure $\lambda \in \mathcal{M}_1(\mathcal{S})$. Let $P = \lambda^{\otimes\mathbb{Z}^d}$ be the product measure on Ω. If we define the coordinate process $(X_i)_{i\in\mathbb{Z}^d}$ as $X_i(\omega) = \omega_i$, then this process is an i.i.d. random field on $(\Omega, \mathscr{F}, P)$. With our statistical mechanics applications in mind, we sometimes refer to the X_i's as *spins*.

The *shift group* $\{\theta_i : i \in \mathbb{Z}^d\}$ on Ω is the group of homeomorphisms defined by $(\theta_i\omega)_j = \omega_{i+j}$ for $\omega \in \Omega$ and $i, j \in \mathbb{Z}^d$. The set of *shift-invariant probability measures* on Ω is

$$\mathcal{M}_\theta(\Omega) = \{\mu \in \mathcal{M}_1(\Omega) : \mu \circ \theta_i = \mu \ \forall i \in \mathbb{Z}^d\}.$$

Note that $\mathcal{M}_\theta(\Omega)$ is a closed convex subset of $\mathcal{M}_1(\Omega)$. Under a shift-invariant measure the coordinate process $X_i(\omega) = \omega_i$ is a *stationary process.* Shift-invariant measures themselves are also called stationary measures. The σ-algebra of shift-invariant Borel sets in Ω is

$$\mathcal{I} = \{A \in \mathscr{F} : \theta_i A = A \ \forall i \in \mathbb{Z}^d\}. \tag{6.1}$$

Exercise 6.3. Prove that $\mathcal{I}$ is a σ-algebra.

Definition 6.4. A measure $\mu \in \mathcal{M}_\theta(\Omega)$ is *ergodic* if $\mu(A) \in \{0, 1\}$ for all $A \in \mathcal{I}$. We denote the set of ergodic probability measures by $\mathcal{M}_e(\Omega)$.

By definition, *extreme points* of $\mathcal{M}_\theta(\Omega)$ are measures $Q \in \mathcal{M}_\theta(\Omega)$ such that $Q = tP_1 + (1-t)P_2$ with $t \in (0, 1)$ and $P_1, P_2 \in M_\theta(\Omega)$ happens only for $P_1 = P_2 = Q$. It turns out that the set of extreme points of $\mathcal{M}_\theta(\Omega)$ is exactly

$\mathcal{M}_e(\Omega)$. This is a consequence of the ergodic theorem. A related, deeper fact is the ergodic decomposition theorem: every shift-invariant probability measure is a mixture of ergodic measures. See Section B.6 for more.

Denote points of the indexing lattice $\mathbb{Z}^d$ by $i = (i_1, i_2, \ldots, i_d)$, and fix an increasing sequence of cubes $V_n = \{i \in \mathbb{Z}^d : -n < i_1, \ldots, i_d < n\}$ whose union exhausts $\mathbb{Z}^d$. The *empirical fields* $R_n : \Omega \to \mathcal{M}_1(\Omega)$ are defined by

$$R_n(\omega) = \frac{1}{|V_n|} \sum_{i \in V_n} \delta_{\theta_i \omega}.$$

For the expectation of a function f under the empirical measure, we use the equivalent notations $R_n(f) = R_n(\omega, f) = E^{R_n(\omega)}[f] = |V_n|^{-1} \sum_{i \in V_n} f(\theta_i \omega)$. By the multidimensional ergodic theorem (page 283), $R_n(\omega, g) \to E^P[g]$ P-a.s. for any $g \in L^1(P)$. By the argument given for (5.5), this implies the limit

$$R_n(\omega) \to P \text{ weakly in } \mathcal{M}_1(\Omega) \text{ for } P\text{-a.e. } \omega. \tag{6.2}$$

Our goal is to study large deviations from this law of large numbers.

For $\omega \in \Omega$, the *periodized configuration* $\omega^{(n)}$ is defined by $\omega_i^{(n)} = \omega_i$ for $i \in V_n$, and $\omega_i^{(n)} = \omega_j^{(n)}$ whenever $i_k = j_k \mod (2n-1)$ for all $k = 1, \ldots, d$. The *periodic empirical fields* $\widetilde{R}_n : \Omega \to \mathcal{M}_1(\Omega)$ are defined by

$$\widetilde{R}_n(\omega) = \frac{1}{|V_n|} \sum_{i \in V_n} \delta_{\theta_i \omega^{(n)}}. \tag{6.3}$$

Due to periodization, $\widetilde{R}_n$ is measurable with respect to $\mathscr{F}_{V_n}$, the σ-algebra generated by $\{\omega_i : i \in V_n\}$. It is therefore easier to work with $\widetilde{R}_n$ than with R_n, which depends on the whole configuration ω. However, $\widetilde{R}_n$ and R_n come asymptotically close together.

***Exercise 6.5.** Prove that for every bounded local measurable function g

$$\sup_\omega \bigl|\widetilde{R}_n(\omega, g) - R_n(\omega, g)\bigr| \to 0.$$

As a consequence, $\widetilde{R}_n$ also converges weakly to P. By Exercise 2.16 the large deviation principle for $\widetilde{R}_n$ transfers easily to one for R_n if the rate function is tight.

6.2. Specific relative entropy

The rate function for the large deviations of $\widetilde{R}_n$ and R_n will be a relative entropy. However, a complication appears.

Exercise 6.6. Prove that if $Q \in \mathcal{M}_\theta(\Omega)$ is absolutely continuous relative to $P \in \mathcal{M}_e(\Omega)$, then $Q = P$.

Hint: Prove that $Q \in \mathcal{M}_e(\Omega)$ and use Exercise B.20.

The difficulty comes from having an infinite product space. To get a nontrivial entropy we take a limit of finite-dimensional entropies normalized by volume.

For $Q \in \mathcal{M}_1(\Omega)$ and $\Lambda \subset \mathbb{Z}^d$, let Q_Λ be the restriction of Q to $\mathscr{F}_\Lambda$, the σ-algebra generated by $\omega_\Lambda = (\omega_i)_{i\in\Lambda}$. Let $H_\Lambda(Q\,|\,P)$ be the entropy of Q_Λ relative to P_Λ. For $\Lambda = V_n$ we abbreviate and write Q_n and H_n.

Theorem 6.7. *For each $Q \in \mathcal{M}_\theta(\Omega)$, the specific relative entropy*

$$h(Q\,|\,P) = \lim_{n\to\infty} \frac{1}{|V_n|} H_n(Q\,|\,P)$$

exists and is also given by

$$h(Q\,|\,P) = \sup_{\Lambda\in\mathscr{R}} \frac{1}{|\Lambda|} H_\Lambda(Q\,|\,P), \tag{6.4}$$

where $\mathscr{R}$ is the collection of finite rectangles in $\mathbb{Z}^d$.

Here we can use the multidimensional generalization of subadditivity and superadditivity.

Fekete's lemma. *Suppose that for $\Lambda \in \mathscr{R}$ we have $a_\Lambda \in [0,\infty]$ that satisfy*

(a) $a_\Lambda + a_\Delta \le a_{\Lambda\cup\Delta}$ *whenever* $\Lambda \cap \Delta = \varnothing$ *and* $\Lambda \cup \Delta \in \mathscr{R}$,

(b) $a_\Lambda = a_{i+\Lambda}$ *for all* $i \in \mathbb{Z}^d$ *and* $\Lambda \in \mathscr{R}$.

Then

$$\lim_{n\to\infty} \frac{a_{V_n}}{|V_n|} = \sup_\Lambda \frac{a_\Lambda}{|\Lambda|}.$$

Proof. Fix Λ. Let $\{k+\Lambda : k \in \mathcal{K}\}$ be a tiling of $\mathbb{Z}^d$ by disjoint shifted copies of Λ. Let $b(n)$ be the number of these copies that are contained in V_n. Let $\{k+\Lambda : k \in \mathcal{K}'\}$ be the copies that intersect V_n but do not lie entirely inside V_n. Set $\Lambda_k^n = V_n \cap (k+\Lambda)$. Apply (a) and (b) repeatedly to get the inequality

$$a_{V_n} \ge b(n)a_\Lambda + \sum_{k\in\mathcal{K}'} a_{\Lambda_k^n} \ge b(n)a_\Lambda.$$

Thus,

$$\varliminf_{n\to\infty} \frac{a_{V_n}}{|V_n|} \ge a_\Lambda \lim_{n\to\infty} \frac{b(n)}{|V_n|} = \frac{a_\Lambda}{|\Lambda|}.$$

The other direction requires no proof. ∎

Proof of Theorem 6.7. We verify (a) and (b) of the lemma above for $a_\Lambda = H_\Lambda(Q \mid P)$. By the variational characterization (5.2) of relative entropy

$$H_\Lambda(Q \mid P) = \sup_{f \in \mathscr{C}_b(\mathcal{S}^\Lambda)} \{E^{Q_\Lambda}[f] - \log E^{P_\Lambda}[e^f]\}.$$

We make no notational distinction between Q_Λ as a measure on the σ-algebra $\mathscr{F}_\Lambda$ on the space Ω and as a measure on the space $\mathcal{S}^\Lambda$. Similarly, a function $f : \mathcal{S}^\Lambda \to \mathbb{R}$ can be composed with a projection to give an $\mathscr{F}_\Lambda$-measurable function $\omega \mapsto f(\omega_\Lambda)$ on Ω, and it is not necessary to introduce extra notation to distinguish between the two roles of f.

Let $\Lambda \cap \Delta = \varnothing$, $f \in \mathscr{C}_b(\mathcal{S}^\Lambda)$, and $g \in \mathscr{C}_b(\mathcal{S}^\Delta)$. As local functions on Ω, f and g are independent under P, and so

$$\begin{aligned} H_{\Lambda\cup\Delta}(Q \mid P) &\ge E^Q[f+g] - \log E^P[e^{f+g}] \\ &= (E^Q[f] - \log E^P[e^f]) + (E^Q[g] - \log E^P[e^g]). \end{aligned}$$

Taking the sup over such f and g gives

$$H_{\Lambda\cup\Delta}(Q \mid P) \ge H_\Lambda(Q \mid P) + H_\Delta(Q \mid P).$$

The shift invariance part (b) follows also from the variational characterization because, for $Q \in \mathcal{M}_\theta(\Omega)$, Q_Λ and $Q_{i+\Lambda}$ coincide as measures on $\mathcal{S}^\Lambda$. ☐

Proposition 6.8. *As a function on $\mathcal{M}_\theta(\Omega)$, $h(\,\cdot \mid P)$ is affine, lower semicontinuous, and for $c \in \mathbb{R}$ the sublevel sets $\{Q \in \mathcal{M}_\theta(\Omega) : h(Q \mid P) \le c\}$ are compact in the weak topology generated by $\mathscr{C}_b(\Omega)$. Furthermore, $h(Q \mid P) = 0$ if and only if $Q = P$.*

Proof. Since each $|\Lambda|^{-1} H_\Lambda$ is convex and lower semicontinuous their supremum h inherits these properties. To show that h is affine, we need $h(Q \mid P) \ge th(Q^1 \mid P) + (1-t)h(Q^2 \mid P)$ for $Q = tQ^1 + (1-t)Q^2$ and $0 < t < 1$.

Assume $Q_n \ll P_n$ for all n. (Otherwise $h(Q \mid P) = \infty$ and there is nothing to prove.) Let $f_n^1 = \frac{dQ_n^1}{dP_n}$ and $f_n^2 = \frac{dQ_n^2}{dP_n}$. Then

$$\begin{aligned} H_n(Q \mid P) &= E^Q\Big[\log \frac{dQ_n}{dP_n}\Big] \\ &= tE^{Q^1}\big[\log(tf_n^1 + (1-t)f_n^2)\big] + (1-t)E^{Q^2}\big[\log(tf_n^1 + (1-t)f_n^2)\big] \\ &\ge tE^{Q^1}\big[\log(tf_n^1)\big] + (1-t)E^{Q^2}\big[\log((1-t)f_n^2)\big] \\ &= tH_n(Q^1 \mid P) + (1-t)H_n(Q^2 \mid P) + t\log t + (1-t)\log(1-t). \end{aligned}$$

Divide by $|V_n|$ and let $n \nearrow \infty$.

Since the weak topology is metric, compactness is equivalent to sequential compactness. Let $\{Q^\ell\}_{\ell\in\mathbb{N}}$ satisfy $h(Q^\ell \,|\, P) \le c$. We show the existence of a convergent subsequence $Q^{\ell_j} \to Q$ via a diagonal argument.

By (6.4), $H_1(Q^\ell \,|\, P) \le h(Q^\ell \,|\, P) \le c$, and Proposition 5.7 implies the existence of a subsequence $\{Q^{\ell(j,1)}\}_{j\in\mathbb{N}}$ whose marginals on $\mathcal{S}^{V_1}$ converge weakly to a probability measure $\rho_1 \in \mathcal{M}_1(\mathcal{S}^{V_1})$.

Next, $H_2(Q^{\ell(j,1)} \,|\, P) \le |V_2|\, h(Q^{\ell(j,1)} \,|\, P) \le |V_2|\, c$ implies the existence of a further subsequence $\{Q^{\ell(j,2)}\}_{j\in\mathbb{N}} \subset \{Q^{\ell(j,1)}\}_{j\in\mathbb{N}}$ whose marginals on $\mathcal{S}^{V_2}$ converge weakly to a probability measure $\rho_2 \in \mathcal{M}_1(\mathcal{S}^{V_2})$.

Inductively, given $\{Q^{\ell(j,n-1)}\}_{j\in\mathbb{N}}$, compact sublevel sets of relative entropy give a further subsequence $\{Q^{\ell(j,n)}\}_{j\in\mathbb{N}} \subset \{Q^{\ell(j,n-1)}\}_{j\in\mathbb{N}}$ whose marginals on $\mathcal{S}^{V_n}$ converge to $\rho_n \in \mathcal{M}_1(\mathcal{S}^{V_n})$.

Let $\ell_j = \ell(j,j)$. We argue that the subsequence $Q^{\ell_j} = Q^{\ell(j,j)}$ converges weakly as $j \to \infty$.

For each fixed n, $\{Q^{\ell(j,j)}\}_{j\ge n}$ is a subsequence of $\{Q^{\ell(j,n)}\}_{j\in\mathbb{N}}$, and consequently the S^{V_n}-marginals $\{Q_n^{\ell(j,j)}\}_{j\in\mathbb{N}}$ converge weakly to ρ_n. It follows from the limits that the $\{\rho_n\}$ form a consistent family of finite-dimensional distributions. This means that for $m < n$ the restriction of ρ_n to $\mathcal{S}^{V_m}$ is precisely ρ_m. By Kolmogorov's extension theorem there exists a probability measure $Q \in \mathcal{M}_1(\Omega)$ such that $Q_n = \rho_n$ for all n. Then the finite-dimensional marginals of $Q^{\ell(j,j)}$ converge to those of Q and by Exercise 6.1 the sequence $Q^{\ell(j,j)}$ itself converges weakly to Q. This proves the sequential compactness (and hence the compactness) of $\{h \le c\}$.

The statement about the zeroes of h follows directly from Lemma 5.3 and (6.4). □

Recall that $\mathcal{M}_e(\Omega)$ is the set of extreme points of the convex set $\mathcal{M}_\theta(\Omega)$. An interesting phenomenon happens: $\mathcal{M}_e(\Omega)$ is in fact dense in $\mathcal{M}_\theta(\Omega)$. This denseness allows us to apply the ergodic theorem to prove the lower bound in the LDP we are after, quite similarly to the way the ergodic theorem was used to prove the lower bound in Sanov's theorem.

Lemma 6.9. *For $Q \in \mathcal{M}_\theta(\Omega)$, there are measures $Q^k \in \mathcal{M}_e(\Omega)$ that converge weakly to Q and satisfy $h(Q^k \,|\, P) \to h(Q \,|\, P)$.*

Proof. Let $\{j + V_n : j \in I_n\}$, $I_n = (2n-1)\mathbb{Z}^d$, be a covering of $\mathbb{Z}^d$ by disjoint shifted copies of V_n. Let $Q^{n,\otimes}$ be the measure on Ω which makes the σ-algebras $\mathscr{F}_{j+V_n}$, $j \in I_n$, independent, but which coincides with Q on each such σ-algebra. More precisely, $Q^{n,\otimes}(d\omega) = \bigotimes_{j\in I_n} Q_{j+V_n}(d\omega_{j+V_n})$. Then

$$Q^n = \frac{1}{|V_n|} \sum_{i\in V_n} Q^{n,\otimes} \circ \theta_{-i}$$

is shift invariant. (Note that $Q^{n,\otimes} \circ \theta_j = Q^{n,\otimes}$ if $j \in I_n$.) We prove that this sequence satisfies the claim of the lemma.

First, we prove the weak convergence to Q. To this end, let g be any bounded $\mathscr{F}_m$-measurable function and take $n > m$. Then

$$E^{Q^n}[g] = \frac{1}{|V_n|} \sum_{\substack{i \in V_n \\ i+V_m \subset V_n}} E^{Q^{n,\otimes}}[g \circ \theta_i] + \mathcal{O}(c_{m,n}),$$

where $c_{m,n} = |\{i \in V_n : i + V_m \not\subset V_n\}|/|V_n| \to 0$ as $n \to \infty$. But if $i + V_m \subset V_n$, then $E^{Q^{n,\otimes}}[g \circ \theta_i] = E^Q[g]$. Thus,

$$E^{Q^n}[g] = (1 - c_{m\,n})E^Q[g] + \mathcal{O}(c_{m,n}) \underset{n\to\infty}{\longrightarrow} E^Q[g].$$

Next, we show that the measures Q^n are ergodic. Let A be in the shift-invariant σ-algebra. Then

$$Q^n(A) = \frac{1}{|V_n|} \sum_{i \in V_n} Q^{n,\otimes}(\theta_{-i}A) = Q^{n,\otimes}(A).$$

Exercise B.23 gives a tail measurable event B such that $0 = Q^n(A \triangle B) \geq \frac{1}{|V_n|} Q^{n,\otimes}(A \triangle B)$. Thus, $Q^{n,\otimes}(A) = Q^{n,\otimes}(B)$. By Kolmogorov's 0-1 law (Exercise B.22), $Q^{n,\otimes}(B) \in \{0, 1\}$. The ergodicity of Q^n has been checked.

Finally, we need to prove the convergence of specific entropies. By the lower semicontinuity of h we know that $h(Q \,|\, P) \leq \underline{\lim}_{n\to\infty} h(Q^n \,|\, P)$. To prove the other direction we first use convexity to write

$$h(Q^k \,|\, P) = \lim_{n\to\infty} \frac{1}{|V_n|} H_n(Q^k \,|\, P) \leq \overline{\lim_{n\to\infty}} \frac{1}{|V_n|} \frac{1}{|V_k|} \sum_{i \in V_k} H_n(Q^{k,\otimes} \circ \theta_{-i} \,|\, P).$$

Note that the shift invariance of P implies that

$$H_n(Q^{k,\otimes} \circ \theta_{-i} \,|\, P) = H_{i+V_n}(Q^{k,\otimes} \,|\, P) \leq H_{W^i_{n,k}}(Q^{k,\otimes} \,|\, P),$$

where $W^i_{n,k}$ is the smallest cube that contains $i + V_n$ and is made up of V_k and disjoint shifted copies of it. The last inequality follows from the variational formulation of relative entropy.

***Exercise 6.10.** Prove the following.

(a) Let $\mathcal{X}$ and $\mathcal{Y}$ be two Polish spaces. Let $\alpha, \mu \in \mathcal{M}_1(\mathcal{X})$ and $\beta, \nu \in \mathcal{M}_1(\mathcal{Y})$. Then $H(\alpha \otimes \beta \,|\, \mu \otimes \nu) = H(\alpha \,|\, \mu) + H(\beta \,|\, \nu)$.

(b) Let $\mathcal{X}$ be a Polish space and let $\lambda, \mu \in \mathcal{M}_1(\mathcal{X})$. Then $h(\mu^{\otimes \mathbb{Z}^d} \,|\, \lambda^{\otimes \mathbb{Z}^d}) = H(\mu \,|\, \lambda)$.

From (a) in the exercise above $H_{W^i_{n,k}}(Q^{k,\otimes}\,|\,P) = \frac{|W^i_{n,k}|}{|V_k|}H_k(Q\,|\,P)$. Since $|W^i_{n,k}| \le (2n+4k)^d$ and $|V_n| = (2n-1)^d$, one has

$$h(Q^k\,|\,P) \le \varlimsup_{n\to\infty} \frac{1}{|V_n|}\frac{1}{|V_k|}\sum_{i\in V_k} H_n(Q^{k,\otimes}\circ\theta_{-i}\,|\,P)$$

$$(6.5) \qquad \le \frac{1}{|V_k|}H_k(Q\,|\,P) \le h(Q\,|\,P).$$

This completes the proof of Lemma 6.9. □

***Exercise 6.11.** Suppose $\mathscr{D}_1 \subset \mathscr{D}_2 \subset \cdots$ are σ-algebras generating a σ-algebra $\mathscr{D}$. Let μ and ν be two probability measures on $\mathscr{D}$. Prove that $H_{\mathscr{D}_n}(\mu\,|\,\nu)$ increases to $H(\mu\,|\,\nu)$ as $n\to\infty$. Here $H_{\mathscr{D}_n}$ is the relative entropy of the restrictions to $\mathscr{D}_n$.

Hint: Use the variational formulation of relative entropy to see that the limit exists and equals $\sup_n H_{\mathscr{D}_n}(\mu\,|\,\nu)$. To identify the limit as $H(\mu\,|\,\nu)$, prove a suitable analogue of Lemma A.11. Note that it is easier to solve the exercise in the special case $\mathscr{D}_n = \mathscr{F}_{V_n}$ than for general σ-algebras because then one can apply Lemma A.11.

We end this section with an exercise that shows how specific relative entropy can in fact be seen as a relative entropy.

Exercise 6.12. Let $Q \in \mathcal{M}_\theta(\Omega)$. Order $\mathbb{Z}^d$ lexicographically. (In $d=1$ this is the usual ordering of integers. In $d=2$, $(a,s)\le(b,t)$ if $a<b$, or $a=b$ and $s<t$. The higher-dimensional generalization should be clear.) Let $U_i = \{k\in\mathbb{Z}^d : k<i\}$ be the lexicographic *past* of the site i. Let $Q(\,\cdot\,|\,\mathscr{F}_{U_0})$ be a conditional probability measure for Q, given the past of 0. Restrict this measure to $\mathscr{F}_0$ to get $Q_0(\,\cdot\,|\,\mathscr{F}_{U_0})$, the conditional distribution of X_0 under Q, given the past of 0. Let $U_0^* = U_0\cup\{0\}$. Think of $Q_{U_0}\otimes\lambda$ as a probability measure on $\mathscr{F}_{U_0^*} = \mathscr{F}_{U_0}\otimes\mathscr{F}_0$ in the obvious way. Prove that

$$h(Q\,|\,P) = H(Q_{U_0^*}\,|\,Q_{U_0}\otimes\lambda).$$

Hint: Fix $V_n = \{i^{(1)} > i^{(2)} > \cdots > i^{(s)}\}$, with $s=|V_n|$. Let $V_n^k = \{i^{(\ell)} : k\le \ell\le s\}$. By telescoping and shift invariance

$$H_n(Q\,|\,P) = H_0(Q\,|\,P) + \sum_{k=1}^{s-1}\{H_{V_n^k - i^{(k)}}(Q\,|\,P) - H_{V_n^{k+1}-i^{(k)}}(Q\,|\,P)\}.$$

By the conditional entropy formula (Exercise 5.13), the summation term is equal to $H_{V_n^k - i^{(k)}}(Q_{U_0^*}\,|\,Q_{U_0}\otimes\lambda)$. Now any fixed finite subset of U_0^* is contained in $V_n^k - i^{(k)}$ for all but an asymptotically vanishing fraction of $i^{(k)}\in V_n$.

6.3. Pressure and the large deviation principle

We are now ready to state the main theorem of this chapter. Throughout this section, the setting is the one defined in Section 6.1. For $Q \in \mathcal{M}_1(\Omega)$ define $I(Q) = h(Q \mid P)$ if $Q \in \mathcal{M}_\theta(\Omega)$ and $I(Q) = \infty$ otherwise.

Theorem 6.13. *Let μ_n be the distribution of the empirical field R_n under the i.i.d. product measure P. Then* $\mathrm{LDP}(\mu_n, |V_n|, I)$ *holds and I is tight. The same holds for the periodized empirical fields $\widetilde{R}_n$.*

The rest of this section is dedicated to the proof of this theorem. The proof is similar to the proof of Sanov's theorems but the lower bound uses the ergodic theorem instead of the law of large numbers. The upper bound will follow from the general Theorem 4.24 combined with exponential tightness. The first step is to prove that the limit that defines the pressure in (4.2) exists.

Proposition 6.14. *Assume the setting as in the statement of Theorem 6.13. Let f be a bounded measurable local function. Then the limit defining the pressure exists and is the same for R_n and the periodized empirical process $\widetilde{R}_n$:*

$$\begin{aligned} p(f) &= \lim_{n\to\infty} \frac{1}{|V_n|} \log E\left[e^{\sum_{i\in V_n} f\circ\theta_i}\right] \\ &= \lim_{n\to\infty} \frac{1}{|V_n|} \log E\left[e^{|V_n| R_n(\omega,f)}\right] = \lim_{n\to\infty} \frac{1}{|V_n|} \log E\left[e^{|V_n| \widetilde{R}_n(\omega,f)}\right]. \end{aligned}$$

In particular, this holds for all $f \in \mathscr{C}_{b,\mathrm{loc}}(\Omega)$.

Proof. Pick r so that f is $\mathscr{F}_{V_r}$-measurable. Take two integers $m < n$ and let $V_m^{(\ell)} \subset V_n$, $\ell = 1, \dots, k^d$, be k^d shifted copies of V_m arranged so that there is a distance of $2r$ between each adjacent pair in each coordinate direction (Figure 6.1). We can take $k = \lfloor \frac{2n-1}{2m+2r-1} \rfloor$.

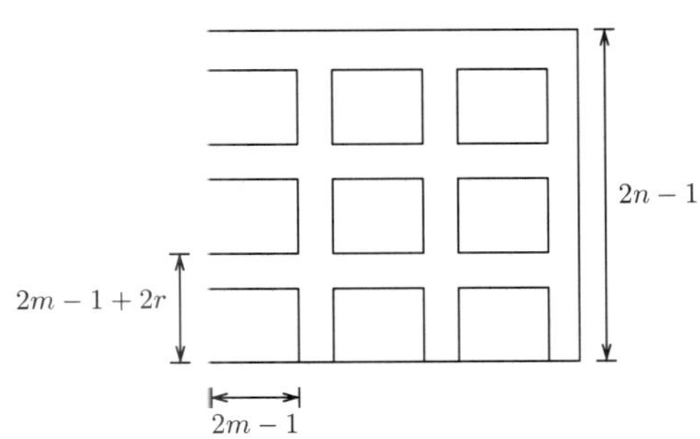

Figure 6.1. Cubes $V_m^{(\ell)}$.

The volume of V_n not covered by the copies of V_m is

$$|V_n| - k^d|V_m| \le |V_n| - \Big(\frac{n-m-r}{m+r}\Big)^d |V_m| = |V_n|\,\kappa_{n,m}$$

with

$$\lim_{m\to\infty}\lim_{n\to\infty}\kappa_{n,m} = \lim_{m\to\infty}\Big(1-\frac{|V_m|}{(2m+2r)^d}\Big) = 0.$$

Since f is $\mathscr{F}_{V_r}$-measurable, the collections $\{f\circ\theta_i : i\in V_m^{(\ell)}\}$ are independent for distinct ℓ. Write

$$\begin{aligned}
(6.6)\qquad p_n(f) &= \frac{1}{|V_n|}\log E\Big[e^{\sum_{i\in V_n} f\circ\theta_i}\Big]\\
&\le \frac{1}{|V_n|}\log E\Big[e^{\sum_{\ell=1}^{k^d}\sum_{i\in V_m^{(\ell)}} f\circ\theta_i}\,e^{\kappa_{n,m}|V_n|\|f\|_\infty}\Big]\\
&= \kappa_{n,m}\,\|f\|_\infty + \frac{k^d}{|V_n|}\log E\Big[e^{\sum_{i\in V_m} f\circ\theta_i}\Big]\\
&\le \kappa_{n,m}\,\|f\|_\infty + \frac{|V_m|}{(2m+2r-1)^d}\,p_m(f).
\end{aligned}$$

Taking $n\to\infty$ and then $m\to\infty$ one has

$$\overline{\lim_{n\to\infty}}\,p_n(f) \le \underline{\lim_{m\to\infty}}\,p_m(f).$$

This proves the existence of the limit $p(f)$. To prove the second claim, write

$$\begin{aligned}
e^{-|V_n|\varepsilon_n}E\Big[\exp\big\{\,|V_n|\,R_n(\omega,f)\big\}\Big] &\le E\Big[\exp\big\{\,|V_n|\,\widetilde{R}_n(\omega,f)\big\}\Big]\\
&\le E\Big[\exp\big\{\,|V_n|\,R_n(\omega,f)\big\}\Big]e^{|V_n|\varepsilon_n},
\end{aligned}$$

where $\varepsilon_n = \sup_\omega|\widetilde{R}_n(\omega,f)-R_n(\omega,f)|\to 0$ by Exercise 6.5. ☐

Next we give a simple upper bound for the pressure p.

Lemma 6.15. *If f is a bounded $\mathscr{F}_{V_m}$-measurable function, then*

$$p(f) \le \frac{1}{|V_m|}\log E[e^{|V_m|f}].$$

Proof. Pick n so that V_n is a union of $r = |V_n|/|V_m|$ disjoint shifted copies of V_m. Let $x_1,\dots,x_r$ be the centers of these copies, so $V_n = \bigcup_{k=1}^r(x_k + V_m)$. Then, use the generalized version of Hölder's inequality (proved by induction) and then independence to write

$$E\Big[e^{\sum_{i\in V_n} f\circ\theta_i}\Big] = E\Big[\prod_{j\in V_m} e^{\sum_{k=1}^r f\circ\theta_{x_k+j}}\Big] \le \prod_{j\in V_m} E\Big[e^{|V_m|\sum_{k=1}^r f\circ\theta_{x_k+j}}\Big]^{1/|V_m|}$$

$$= \prod_{j\in V_m}\prod_{k=1}^r E\big[e^{|V_m|f}\big]^{1/|V_m|} = E\big[e^{|V_m|f}\big]^{|V_n|/|V_m|}.$$

The claim follows. □

Put the space $\mathcal{M}(\Omega)$ of finite Borel measures in duality with $\mathscr{C}_{b,\mathrm{loc}}(\Omega)$. We identify the convex conjugate of p in this duality.

Lemma 6.16. *For $Q \in \mathcal{M}_\theta(\Omega)$, $p^*(Q) = h(Q\,|\,P)$.*

Proof. Let f be a bounded continuous $\mathscr{F}_{V_m}$-measurable function. Then

$$p^*(Q) \ge E^Q[f/|V_m|] - p(f/|V_m|) \ge (E^Q[f] - \log E[e^f])/|V_m|.$$

Taking the sup over such f and using the variational characterization of entropy implies $p^*(Q) \ge H_m(Q\,|\,P)/|V_m|$, for all m, and thus $p^*(Q) \ge h(Q\,|\,P)$.

Conversely, $\bar f = \sum_{i\in V_n} f\circ\theta_i$ is $\mathscr{F}_{V_{m+n}}$-measurable; hence

$$H_{m+n}(Q\,|\,P) \ge E^Q[\bar f] - \log E\Big[e^{\sum_{i\in V_n} f\circ\theta_i}\Big] = |V_n|\,(E^Q[f] - p_n(f)),$$

where $p_n(f)$ was defined in (6.6). Dividing by $|V_n|$ and taking $n\to\infty$ one finds that $h(Q\,|\,P) \ge E^Q[f] - p(f)$ and hence $h(Q\,|\,P) \ge p^*(Q)$. □

***Exercise 6.17.** Prove that if $Q \in \mathcal{M}(\Omega)\setminus\mathcal{M}_\theta(\Omega)$, then $p^*(Q) = \infty$.

Next we take care of exponential tightness.

Lemma 6.18. *For each $b > 0$ there exists a compact set $K_b \subset \mathcal{M}_1(\Omega)$ such that $P\{\widetilde R_n \notin K_b\} \le e^{-|V_n|b}$.*

Proof. By Sanov's theorem (page 73), the distributions of the empirical measures $L_n = |V_n|^{-1}\sum_{i\in V_n}\delta_{X_i}$ form an exponentially tight family of measures on $\mathcal{M}_1(\mathcal{S})$. Thus, for each $m\in\mathbb{N}$ and $j\in\mathbb{Z}^d$, there is a compact $A_{m,j}\subset\mathcal{M}_1(\mathcal{S})$ such that

$$P\{L_n \notin A_{m,j}\} \le e^{-|V_n|(m+|j|)}.$$

Since $A_{m,j}$ is compact, Prohorov's theorem (page 282) implies it is a tight family of probability measures. Hence $\forall m,j$ there exists a compact $U_{m,j}\subset\mathcal{S}$ such that $\mu(U_{m,j}^c) < e^{-(m+|j|)}$ for all $\mu\in A_{m,j}$. Define

$$D_m = \{Q\in\mathcal{M}_1(\Omega) : Q(\omega : \omega_j \in U_{m,j}^c) \le e^{-(m+|j|)}\ \forall j\in\mathbb{Z}^d\}$$

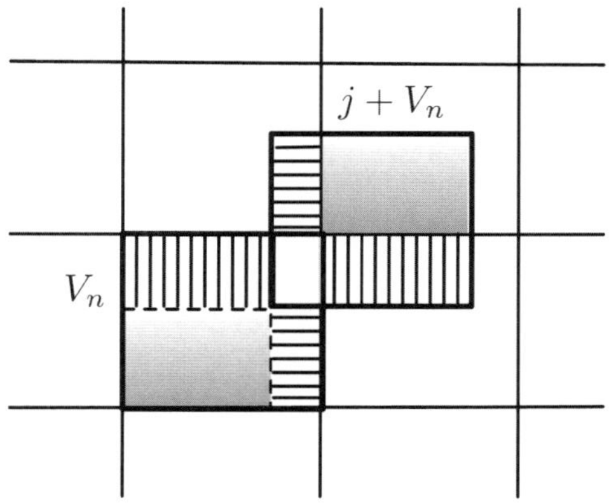

Figure 6.2. The bijection from $j + V_n$ onto V_n.

and observe that if $Q_\ell \in D_m$ converge weakly to Q, then the portmanteau theorem (page 279) implies

$$Q(\omega_j \in U^c_{m,j}) \le \varliminf_{\ell\to\infty} Q_\ell(\omega_j \in U^c_{m,j}) \le e^{-(m+|j|)}.$$

Thus, D_m is closed. Take ℓ large enough such that $\sum_{m\ge\ell-b}\sum_j e^{-(m+|j|)} \le 1$ and define $K_b = \bigcap_{m\ge\ell} D_m$. Then, K_b is a closed tight family of measures and hence, by Prohorov's theorem, is compact. Indeed, for $\varepsilon > 0$ take $m \ge \ell$ such that $\sum_j e^{-(m+|j|)} < \varepsilon$. By Tychonov's theorem (Theorem A3 in [**69**]), $U_m = \prod_{j\in\mathbb{Z}^d} U_{m,j}$ is compact and, for $Q \in K_b$,

$$Q(U^c_m) \le \sum_j Q(\omega_j \in U^c_{m,j}) \le \sum_j e^{-(m+|j|)} < \varepsilon.$$

Finally,

$$\begin{aligned} P\{\widetilde R_n \notin K_b\} &\le \sum_{m\ge\ell} P\{\widetilde R_n \notin D_m\} \\ &\le \sum_{m\ge\ell}\sum_j P\Big\{\widetilde R_n(\omega_j \in U^c_{m,j}) > e^{-(m+|j|)}\Big\}. \end{aligned}$$

But

$$\begin{aligned} \widetilde R_n(\omega_j \in U^c_{m,j}) &= \frac{1}{|V_n|}\sum_{i\in V_n} \delta_{\theta_i\omega^{(n)}}(\omega_j \in U^c_{m,j}) = \frac{1}{|V_n|}\sum_{i\in V_n} \mathbb{1}_{U^c_{m,j}}(\omega^{(n)}_{i+j}) \\ &= \frac{1}{|V_n|}\sum_{\tilde\imath\in V_n} \mathbb{1}_{U^c_{m,j}}(\omega_{\tilde\imath}) = L_n(U^c_{m,j}), \end{aligned}$$

where we observe that there is a bijection $j + i \mapsto \tilde\imath$ from $j + V_n$ onto V_n such that $\omega^{(n)}_{i+j} = \omega_{\tilde\imath}$. See Figure 6.2.

Since $L_n(U^c_{m,j}) > e^{-(m+|j|)}$ implies $L_n \in A^c_{m,j}$, we have

$$P\{\widetilde{R}_n \notin K_b\} \leq \sum_{m\geq \ell}\sum_{j} e^{-|V_n|(m+|j|)} \leq e^{-|V_n|b} \sum_{m\geq \ell-b}\sum_{j} e^{-(m+|j|)} \leq e^{-|V_n|b}.$$

The lemma is proved. □

We are ready to complete the proof of the process level LDP.

Proof of Theorem 6.13. We argue first the LDP for the distributions of $\widetilde{R}_n$ and at the end extend the result to R_n.

Recall that the weak topology on $\mathcal{M}_1(\Omega)$ generated by $\mathscr{C}_b(\Omega)$ is the same as that generated by $\mathscr{C}_{b,\mathrm{loc}}(\Omega)$ (Exercise 6.2). The upper large deviation bound (2.3) for compact sets follows from Theorem 4.24 (with $\mathcal{E} = \mathcal{M}_1(\Omega)$), Proposition 6.14, Lemma 6.16, and Exercise 6.17. Then, by Theorem 2.19, exponential tightness implies the upper bound for all closed sets.

By Lemma 6.9, for the lower bound it suffices to show that for open $G \subset \mathcal{M}_\theta(\Omega)$,

$$\varliminf_{n\to\infty} \frac{1}{|V_n|} \log P\{\widetilde{R}_n \in G\} \geq - \inf_{Q\in G\cap\mathcal{M}_e(\Omega)} h(Q \,|\, P).$$

Thus let $Q \in G$ be ergodic. We may assume $h(Q \,|\, P) < \infty$; otherwise there is nothing to prove. Then we have for all n a Radon-Nikodym derivative $f_n = \frac{dQ_n}{dP_n}$ on $\mathscr{F}_n$. The $\mathscr{F}_n$-measurability of $\widetilde{R}_n$ enables us to perform the change of measure below. Using Jensen's inequality in the third line,

$$\begin{aligned}
\frac{1}{|V_n|}\log P\{\widetilde{R}_n \in G\} &\geq \frac{1}{|V_n|}\log \int \mathbb{1}_G(\widetilde{R}_n) f_n^{-1}\, dQ_n \\
&= \frac{1}{|V_n|}\log Q\{\widetilde{R}_n \in G\} + \frac{1}{|V_n|}\log\Big[\frac{1}{Q\{\widetilde{R}_n \in G\}}\int_{\widetilde{R}_n\in G} f_n^{-1}\, dQ_n\Big] \\
&\geq \frac{1}{|V_n|}\log Q\{\widetilde{R}_n \in G\} - \frac{1}{|V_n|\, Q\{\widetilde{R}_n \in G\}}\int_{\widetilde{R}_n\in G} \log f_n\, dQ_n. \qquad (6.7)
\end{aligned}$$

Use $x\log x \geq -1/e$ for $x > 0$ to write

$$\begin{aligned}
\int_{\widetilde{R}_n\in G} \log f_n\, dQ_n &= \int \log f_n\, dQ_n - \int_{\widetilde{R}_n\in G^c} f_n \log f_n\, dP \\
&\leq H_n(Q \,|\, P) + 1/e.
\end{aligned}$$

Now line (6.7) is bounded below by the expression

$$\frac{1}{|V_n|}\log Q\{\widetilde{R}_n \in G\} - \frac{1}{Q\{\widetilde{R}_n \in G\}}\frac{1}{|V_n|}H_n(Q \,|\, P) - \frac{1}{e\,|V_n|\, Q\{\widetilde{R}_n \in G\}}.$$

By the construction of the weak topology (part (a) of Proposition 4.6) there exists an open neighborhood of Q inside G that is determined by finitely many functions in $\mathscr{C}_{b,\mathrm{loc}}(\Omega)$. By the ergodic theorem $Q\{|\widetilde{R}_n(f) - E^Q[f]| < \varepsilon\} \to 1$ for all such functions f and any $\varepsilon > 0$. Hence $Q\{\widetilde{R}_n \in G\} \to 1$ as $n \to \infty$ and in the limit we get the inequality

$$\varliminf_{n\to\infty} \frac{1}{|V_n|} \log P\{\widetilde{R}_n \in G\} \geq -h(Q \,|\, P).$$

The lower large deviation bound follows by taking the supremum over $Q \in G \cap \mathcal{M}_e(\Omega)$ on the right.

The LDP extends to R_n by Exercises 6.5 and 2.16. The theorem is proved. □

We end this chapter with an exercise on a phenomenon related to the one observed in Section 5.3. For coin tossing this result is elementary, as suggested in Exercise 2.38, but here we do it with large deviation theory. The reader may prefer to state and prove a more general result for i.i.d. sequences.

Exercise 6.19. This exercise shows that if a coin tossing process is conditioned to yield an abnormally large fraction of 1's, the entire process behaves (in the limit) like a sequence of tosses from a biased coin.

Let $\Omega = \{0,1\}^{\mathbb{N}}$ be the sample space for coin tosses, with sample points $\omega = (x_k)_{k\in\mathbb{N}}$ and coordinate variables $X_k(\omega) = x_k$. For $s \in [0,1]$ let $\nu_s = (1-s)\delta_0 + s\delta_1$ be the probability measure on $\{0,1\}$ that gives a 1 with probability s, and let $Q_s = \nu_s^{\otimes\mathbb{N}}$ be the probability measure on Ω under which the $\{X_k\}$ are i.i.d. ν_s-distributed. Let $P = Q_{1/2}$ be the fair coin tossing measure, and let $S_n = X_1 + \cdots + X_n$ be the number of 1's in the first n observations.

Let $1/2 < s \leq 1$. Use large deviation theory to show that $P\{\,\cdot \,|\, S_n \geq ns\}$ converges weakly to Q_s.

Hint: Adapt the proof of the maximum entropy principle (page 79). Observe that if f is a bounded local function on Ω, then

$$E^P[f \,|\, S_n \geq ns] = E^P[R_n(f) \,|\, S_n \geq ns] + \text{a small error}.$$

Use Exercise 6.12 to show that Q_s is the unique minimizer of $h(Q \,|\, P)$ subject to $E^Q[X_0] \geq s$.

Literature notes. Large deviations on levels 1–3 for $\mathbb{R}^d$-valued i.i.d. random vectors are covered in [**32**]. Results on levels 2 or 3 under certain mixing conditions can be found in [**15**] and [**18**].

Part II

Statistical mechanics

Chapter 7

Formalism for classical lattice systems

In Part I of the book we discussed mainly i.i.d. random variables. We turn to constructing and studying models that allow for interaction between spins σ_i. First a product probability measure $\lambda = \lambda_0^{\otimes \mathbb{Z}^d}$ is put on a configuration space $\Omega = \mathcal{S}^{\mathbb{Z}^d}$. This *a priori* or *reference measure* λ represents the absence of interaction. Gibbs measures will be defined by superimposing interaction on λ. Because $\mathbb{Z}^d$ is infinite this task will not be technically trivial.

The point of view is that a probability measure on the configuration space is a natural notion of a *macroscopic state* of the system. The *microscopic state* $\sigma \in \Omega$ involves impossibly many variables to describe, and moreover, these variables are subject to statistical fluctuations. However, macroscopic behavior can still be deterministic. Over the full infinite system, the laws of large numbers apply. Statistical fluctuations average out and yield well-defined limiting averages that can be computed as expectations under the probability measure.

7.1. Finite-volume model

In a finite volume $\Lambda \subset \mathbb{Z}^d$ we can simply modify a product measure with a Radon-Nikodym factor to define a model of interacting spins. Let $\mathcal{S}$ be a Polish space and let $\Omega_\Lambda = \mathcal{S}^\Lambda$, with Borel σ-algebra $\mathscr{F}_\Lambda$ and reference measure $\lambda_\Lambda = \lambda_0^{\otimes\Lambda}$. We no longer distinguish notationally between the spins (coordinates) σ_i of a configuration $\sigma_\Lambda = (\sigma_i)_{i\in\Lambda} \in \Omega_\Lambda$ and the $\mathcal{S}$-valued random variable σ_i on the probability space $(\Omega_\Lambda, \mathscr{F}_\Lambda, \lambda_\Lambda)$. Under λ_Λ the spins $\{\sigma_i\}_{i\in\Lambda}$ are i.i.d. random variables and each σ_i is λ_0-distributed.

A *Hamiltonian* $\mathcal{H}_\Lambda$ is a bounded $\mathscr{F}_\Lambda$-measurable function that gives the energy $\mathcal{H}_\Lambda(\sigma_\Lambda)$ of a configuration $\sigma_\Lambda \in \Omega_\Lambda$. The equilibrium of the interacting spins in the volume Λ is described by the *Gibbs measure* μ_Λ defined by

$$d\mu_\Lambda = \frac{e^{-\beta\mathcal{H}_\Lambda}}{Z_\Lambda}\, d\lambda_\Lambda. \tag{7.1}$$

The normalization constant $Z_\Lambda = E^{\lambda_\Lambda}[e^{-\beta\mathcal{H}_\Lambda}]$ is called the *partition function*, and the inverse temperature is $\beta = 1/T > 0$.

That a measure of the form (7.1) should be the right description of equilibrium is justified by the *equivalence of ensembles*, which states the following. Condition the reference measure on the event $|\mathcal{H}_\Lambda - z| \le \delta$ for an energy value z. Then in a limit where first $\Lambda \nearrow \mathbb{Z}^d$ and then $\delta \searrow 0$, this conditioned measure converges to a Gibbs measure at inverse temperature β determined by z. In the jargon of the field, the *microcanonical ensemble* converges to the *canonical ensemble*. This can be proved rigorously, but we shall not address it in the text. (See the literature notes at the end of the chapter.) Exercise 5.21 gave us a simple example of this principle.

Example 7.1 (Ising model in a finite volume)**.** Consider the Ising ferromagnet $\mathcal{S} = \{-1, +1\}$. The Hamiltonian in a finite-volume Λ is

$$\mathcal{H}_\Lambda(\sigma_\Lambda) = -J \sum_{\{x,y\}\subset\Lambda:|x-y|_1=1} \sigma_x\sigma_y - h\sum_{x\in\Lambda}\sigma_x \tag{7.2}$$

where the first sum is over the nearest-neighbor pairs of spins in Λ, $J > 0$ is the coupling constant, and $h \in \mathbb{R}$ is the external magnetic field.

The equilibrium of the interacting spins is also defined by a variational principle: the *free energy* $-\beta^{-1}\log Z_\Lambda$ (from statistical mechanics) minimizes the Helmholtz free energy (from thermodynamics),

$$-\beta^{-1}\log Z_\Lambda = \inf_{\nu\in\mathcal{M}_1(\Omega_\Lambda)}\{E^\nu[\mathcal{H}_\Lambda] + \beta^{-1}H(\nu\,|\,\lambda_\Lambda)\}. \tag{7.3}$$

As was observed in Exercise 5.19, the infimum is uniquely attained at the Gibbs measure $\nu = \mu_\Lambda$, also called the *equilibrium measure*.

The variational principle reflects a balance between order and randomness. Minimizing the energy term $E^\nu[\mathcal{H}_\Lambda]$ alone leads to measures supported on the minima of $\mathcal{H}_\Lambda$, called the *ground states*. This is extreme order. The entropy $H(\nu\,|\,\lambda_\Lambda)$ is minimized by $\nu = \lambda_\Lambda$, which represents randomness. The Gibbs measure strikes the right balance between these two tendencies. Note also the role of β in (7.3): the higher β, the weaker the influence of entropy. In other words, at lower temperature order dominates. We observed this effect in the Curie-Weiss model (Section 3.4): spontaneous magnetization occurs at low temperature but not at high temperature.

7.2. Potentials and Hamiltonians

In order to observe a phase transition we develop Gibbs measures for infinite volume. Formula (7.1) does not work directly for $\Lambda = \mathbb{Z}^d$ because $\mathcal{H}_{\mathbb{Z}^d}$ can be ill-defined (as in (7.2)). Infinite-volume Gibbs measures are defined indirectly through their conditional probabilities. On the full configuration space $\Omega = \mathcal{S}^{\mathbb{Z}^d}$ let $\mathscr{F}_\Lambda$ denote the σ-algebra generated by the spins $\sigma_\Lambda = (\sigma_i)_{i\in\Lambda}$. The Borel, or product. σ-algebra of Ω is $\mathscr{F}$. Shift maps are defined as before: $(\theta_i\sigma)_j = \sigma_{i+j}$ for a configuration $\sigma = (\sigma_i)_{i\in\mathbb{Z}^d} \in \Omega$ and $i, j \in \mathbb{Z}^d$.

Let a function $\Phi_A(\sigma)$ represent the energy of the interaction that simultaneously involves all the spins in a particular finite set A. Together these energies make up an interaction potential.

Definition 7.2. $\Phi = \{\Phi_A : A \subset \mathbb{Z}^d \text{ finite}, \ \Phi_A : \Omega \to \mathbb{R}\}$ is an *absolutely summable, shift-invariant interaction potential* if for all $i \in \mathbb{Z}^d$ and finite $A \subset \mathbb{Z}^d$ the following hold:

(a) Φ_A is $\mathscr{F}_A$-measurable

(b) $\Phi_{A+i} = \Phi_A \circ \theta_i$,

(c) $\|\Phi\| = \sum_{A:0\in A} \|\Phi_A\|_\infty < \infty$, with $\|\Phi_A\|_\infty = \sup_{\sigma\in\Omega} |\Phi_A(\sigma)|$.

The summability in (c) makes all the infinite sums in the sequel well-defined. Φ is a *continuous* interaction potential if all functions Φ_A are continuous. After some preliminary work this will be a standing assumption. This is general enough and also necessary because the large deviation theory we have developed is for the weak topology of probability measures.

Potential Φ has *finite range* R if $\operatorname{diam}(A) > R$ implies $\Phi_A = 0$. If $\Phi_A = 0$ when A has more than two points ($|A| > 2$), then Φ is a *two-body*, or *pair potential.* A pair potential with range 1 is called a *nearest-neighbor potential.*

The *one-body*, or *self-potential,* $\Phi_{\{i\}}$, for $i \in \mathbb{Z}^d$, does not contribute to interaction and could be subsumed in the a priori product measure $\lambda = \lambda_0^{\otimes\mathbb{Z}^d}$. However, it is conceptually clear to keep it separate. The a priori measure is often canonically associated to the single spin space $\mathcal{S}$, while the self-potential describes the effect of external forces such as a magnetic field.

Let $\Lambda \subset \mathbb{Z}^d$ be finite. The *Hamiltonian* $\mathcal{H}_\Lambda^{\text{free}} = \mathcal{H}_\Lambda^{\Phi,\text{free}}$ for the volume Λ with *free boundary condition* (no influence from Λ^c) is the bounded $\mathscr{F}_\Lambda$-measurable function

$$\mathcal{H}_\Lambda^{\text{free}} = \sum_{A:A\subset\Lambda} \Phi_A.$$

The interaction between spins in Λ and Λ^c is described by

$$W_{\Lambda,\Lambda^c} = W_{\Lambda,\Lambda^c}^{\Phi} = \sum_{\substack{A:A\cap\Lambda\neq\varnothing \\ A\cap\Lambda^c\neq\varnothing}} \Phi_A.$$

Now fix a configuration $\tau_{\Lambda^c} \in \Omega_{\Lambda^c}$ on the sites outside Λ. The Hamiltonian $\mathcal{H}_\Lambda^{\tau_{\Lambda^c}} = \mathcal{H}_\Lambda^{\Phi,\tau_{\Lambda^c}}$ for Λ with boundary condition τ_{Λ^c} is

$$\mathcal{H}_\Lambda^{\tau_{\Lambda^c}}(\sigma_\Lambda) = \mathcal{H}_\Lambda^{\mathrm{free}}(\sigma_\Lambda) + W_{\Lambda,\Lambda^c}(\sigma_\Lambda, \tau_{\Lambda^c}) = \sum_{A:A\cap\Lambda\neq\varnothing} \Phi_A(\sigma_\Lambda, \tau_{\Lambda^c}).$$

When a boundary condition τ_{Λ^c} has been specified, the Gibbs measure $\pi_\Lambda(\tau, \cdot)$ of the spins in Λ is defined in terms of the integral of a bounded test function f on Ω_Λ by

$$\int_{\Omega_\Lambda} f(\sigma_\Lambda)\, \pi_\Lambda(\tau, d\sigma_\Lambda) = \frac{1}{Z_\Lambda^{\tau_{\Lambda^c}}} \int_{\Omega_\Lambda} f(\sigma_\Lambda) e^{-\beta \mathcal{H}_\Lambda^{\tau_{\Lambda^c}}(\sigma_\Lambda)}\, \lambda_\Lambda(d\sigma_\Lambda). \tag{7.4}$$

We can equally well think of $\pi_\Lambda(\tau, \cdot)$ as a measure on the full space Ω, in which case the integral of a test function $f : \Omega \to \mathbb{R}$ is

$$\int_{\Omega} f(\sigma)\, \pi_\Lambda(\tau, d\sigma) = \frac{1}{Z_\Lambda^{\tau_{\Lambda^c}}} \int_{\Omega_\Lambda} f(\sigma_\Lambda, \tau_{\Lambda^c}) e^{-\beta \mathcal{H}_\Lambda^{\tau_{\Lambda^c}}(\sigma_\Lambda)}\, \lambda_\Lambda(d\sigma_\Lambda). \tag{7.5}$$

As before, $Z_\Lambda^{\tau_{\Lambda^c}} = \int_{\Omega_\Lambda} e^{-\beta \mathcal{H}_\Lambda^{\tau_{\Lambda^c}}(\sigma_\Lambda)}\, \lambda_\Lambda(d\sigma_\Lambda)$ is the normalization constant called the *partition function.*

Example 7.3. Here are several standard examples of statistical mechanical models that fall under this general definition.

Nearest-neighbor Ising model. The single spin space is $\mathcal{S} = \{-1, +1\}$ with uniform reference measure $\lambda_0(\pm 1) = 1/2$. The potential is given by

$$\Phi_A(\sigma) = \begin{cases} -J\sigma_i\sigma_j, & A = \{i, j\} \text{ with } |i - j|_1 = 1, \\ -h\sigma_i, & A = \{i\}, \\ 0, & \text{otherwise.} \end{cases}$$

Here $|i|_1$ is the ℓ^1 norm and thus $|i - j|_1 = 1$ means that i and j are nearest neighbors. This is also expressed by writing $i \sim j$. Some texts denote nearest-neighbor pairs by $\langle i, j\rangle$. If the coupling constant $J > 0$ this is the Ising ferromagnet, and if $J < 0$, this is the Ising antiferromagnet.

Long-range Ising model. More generally, we can allow couplings over arbitrarily long distances, as long as the summability condition is satisfied. The pair potential is $\Phi_{\{i,j\}}(\sigma) = -J_{j-i}\sigma_i\sigma_j$ for $i \neq j$ in $\mathbb{Z}^d$ and satisfies

$$\sum_{j\in\mathbb{Z}^d} |J_j| < \infty.$$

Potts model. This is a generalization of the Ising model to q spin values in the space $\mathcal{S} = \{1, 2, \ldots, q\}$. The nearest-neighbor pair interaction is $\Phi_{\{i,j\}}(\sigma) = -J\mathbb{1}_{\sigma_i=\sigma_j}$.

Classical Heisenberg model. This is analogous to the Ising model but with $\mathcal{S} = \mathbb{S}^2$, the two-dimensional sphere in $\mathbb{R}^3$ (surface of the three-dimensional unit ball). The nearest-neighbor couplings are $\Phi_{\{i,j\}}(\sigma) = -J\sigma_i \cdot \sigma_j$ (scalar product of 3-vectors). If the spins are two-dimensional unit vectors $\sigma_i = (\cos\theta_i, \sin\theta_i)$, then this is called the *XY model*, or the *rotor model*. The general case of n-dimensional unit vectors is the *n-vector model*.

Lattice gas. The Ising model can be used to model atoms on a lattice. Take $\mathcal{S} = \{0, 1\}$ where 0 stands for an empty site and 1 stands for a site occupied by an atom. The interaction is $\Phi_A(\sigma) = -J(A)\prod_{i\in A}\sigma_i$ for positive coupling constants J that satisfy $J(\theta_i A) = J(A)$. For example, taking $J(\{i,j\}) = J > 0$ for nearest-neighbor pairs $\{i,j\}$ and $J(A) = 0$ otherwise would model particles that are attracted to each other.

***Exercise 7.4.** Show that the Gibbs kernels defined by (7.5) for a potential satisfying Definition 7.2 have the following shift invariance: for $f \in b\mathscr{F}$ on Ω,

$$\int_\Omega f(\theta_{-i}\sigma)\,\pi_\Lambda(\tau, d\sigma) = \int_\Omega f(\sigma)\,\pi_{\Lambda+i}(\theta_{-i}\tau, d\sigma).$$

7.3. Specifications

The finite-volume Gibbs measures with boundary conditions in (7.5) are a special case of a more general family of stochastic kernels called specifications. We discuss these next.

Given two measurable spaces $(\mathcal{Y}, \mathscr{C})$ and $(\mathcal{Z}, \mathscr{D})$, a *stochastic kernel* π from $(\mathcal{Y}, \mathscr{C})$ to $(\mathcal{Z}, \mathscr{D})$ is a map $\pi : \mathcal{Y} \times \mathscr{D} \to [0, 1]$ such that

(a) $\pi(y, \cdot) \in \mathcal{M}_1(\mathcal{Z}, \mathscr{D})$ for each $y \in \mathcal{Y}$,

(b) $y \mapsto \pi(y, D)$ is $\mathscr{C}$-measurable for all $D \in \mathscr{D}$.

We will often write $\pi(y)$ for the probability measure $\pi(y, \cdot)$.

The physical perspective would be that the stochastic kernel describes the equilibrium in $\mathcal{Z}$, given a point $y \in \mathcal{Y}$. From a dynamical systems point of view, a stochastic kernel is a generalization of a deterministic map. A stochastic kernel acts as a linear transformation that pulls functions back and pushes measures forward. The $b\mathscr{D} \to b\mathscr{C}$ map $f \mapsto \pi f$ is defined by

$$\pi f(y) = \int_{\mathcal{Z}} f(z)\pi(y, dz), \quad y \in \mathcal{Y},$$

and the $\mathcal{M}(\mathcal{Y}, \mathscr{C}) \to \mathcal{M}(\mathcal{Z}, \mathscr{D})$ map $\mu \mapsto \mu\pi$ is defined by

$$\mu\pi(D) = \int_{\mathcal{Y}} \pi(y, D)\mu(dy), \quad D \in \mathscr{D}.$$

Sometimes we will write $\pi^y(D)$, $\pi^y f$, and $\pi(f)$ instead of $\pi(y, D)$, $\pi f(y)$, and πf, respectively.

Basic examples of stochastic kernels in probability theory are Markov transition probabilities and conditional probability measures.

Example 7.5. The transition probability $P(x, dy)$ of a time-homogeneous Markov chain is a stochastic kernel on the state space of the chain.

Example 7.6. Let $\mathcal{X}$ be a Polish space, let $\mathscr{B}_{\mathcal{X}}$ be its Borel σ-algebra, and let $\mu \in \mathcal{M}_1(\mathcal{X})$. Let $\mathscr{A} \subset \mathscr{B}_{\mathcal{X}}$ be a sub-σ-algebra of $\mathscr{B}_{\mathcal{X}}$. Then there exists a *conditional probability measure* for μ, given $\mathscr{A}$, which is a stochastic kernel π from $(\mathcal{X}, \mathscr{A})$ to $(\mathcal{X}, \mathscr{B}_{\mathcal{X}})$ such that $E^\mu[fg] = E^\mu[f\,\pi g]$ for all bounded $\mathscr{A}$-measurable functions f and all bounded $\mathscr{B}_{\mathcal{X}}$-measurable functions g. See Theorems B.4 and B.5.

Stochastic kernels can be composed. If π is a kernel from $(\mathcal{Y}, \mathscr{C})$ to $(\mathcal{Z}, \mathscr{D})$ and ρ is a kernel from $(\mathcal{Z}, \mathscr{D})$ to $(\mathcal{W}, \mathscr{E})$, then

$$\pi\rho(y, E) = \int \rho(z, E)\pi(y, dz), \quad E \in \mathscr{E},$$

is a kernel from $(\mathcal{Y}, \mathscr{C})$ to $(\mathcal{W}, \mathscr{E})$.

***Exercise 7.7.** Check that the composition gives a stochastic kernel and the properties $(\pi\rho)f = \pi(\rho f)$ and $\mu(\pi\rho) = (\mu\pi)\rho$.

Definition 7.8. A *specification* is a family of stochastic kernels $\Pi = \{\pi_\Lambda : \Lambda \subset \mathbb{Z}^d \text{ finite}\}$ such that

(a) π_Λ is a stochastic kernel from $(\Omega, \mathscr{F}_{\Lambda^c})$ to $(\Omega, \mathscr{F})$,

(b) π_Λ is $\mathscr{F}_{\Lambda^c}$-proper: if $B \in \mathscr{F}_{\Lambda^c}$, then $\pi_\Lambda(\sigma, B) = \mathbb{1}_B(\sigma)$ for all $\sigma \in \Omega$, or succinctly $\pi_\Lambda \mathbb{1}_B = \mathbb{1}_B$,

(c) if $\Delta \subset \Lambda \subset \mathbb{Z}^d$ are finite, then $\pi_\Lambda \pi_\Delta = \pi_\Lambda$.

The restriction of $\pi_\Lambda(\sigma, \cdot)$ to $\mathscr{F}_\Lambda$ can be thought of as the equilibrium probability for the volume Λ, given the boundary condition σ_{Λ^c} outside Λ. Then (c) means that equilibrium in a volume Λ is compatible with the equilibria of all its subvolumes Δ.

***Exercise 7.9.** (a) Show that part (b) of Definition 7.8 implies $\pi_\Lambda(fg) = g\pi_\Lambda f$ for $g \in b\mathscr{F}_{\Lambda^c}$ and $f \in b\mathscr{F}$.

(b) Let $\Delta \subset \Lambda \subset \mathbb{Z}^d$ be finite. Show that $\pi_\Delta \pi_\Lambda = \pi_\Lambda$. (Compare with part (c) of Definition 7.8.)

Hint: For (a) it is enough to consider $f = \mathbb{1}_A$, $g = \mathbb{1}_B$. Start with $\pi(\mathbb{1}_A \mathbb{1}_B) \leq \pi\mathbb{1}_A \wedge \pi\mathbb{1}_B$.

Exercise 7.10. Suppose $\pi_\Lambda f = E^P[f \,|\, \mathscr{F}_{\Lambda^c}]$ for a probability measure P and all finite Λ. Show that condition (c) of Definition 7.8 is then the familiar consistency condition for conditional expectations, namely for finite $\Delta \subset \Lambda$,

$$E^P[f \,|\, \mathscr{F}_{\Lambda^c}] = E^P\big(E^P[f \,|\, \mathscr{F}_{\Delta^c}] \,\big|\, \mathscr{F}_{\Lambda^c}\big).$$

***Exercise 7.11** (Gibbs specifications). Let Φ be an absolutely summable shift-invariant interaction potential that defines the finite-volume Gibbs kernels $\pi_\Lambda : (\Omega, \mathscr{F}_{\Lambda^c}) \to (\Omega, \mathscr{F})$ in (7.5). Prove that $\Pi^\Phi = \{\pi_\Lambda : \Lambda \subset \mathbb{Z}^d \text{ finite}\}$ is a specification. Specifications defined by interaction potentials are called *Gibbs specifications.*

We come to a key definition. The criterion in the definition makes sense because $\pi_\Lambda f$ is $\mathscr{F}_{\Lambda^c}$-measurable. See Section B.3 for the definition and a review of conditional expectation.

Definition 7.12. The set of *Gibbs measures* of a specification Π is

$$\mathscr{G}^\Pi = \{\mu \in \mathcal{M}_1(\Omega) : E^\mu[f|\mathscr{F}_{\Lambda^c}] = \pi_\Lambda f \ \mu\text{-a.s.}, \ \forall f \in b\mathscr{F}, \ \forall \text{ finite } \Lambda \subset \mathbb{Z}^d\}.$$

There is a small inconsistency in terminology: a Gibbs specification is a special type of specification, but a Gibbs measure is a notion relevant for general specifications.

The existence and uniqueness of Gibbs measures are the central questions. To work towards some answers we weaken the conditions for membership in $\mathscr{G}^\Pi$. Recall the definition of the rectangles $V_n = \{i \in \mathbb{Z}^d : -n < i_1, \dots, i_d < n\}$. If Π is defined by a potential Φ, we may write $\mathscr{G}^\Phi$ for $\mathscr{G}^\Pi$.

Lemma 7.13. *The following statements are equivalent.*

(a) $\mu \in \mathscr{G}^\Pi$.

(b) $\mu\pi_\Lambda = \mu \ \forall$ *finite* $\Lambda \subset \mathbb{Z}^d$.

(c) $\mu\pi_{V_n} = \mu \ \forall n$.

Proof. The implication (a) $\Rightarrow$ (b) follows from $E^\mu(E^\mu[f|\mathscr{F}_{\Lambda^c}]) = E^\mu[f]$. (b) trivially implies (c). To prove that (c) implies (b) suppose $\Lambda \subset V_n$. Then

$$\mu\pi_\Lambda = \mu\pi_{V_n}\pi_\Lambda = \mu(\pi_{V_n}\pi_\Lambda) = \mu\pi_{V_n} = \mu.$$

Finally, (b) implies (a) because if $g \in b\mathscr{F}_{\Lambda^c}$ and $f \in b\mathscr{F}$, then by Exercise 7.9,

$$E^\mu[fg] = E^\mu[\pi_\Lambda(fg)] = E^\mu[g\,\pi_\Lambda f].$$ □

Exercise 7.14. Show that $\mathscr{G}^\Pi$ is a convex set. Prove additionally that

$$\begin{aligned}\mathscr{G}^\Pi &= \{\mu \in \mathcal{M}_1(\Omega) : E^\mu[f|\mathscr{F}_{\Lambda^c}] = \pi_\Lambda f, \ \forall f \in \mathscr{C}_b(\Omega), \ \forall \text{ finite } \Lambda \subset \mathbb{Z}^d\} \\ &= \{\mu \in \mathcal{M}_1(\Omega) : E^\mu[f|\mathscr{F}_{\Lambda^c}] = \pi_\Lambda f, \ \forall f \in \mathscr{C}_{b,\text{loc}}(\Omega), \ \forall \text{ finite } \Lambda \subset \mathbb{Z}^d\}.\end{aligned}$$

Hint: The proof of the lemma above works word for word with these alternate definitions of $\mathscr{G}^\Pi$. You can also use Lemma A.11.

Definition 7.15. A specification Π is *Feller-continuous* if $\pi_\Lambda f \in \mathscr{C}_b(\Omega)$ for all $f \in \mathscr{C}_b(\Omega)$ and all finite $\Lambda \subset \mathbb{Z}^d$.

Exercise 7.16. Continuing from Exercise 7.11, show that if $\Phi \subset \mathscr{C}_b(\Omega)$, then Π^Φ is a Feller-continuous specification.

Theorem 7.17. *Suppose Π is a Feller-continuous specification. Let Γ_n be any sequence of finite subsets of $\mathbb{Z}^d$ that exhausts $\mathbb{Z}^d$, in the sense that any finite $\Lambda \subset \mathbb{Z}^d$ is included in Γ_n for all large enough n. Let $\nu_n \in \mathcal{M}_1(\Omega)$ be any sequence. If $\nu_n \pi_{\Gamma_n}$ converges weakly to some $\mu \in \mathcal{M}_1(\Omega)$, then $\mu \in \mathscr{G}^\Pi$.*

Proof. Fix a finite $\Lambda \subset \mathbb{Z}^d$ and $f \in \mathscr{C}_b(\Omega)$. Then $\pi_\Lambda f \in \mathscr{C}_b(\Omega)$ and

$$E^\mu[\pi_\Lambda f] = \lim_{n\to\infty} E^{\nu_n}[\pi_{\Gamma_n}\pi_\Lambda f] = \lim_{n\to\infty} E^{\nu_n}[\pi_{\Gamma_n} f] = E^\mu[f]. \qquad \square$$

Corollary 7.18. *If $\mathcal{S}$ is compact and Π is a Feller-continuous specification, $\mathscr{G}^\Pi$ is not empty.*

Proof. If $\mathcal{S}$ is compact, so is $\mathcal{M}_1(\Omega)$, and any sequence $\nu_n \pi_{\Gamma_n}$ has a convergent subsequence. □

The occurrence of multiple Gibbs measures is significant. It means the system has several possible equilibria to choose from. The following definition was introduced by Dobrushin [**23**].

Definition 7.19. A *phase transition* is said to occur if $\mathscr{G}^\Pi$ has more than one element.

We give two proofs of the occurrence of phase transition in the low temperature Ising model: in Chapter 9 with the classical Peierls argument and in Chapter 10 with a percolation argument. The next section sheds some light on the physical significance of this definition.

We record the simplest examples of Gibbs measures, namely product measures (where interaction is absent) and Markov chains.

Exercise 7.20 (Product measure as Gibbs measure)**.** Let $\nu_i \in \mathcal{M}_1(\mathcal{S})$ for $i \in \mathbb{Z}^d$ be given. Define the product measures $\nu = \bigotimes_{i\in\mathbb{Z}^d} \nu_i$ on Ω and $\nu_\Lambda = \bigotimes_{i\in\Lambda} \nu_i$ on Ω_Λ. For $\tau \in \Omega$ define probability measures $\pi_\Lambda(\tau, \cdot\,)$ on Ω through the integrals

$$\pi_\Lambda(\tau, f) = \int_{\Omega_\Lambda} f(\omega_\Lambda, \tau_{\Lambda^c})\, \nu_\Lambda(d\omega_\Lambda) \qquad \text{for } f \in b\mathscr{F}.$$

Show that $\{\pi_\Lambda : \Lambda \subset \mathbb{Z}^d \text{ finite}\}$ is a specification whose unique Gibbs measure is ν.

Example 7.21 (Finite Markov chain)**.** Another simple example of a unique Gibbs measure is a stationary Markov chain on a finite state space with a strictly positive transition matrix. Let $\mathcal{S} = \{1, 2, \dots, s\}$ and consider a transition probability matrix $P = (p(x,y))_{x,y\in\mathcal{S}}$ that satisfies $p(x,y) > 0$. Such a Markov chain has a unique invariant distribution $\alpha \in \mathcal{M}_1(\mathcal{S})$. Let

$\nu_\alpha \in \mathcal{M}_1(\mathcal{S}^{\mathbb{Z}})$ be the distribution of the stationary Markov chain, characterized by

$$\text{(7.6)} \qquad \nu_\alpha\{\sigma_k = x_k, \sigma_{k+1} = x_{k+1}, \dots, \sigma_m = x_m\} = \alpha(x_k) \prod_{i=k}^{m-1} p(x_i, x_{i+1})$$

for $k < m$ in $\mathbb{Z}$ and $x_k, \dots, x_m \in \mathcal{S}$. (In stochastic process texts a Markov chain is typically restricted to nonnegative time indices, but a stationary process on $\mathbb{Z}_+$ can be uniquely extended to $\mathbb{Z}$.) To find a potential that characterizes ν_α, let us look at conditional probabilities for guidance. $P^k = (p^k(x,y))_{x,y\in\mathcal{S}}$ denotes the k-step transition matrix. For $x, y, z \in \mathcal{S}$

$$\nu_\alpha(\sigma_k = y \,|\, \sigma_{k-1} = x, \sigma_{k+1} = z) = \frac{\alpha(x)p(x,y)p(y,z)}{\alpha(x)p^2(x,z)}$$

$$= \frac{p(x,y)p(y,z)}{p^2(x,z)} = \frac{e^{\Phi_{k-1,k}(x,y)+\Phi_{k,k+1}(y,z)}\lambda_0(y)}{\sum_{u\in\mathcal{S}} e^{\Phi_{k-1,k}(x,u)+\Phi_{k,k+1}(u,z)}\lambda_0(u)}$$

with uniform reference measure $\lambda_0(x) = 1/s$ and

$$\text{(7.7)} \qquad \Phi_{k,k+1}(\sigma) = \log p(\sigma_k, \sigma_{k+1}).$$

Now let us take the potential (7.7) as given and prove that $\mathscr{G}^\Phi = \{\nu_\alpha\}$. The existence of $\mu \in \mathscr{G}^\Phi$ is guaranteed by Corollary 7.18. Consider the event in (7.6). Let $n \in \mathbb{N}$. Then

$$\mu\{(\sigma_k, \dots, \sigma_m) = (x_k, \dots, x_m)\}$$
$$= \int \mu\{(\sigma_k, \dots, \sigma_m) = (x_k, \dots, x_m) \,|\, \mathscr{F}_{\{k-n,\dots,m+n\}^c}\}(\tau)\, \mu(d\tau)$$
$$= \int \frac{p^{n+1}(\tau_{k-n-1}, x_k) \cdot \prod_{i=k}^{m-1} p(x_i, x_{i+1}) \cdot p^{n+1}(x_m, \tau_{m+n+1})}{p^{m-k+2n+2}(\tau_{k-n-1}, \tau_{m+n+1})}\, \mu(d\tau).$$

According to the convergence theorem of irreducible aperiodic finite Markov chains (Exercise 5.16 or [**27**, Theorem 6.6.4]), $p^n(x,y) \to \alpha(y)$. Let $n \to \infty$ above. The ratio

$$\frac{p^{n+1}(x_m, \tau_{m+n+1})}{p^{m-k+2n-2}(\tau_{k-n-1}, \tau_{m+n+1})} \to 1$$

and we conclude that

$$\mu\{(\sigma_k, \dots, \sigma_m) = (x_k, \dots, x_m)\} = \alpha(x_k) \prod_{i=k}^{m-1} p(x_i, x_{i+1}).$$

Thus $\mu = \nu_\alpha$ is the unique Gibbs measure.

Remark 7.22 (Markov random field). Suppose Φ is a nearest-neighbor pair potential; that is, $\Phi_A = 0$ unless A is a singleton or a nearest-neighbor pair. Let $\Lambda \subset \mathbb{Z}^d$ be finite and let $f \in b\mathscr{F}_\Lambda$. Then a Gibbs measure μ satisfies $E^\mu[f|\mathscr{F}_{\Lambda^c}] = E^\mu[f|\mathscr{F}_{\partial\Lambda}]$ where $\partial\Lambda = \{i \in \mathbb{Z}^d \smallsetminus \Lambda : \exists j \in \Lambda \text{ such that } i \sim j\}$

is the outer boundary of Λ. This property of μ can be regarded as a spatial Markov property, and a measure with this property is called a *Markov field*.

7.4. Phase transition

This section is an interlude that explains some of the physical meaning of the mathematics introduced thus far. It can be skipped without loss of continuity.

Consider a Gibbs specification $\Pi = \{\pi_\Lambda : \Lambda \subset \mathbb{Z}^d \text{ finite}\}$ defined as in (7.5) by an interaction potential Φ and an inverse temperature $\beta > 0$. Elements of $\mathscr{G}^\Pi$ are *Gibbs measures* for the potential Φ at inverse temperature β. The equations $\mu\pi_\Lambda = \mu$ defining $\mathscr{G}^\Pi$ are called the *DLR equations*. This idea was introduced in 1968 by Dobrushin [**23**] and in 1969 by Lanford and Ruelle [**57**].

What makes these models interesting is that they are realistic enough to account for the physical phenomenon known as phase transition. Physically, a *phase transition* is an abrupt change in the physical properties of a system, like water boiling into steam.

Imagine a family of absolutely summable shift-invariant continuous interaction potentials $\{\Phi_\alpha\}$. The index α refers to thermodynamic variables other than temperature, such as the external magnetic field (denoted by h) or the pressure (denoted by P). Since all thermodynamic quantities can be expressed in terms of the free energy and its derivatives, one way to characterize the occurrence of a phase transition at (β, α) is by the formation of singularities in the free energy (that is, one of its derivatives becomes discontinuous).

***Exercise 7.23.** Observe that the finite-volume Curie-Weiss model (Section 3.4) does not have a phase transition at any $\beta > 0$ and $h \in \mathbb{R}$. Prove that the infinite-volume limit of the Curie-Weiss model (Theorem 3.18) exhibits a phase transition at points (β, h) with $h = 0$ and $\beta \geq 1/J$.

Hint: Use the analytic implicit function theorem (page 34 of [**37**]) to show that the magnetization $m(\beta, h)$ is analytic when $h \neq 0$ or $\beta < 1/J$. Also, show that m has a jump when $\beta > 1/J$ and h crosses 0 and that $m(\beta, +) \sim \sqrt{3(J\beta - 1)}$ as $\beta \searrow 1/J$. Next, apply Varadhan's theorem (page 38) to find that the limit of the free energy per spin is given by

$$-p(\beta, h) = -\lim_{n\to\infty} \frac{1}{n\beta} \log \int e^{-\beta \mathcal{H}_n} dP_n = \beta^{-1} I(m) - \frac{J}{2} m^2 - hm. \tag{7.8}$$

Use the equation that m solves to show that $m = \frac{\partial p}{\partial h}$, when $h \neq 0$ or $\beta < 1/J$.

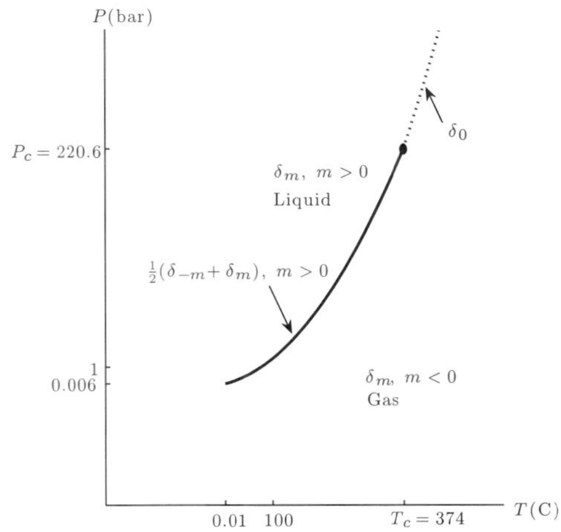

Figure 7.1. The phase diagram for the liquid-gas transition in water. $T = 1/\beta$ and $m = m(\beta, h)$ are as in Theorem 3.18. The solid line represents the values of temperature and pressure at which a phase transition occurs. In the Curie-Weiss and Ising models this corresponds to $\beta > \beta_c$ and $h = 0$.

The free energy of a finite system has no singularities, as was also observed in the last exercise. In the infinite-volume limit, however, singularities can appear. Convergence to the thermodynamic limit implies that the phase behavior is visible already in large but finite systems, even though the singularities are smoothed out by the system's finite size.

Definition 7.19 above defined phase transition as nonuniqueness of Gibbs measures. This does correspond to some degree to the physical notion of phase transition, as we now explain. We refer to the Curie-Weiss and Ising models as examples, although the reader will not see the Ising phase transition until Chapter 9. For simplicity we put the discussion in the context of the more familiar liquid-gas transition instead of the magnetization transition. To this end interpret $\sigma_i = +1$ as "site i is occupied by a water molecule" and $\sigma_i = -1$ as "site i is empty". The empirical mean S_n/n is then linearly related to the density of particles. A positive mean indicates a liquid phase, while a negative mean characterizes the gas phase.

The parameter β is still the inverse temperature, while the parameter h is now called the *chemical potential*. The more familiar physical quantity is pressure. The relation between pressure and chemical potential is complicated, but increasing h corresponds to increasing the pressure. The phase diagram in Figure 3.2 is then transformed into the one in Figure 7.1.

This figure gives the liquid-gas part of the experimental phase diagram for water. (We leave out the liquid-solid phase transition.) It shows that if pressure is held constant and temperature is increased, water will turn into steam at a critical temperature: for example, if the pressure is equal to 1 bar, then water boils at the familiar 100°C.

Qualitatively (ignoring the scales on the axes and the precise equation of the curve) Figure 7.1 corresponds to the phase diagram of both the Curie-Weiss and the two-dimensional Ising model, with the chemical potential h replaced by the pressure. This is why we get a curve instead of a straight line $h = 0$.

Now we can see Definition 7.19 in this picture. Along the line of phase transition the average density takes one of two values with equal probability. In the Ising model, this corresponds to having two Gibbs measures. Away from this line the average density is unique. This corresponds to uniqueness of the Gibbs measure in the Ising model. This behavior of the Ising model will be established in Theorem 9.2.

It is noteworthy that Exercise 7.23 shows that at critical pressure P_c and critical temperature T_c free energy develops a singularity. However, Theorem 3.18 states that the average density is unique at these values. It is also known [**3**] that the Gibbs measure is unique for the Ising model at the critical point when $d \geq 2$. The two definitions of phase transition then seem to coincide everywhere but at this critical point. We could fix this by slightly changing Definition 7.19 to say that there is no phase transition at (β, α) if the Gibbs measure is unique in a whole neighborhood of (β, α).

With this last modification, uniqueness of the Gibbs distribution does coincide with the other definition of phase transition in many important models (like the Ising model). However, it is not always a good definition for phase transition in a common physical way of thinking and will not coincide with the other definition in general. For example, the plane rotator model (Ising spins replaced by vectors on the unit circle) exhibits a Kosterlitz-Thouless transition in which, for β above a certain β_c, exponential decay of correlations is replaced by power-law decay [**39**]. On the other hand, it is also known (at least partly rigorously, but based on older physics arguments) that there is a unique Gibbs state [**11**], [**59**], [**38**].

7.5. Extreme Gibbs measures

A separation of microscopic and macroscopic scales can be described with the mathematical formalism we have introduced. A microscopic observation should involve only finitely many spins, while a macroscopic observation should not depend on any finite collection of spins. Consequently a microscopic observable is a real-valued $\mathscr{F}_\Lambda$-measurable function on Ω for a finite $\Lambda \subset \mathbb{Z}^d$. A macroscopic observable must be measurable with respect to the tail σ-algebra $\mathscr{T} = \bigcap \mathscr{F}_{\Lambda^c}$, where the intersection is over all finite $\Lambda \subset \mathbb{Z}^d$. An alternative that also appears in the literature is to require that a macroscopic observable be shift invariant, that is, $\mathcal{I}$-measurable (the shift-invariant σ-algebra $\mathcal{I}$ was defined in (6.1)).

To illustrate, if f is a microscopic observable, then the limiting average $\lim_{n\to\infty} |V_n|^{-1} \sum_{i\in V_n} f\circ\theta_i$, if it exists, is both $\mathscr{T}$- and $\mathcal{I}$-measurable and so qualifies as a macroscopic observable. Only a constant function can be both a microscopic and a macroscopic observable.

Next consider a probability measure μ that represents a macrostate. We should require that macroscopic observables be deterministic under μ because in equilibrium only microscopic quantities should fluctuate. In other words, μ should satisfy a 0-1 law on either $\mathscr{T}$ or $\mathcal{I}$, depending on the definition we take. This criterion singles out the extreme Gibbs measures as the physically relevant measures. (Extreme points of a convex set were defined on page 84.)

Theorem 7.24. *Let Π be a specification defined by an absolutely summable shift-invariant interaction potential Φ.*

(a) *$\mu \in \mathscr{G}^\Pi$ is $\mathscr{T}$-trivial if and only if μ is an extreme point of $\mathscr{G}^\Pi$.*

(b) *$\mu \in \mathscr{G}^\Pi \cap \mathcal{M}_\theta(\Omega)$ is $\mathcal{I}$-trivial (in other words, ergodic) if and only if μ is an extreme point of $\mathscr{G}^\Pi \cap \mathcal{M}_\theta(\Omega)$.*

Proof. Part (a). Assume first that $\mathscr{T}$ is not trivial under $\mu \in \mathscr{G}^\Pi$. So $0 < \mu(A) < 1$ for some $A \in \mathscr{T}$. Define $\nu_1(B) = \mu(B|A)$ and $\nu_2(B) = \mu(B|A^c)$. Then $\nu_1 \neq \nu_2$ and $\mu = t\nu_1 + (1-t)\nu_2$ for $t = \mu(A) \in (0,1)$. Moreover, ν_1 and ν_2 belong to $\mathscr{G}^\Pi$. Indeed, if $\Lambda \subset \mathbb{Z}^d$ is finite, then

$$\begin{aligned}\nu_1\pi_\Lambda(B) &= \int \pi_\Lambda(\sigma, B)\nu_1(d\sigma) = \mu(A)^{-1}\int \mathbb{1}_A(\sigma)\pi_\Lambda\mathbb{1}_B(\sigma)\mu(d\sigma)\\ &= \mu(A)^{-1}\int \pi_\Lambda(\sigma, A\cap B)\mu(d\sigma) = \mu(B|A) = \nu_1(B).\end{aligned}$$

We used Exercise 7.9 for $\mathbb{1}_A\pi_\Lambda\mathbb{1}_B = \pi_\Lambda\mathbb{1}_{A\cap B}$. A similar computation works for ν_2. Thus μ is not an extreme point of $\mathscr{G}^\Pi$.

Suppose $\mathscr{T}$ is trivial under μ. Let $\mu = t\nu_1 + (1-t)\nu_2$ for ν_1 and ν_2 in $\mathscr{G}^\Pi$ and $t \in (0,1)$. We argue that $\nu_1 = \mu$. Since $\nu_1 \ll \mu$, $\mathscr{T}$ is trivial also under ν_1. Let $f \in b\mathscr{F}$. Since

$$\pi_{V_n} f = E^\mu[f \,|\, \mathscr{F}_{V_n^c}] = E^{\nu_1}[f \,|\, \mathscr{F}_{V_n^c}],$$

the backward-martingale convergence theorem [**27**, Theorem 5.6.3] together with $\mathscr{T}$-triviality implies that

$$\lim_{n\to\infty} \pi_{V_n} f = E^\mu(f) \quad \mu\text{-a.s. and} \quad \lim_{n\to\infty} \pi_{V_n} f = E^{\nu_1}(f) \quad \nu_1\text{-a.s.}$$

But then $\nu_1 \ll \mu$ implies that the limits agree. So $\mu = \nu_1$. Thus μ is an extreme point.

Part (b). Let $\mu \in \mathscr{G}^\Pi \cap \mathcal{M}_\theta(\Omega)$ be such that $0 < \mu(A) < 1$ for some $\mathcal{I}$-measurable set A. By Exercise B.23 there exists a $\mathscr{T}$-measurable set $\widetilde{A}$ such that $\mu(A \triangle \widetilde{A}) = 0$. Thus, defining ν_1 and ν_2 as above we have $\mu = t\nu_1 + (1-t)\nu_2$ with $\nu_1, \nu_2 \in \mathscr{G}^\Pi \cap \mathcal{M}_\theta(\Omega)$ and $t \in (0,1)$. In other words, μ is not extreme. The other direction is a consequence of the more general extremality of ergodic measures. See Exercise B.21. $\square$

Even though only extreme Gibbs measures are physically meaningful, convex combinations of extreme measures may represent ignorance about the actual state of the system. Each Gibbs measure can in fact be written as a convex mixture of extreme ones. One way to see this is via Choquet's theorem [**63**]. The exercises below contain an alternative approach.

Exercise 7.25 (Decomposition of Gibbs measures)**.** Use the limits of π_{Λ_n} as $\Lambda_n \nearrow \mathbb{Z}^d$ to construct a kernel with values in extreme Gibbs measures and that gives versions of $\mu(\,\cdot\,|\,\mathscr{T})$ simultaneously for all $\mu \in \mathscr{G}^\Pi$. Apply Exercise B.30 to prove that every $\mu \in \mathscr{G}^\Pi$ has a unique probability measure Q_μ supported on the extreme points of $\mathscr{G}^\Pi$ such that $\mu = \int \nu \, Q_\mu(d\nu)$.

Exercise 7.26 (Decomposition of shift-invariant Gibbs measures)**.** Let $\mu \in \mathscr{G}^\Pi \cap \mathcal{M}_\theta(\Omega)$. Let κ be the kernel from Exercise B.31. Show that for any finite $\Lambda \subset \mathbb{Z}^d$, $\kappa(\omega)\pi_\Lambda = \kappa(\omega)$ μ-a.s. Use Exercise B.30 to conclude the existence of a unique probability measure Q_μ supported on the extremes of $\mathscr{G}^\Pi \cap \mathcal{M}_\theta(\Omega)$ such that $\mu = \int \nu \, Q_\mu(d\nu)$.

The set of shift-invariant Gibbs measures can be empty even when the set of Gibbs measures is not (see (11.46) in [**41**]). Sufficient conditions for existence of shift-invariant Gibbs measures are given in Section 5.2 of [**41**].

7.6. Uniqueness for small potentials

We prove a uniqueness theorem for Gibbs measures due to Dobrushin [**22**]. There is no spatial structure in this proof. So we can take the index set to be an arbitrary countable set $\mathbb{I}$. Let the spin space $\mathcal{S}$ be a compact metric space. The result is true more generally but this assumption simplifies the proof. Our main application of the theorem will be to the Ising model where $\mathcal{S}$ is a finite set and so compactness is sufficient. The configuration space is $\Omega = \mathcal{S}^{\mathbb{I}}$ with its product (Borel) σ-algebra $\mathscr{F}$. Ω is also compact, and the space $\mathscr{C}(\Omega)$ of continuous functions $f : \Omega \to \mathbb{R}$ consists entirely of bounded, uniformly continuous functions.

Assume given continuous stochastic kernels α_i from $\mathcal{S}^{\mathbb{I}\smallsetminus i}$ to $\mathcal{S}$. (We will abbreviate $\mathbb{I}\smallsetminus i$ for $\mathbb{I}\smallsetminus\{i\}$.) Continuity is meant in the usual sense of weak topology on $\mathcal{M}_1(\mathcal{S})$, so that, for $f \in \mathscr{C}(\mathcal{S})$, $\omega_{\mathbb{I}\smallsetminus i} \mapsto \int_{\mathcal{S}} f(s)\, \alpha_i(\omega_{\mathbb{I}\smallsetminus i}, ds)$ is a continuous function on Ω. To simplify notation we write $\alpha_i(\omega, ds)$ instead of $\alpha_i(\omega_{\mathbb{I}\smallsetminus i}, ds)$, and we think of α_i as a function on Ω that does not depend on the coordinate ω_i.

Probability measure $\alpha_i(\omega, ds)$ represents a distribution of spin ω_i, given the others spins $\omega_{\mathbb{I}\smallsetminus i}$. From these kernels we define kernels ρ_i on Ω by

$$\rho_i(\omega, B) = \int_S \mathbb{1}_B(\omega_{\mathbb{I}\smallsetminus i}, s)\, \alpha_i(\omega, ds), \quad B \in \mathscr{F}. \tag{7.9}$$

The configuration $\eta = (\omega_{\mathbb{I}\smallsetminus i}, s)$ in the integral has $\eta_i = s$ and $\eta_j = \omega_j$ for $j \neq i$.

The *total variation distance* $\|\mu - \nu\|$ of two probability measures can be expressed in terms of a supremum over measurable functions or sets:

$$\begin{aligned} \|\mu - \nu\| &= \sup_{\|f\|_\infty \le 1} |E^\mu(f) - E^\nu(f)| \\ &= 2 \cdot \sup_{A \in \mathscr{B}_S} |\mu(A) - \nu(A)|. \end{aligned} \tag{7.10}$$

With this norm we define a quantity $b_{i,j}$ that measures the strength of the dependence of $\alpha_i(\omega)$ on the spin ω_j:

$$b_{i,j} = \tfrac{1}{2} \sup\{\|\alpha_i(\sigma) - \alpha_i(\omega)\| : \sigma_k = \omega_k \ \forall k \in \mathbb{I} \smallsetminus j\}, \quad i, j \in \mathbb{I}. \tag{7.11}$$

Dobrushin's coefficient is defined by

$$b_* = \sup_{i \in \mathbb{I}} \sum_{j \in \mathbb{I}} b_{i,j}. \tag{7.12}$$

Note that $b_{i,i} = 0$ so the sum can be restricted to $j \neq i$.

Dobrushin's uniqueness theorem. *Assume that the stochastic kernels α_i are continuous and $b_* < 1$. Then there exists at most one $\mu \in \mathcal{M}_1(\Omega)$ such that $\mu\rho_i = \mu$ $\forall i \in \mathbb{I}$.*

The condition $b_* < 1$ is sometimes called *Dobrushin's condition.* Before proving the theorem let us apply it to the lattice systems defined earlier in this chapter. The qualitative meaning of the result is that if a Gibbs specification is a small enough perturbation from product measure, its Gibbs measure must be unique.

Theorem 7.27. *Let S be compact and let Φ be a continuous, absolutely summable shift-invariant interaction potential that satisfies*

$$\sum_{A: 0 \in A} (|A| - 1)\, \|\Phi_A\|_\infty < 1. \tag{7.13}$$

Then there exists exactly one Gibbs measure for Φ (at inverse temperature $\beta = 1$).

Proof. Existence follows from Corollary 7.18. Now

$$\alpha_i(\omega, ds) = \pi_{\{i\}}(\omega, ds) = \left(\int_S e^{-\mathcal{H}^{\omega}_{\{i\}}(s)}\lambda(ds)\right)^{-1} e^{-\mathcal{H}^{\omega}_{\{i\}}(s)}\lambda(ds).$$

By Lemma 7.28 below, for $\sigma_k = \omega_k \ \forall k \neq j$,

$$\|\alpha_i(\sigma) - \alpha_i(\omega)\| \leq \sup_{s\in\mathcal{S}} |\,\mathcal{H}^{\sigma}_{\{i\}}(s) - \mathcal{H}^{\omega}_{\{i\}}(s)\,|$$
$$\leq \sup_{s\in\mathcal{S}} \left| \sum_{A:\{i,j\}\subset A} \left(\Phi_A(\sigma_{A\setminus i}, s) - \Phi_A(\omega_{A\setminus i}, s)\right)\right| \leq 2 \sum_{A:\{i,j\}\subset A} \|\Phi_A\|_\infty$$

from which

$$b_{i,j} \leq \sum_{A:\{i,j\}\subset A} \|\Phi_A\|_\infty .$$

Next, from reordering the sums and shift invariance,

$$\sum_{j\in\mathbb{Z}^d\setminus i} b_{i,j} \leq \sum_{j\in\mathbb{Z}^d\setminus i}\ \sum_{A:A\ni i,j} \|\Phi_A\|_\infty = \sum_{A:i\in A} (|A|-1)\,\|\Phi_A\|_\infty$$
$$= \sum_{A:0\in A} (|A|-1)\,\|\Phi_A\|_\infty . \qquad \square$$

Lemma 7.28. *Let g and h be bounded measurable functions and let λ be a probability measure on an arbitrary measurable space. Then*

$$\left|\frac{E^\lambda(fe^g)}{E^\lambda(e^g)} - \frac{E^\lambda(fe^h)}{E^\lambda(e^h)}\right| \leq \|f\|_\infty\, \|g-h\|_\infty . \tag{7.14}$$

Proof. For $t \in \mathbb{R}$ define the probability measure γ_t by

$$E^{\gamma_t}(f) = \frac{E^\lambda(fe^{h+t(g-h)})}{E^\lambda(e^{h+t(g-h)})}.$$

Then the left-hand side of (7.14) equals

$$
\begin{aligned}
|E^{\gamma_1}(f) - E^{\gamma_0}(f)| &\le \int_0^1 \left| \frac{d}{dt} E^{\gamma_t}(f) \right| dt \\
&= \int_0^1 \left| E^{\gamma_t}[f(g-h)] - E^{\gamma_t}[f] E^{\gamma_t}[g-h] \right| dt \\
&= \int_0^1 \left| E^{\gamma_t}\big[f\big(g-h-E^{\gamma_t}[g-h]\big)\big] \right| dt \\
&\le \|f\|_\infty \int_0^1 E^{\gamma_t} \big| g-h-E^{\gamma_t}[g-h] \big| \, dt \\
&\le \|f\|_\infty \int_0^1 \big\{ E^{\gamma_t}\big[\big(g-h-E^{\gamma_t}[g-h]\big)^2\big] \big\}^{1/2} dt \\
&\le \|f\|_\infty \int_0^1 \big\{ E^{\gamma_t}\big[(g-h)^2\big] \big\}^{1/2} dt \le \|f\|_\infty \, \|g-h\|_\infty \, .
\end{aligned}
$$

The differentiation in the first equality can be readily justified because all functions are bounded. □

We turn to proving Dobrushin's uniqueness theorem (page 113). Measure the oscillation of a function $f : \Omega \to \mathbb{R}$ as spin ω_j varies by

(7.15) $$\operatorname{osc}_j(f) = \sup\{|f(\sigma) - f(\omega)| : \sigma_k = \omega_k \ \forall k \in \mathbb{I} \smallsetminus j\}, \quad j \in \mathbb{I},$$

and then let

(7.16) $$D(f) = \sum_{j \in \mathbb{I}} \operatorname{osc}_j(f).$$

The quantity $D(f)$ controls the overall oscillation of f. Continuous, local functions satisfy $D(f) < \infty$. The class of such functions is rich enough to separate probability measures.

The gist of the proof is that under a Gibbs measure we can iterate the kernels, and Dobrushin's condition will imply that by iterating sufficiently we can damp oscillations and converge to a constant independent of the measure. This thread is developed in a sequence of lemmas. For the rest of the section it will be convenient to take $\mathbb{I} = \mathbb{N}$.

Lemma 7.29. *For $f \in \mathscr{C}(\mathcal{S})$, $\sup f - \inf f \le D(f)$.*

Proof. Fix arbitrary configurations $\sigma, \tau, \omega \in \Omega$. Then

$$\begin{aligned}
&|\, f(\sigma_1, \dots, \sigma_k,\, \omega_{k+1}, \omega_{k+2}, \dots) \;-\; f(\tau_1, \dots, \tau_k,\, \omega_{k+1}, \omega_{k+2}, \dots)\,| \\
&\quad \le \sum_{j=1}^{k} |\, f(\tau_1, \dots, \tau_{j-1}, \sigma_j, \dots, \sigma_k,\, \omega_{k+1}, \omega_{k+2}, \dots) \\
&\qquad\qquad - \; f(\tau_1, \dots, \tau_j, \sigma_{j+1}, \dots, \sigma_k,\, \omega_{k+1}, \omega_{k+2}, \dots)\,| \\
&\quad \le \sum_{j=1}^{k} \operatorname{osc}_j(f) \le D(f).
\end{aligned}$$

By continuity, letting $k \to \infty$ gives $|f(\sigma) - f(\tau)| \le D(f)$ for two arbitrary configurations $\sigma, \tau \in \Omega$. □

Lemma 7.30. *For any two probability measures μ, ν and a bounded measurable function f,*

$$|\, E^\mu(f) \;-\; E^\nu(f)\,| \;\le\; \frac{\sup f - \inf f}{2} \cdot \|\mu - \nu\|\,. \tag{7.17}$$

Proof. Let $g = f - \frac{\sup f + \inf f}{2}$. Then

$$\sup g = \frac{\sup f - \inf f}{2} \quad \text{and} \quad \inf g = \frac{\inf f - \sup f}{2}$$

and so $\|g\|_\infty = \frac{1}{2}(\sup f - \inf f)$. Now write

$$|E^\mu(f) \;-\; E^\nu(f)| = |E^\mu(g) \;-\; E^\nu(g)| \le \|g\|_\infty \, \|\mu - \nu\|\,.$$ □

Lemma 7.31. *For $i, j \in \mathbb{I}$,* $\operatorname{osc}_j(\rho_i f) \le \begin{cases} 0, & i = j, \\ \operatorname{osc}_i(f) b_{i,j} + \operatorname{osc}_j(f), & i \ne j. \end{cases}$

Proof. The case $i = j$ is clear because $\rho_i f(\omega)$ does not depend on ω_i.

Let $\sigma_k = \omega_k \ \forall k \in \mathbb{I} \smallsetminus j$. Abbreviate $g(s) = f(\sigma_{\mathbb{I} \smallsetminus i}, s)$ and $h(s) = f(\omega_{\mathbb{I} \smallsetminus i}, s)$. Then

$$\begin{aligned}
|\rho_i f(\sigma) - \rho_i f(\omega)| &= \left| \int_{\mathcal{S}} g(s)\, \alpha_i(\sigma, ds) \;-\; \int_{\mathcal{S}} h(s)\, \alpha_i(\omega, ds) \right| \\
&\le \left| \int_{\mathcal{S}} g(s)\, \alpha_i(\sigma, ds) \;-\; \int_{\mathcal{S}} g(s)\, \alpha_i(\omega, ds) \right| + \int_{\mathcal{S}} |g(s) - h(s)|\; \alpha_i(\omega, ds) \\
&\le \frac{\sup g - \inf g}{2} \cdot \|\alpha_i(\sigma) - \alpha_i(\omega)\| + \|g - h\|_\infty \\
&\le \operatorname{osc}_i(f) b_{i,j} + \operatorname{osc}_j(f).
\end{aligned}$$

In the last step note that $g(s) - h(s) = f(\sigma_{\mathbb{I}\smallsetminus i}, s) - f(\omega_{\mathbb{I}\smallsetminus i}, s)$ and the arguments in f differ only at site j. □

We claim that a linear operator A on $\mathscr{C}(\Omega)$ can be defined by

$$Af = \lim_{n\to\infty} \rho_1\rho_2\cdots\rho_n f \tag{7.18}$$

where the limit exists in sup norm. We first construct Af for local f and then extend by a density argument. If $f(\omega) = f(\omega_1, \dots, \omega_k)$ depends only on coordinates $(\omega_1, \dots, \omega_k)$, then $\rho_\ell f = f$ for $\ell > k$. Consequently the limit in (7.18) exists and in fact $Af = \rho_1\rho_2\cdots\rho_k f$. The composition $\rho_1\rho_2\cdots\rho_n$ is a stochastic kernel. So, for two given local functions $f, g \in \mathscr{C}(\Omega)$ and for large enough n,

$$\|Af - Ag\|_\infty = \|\rho_1\rho_2\cdots\rho_n f - \rho_1\rho_2\cdots\rho_n g\|_\infty \le \|f - g\|_\infty\,. \tag{7.19}$$

Local functions are dense in $\mathscr{C}(\Omega)$. So we can define Af unambiguously for any $f \in \mathscr{C}(\Omega)$ by $Af = \lim_{k\to\infty} Af_k$ for any sequence $\{f_k\}$ of local functions such that $\|f_k - f\|_\infty \to 0$. (Details in Exercise 7.35.)

Lemma 7.32. *If $b_* \le 1$, then $D(Af) \le b_* D(f)$ for $f \in \mathscr{C}(\Omega)$.*

Proof. To do away with the trivial case, if $b_* = 0$, then $\alpha_i(\omega, ds) = \alpha_i(ds)$ is a fixed measure on $\mathcal{S}$, and $Af = \int f\, d(\bigotimes \alpha_i)$ is a constant (Exercise 7.34).

We claim that

$$D(\rho_1\rho_2\cdots\rho_n f) \le b_* \sum_{j=1}^{n} \operatorname{osc}_j(f) + \sum_{j=n+1}^{\infty} \operatorname{osc}_j(f). \tag{7.20}$$

Case $n = 1$.

$$\begin{aligned} D(\rho_1 f) &= \sum_{j=2}^{\infty} \operatorname{osc}_j(\rho_1 f) \le \sum_{j=2}^{\infty} \big(\operatorname{osc}_1(f) b_{1,j} + \operatorname{osc}_j(f)\big) \\ &\le b_* \operatorname{osc}_1(f) + \sum_{j=2}^{\infty} \operatorname{osc}_j(f). \end{aligned}$$

Induction step: Assume (7.20) is true for $n-1$. Then

$$\begin{aligned}
D(\rho_1\rho_2\cdots\rho_n f) &\le b_* \sum_{j=1}^{n-1} \operatorname{osc}_j(\rho_n f) + \sum_{j=n}^{\infty} \operatorname{osc}_j(\rho_n f)\\
&\le b_* \sum_{j=1}^{n-1} \big(b_{n,j} \operatorname{osc}_n(f) + \operatorname{osc}_j(f)\big) + \sum_{j=n+1}^{\infty} \big(b_{n,j} \operatorname{osc}_n(f) + \operatorname{osc}_j(f)\big)\\
&\le b_* \sum_{j=1}^{n-1} \operatorname{osc}_j(f) + \sum_{j=1}^{\infty} b_{n,j} \operatorname{osc}_n(f) + \sum_{j=n+1}^{\infty} \operatorname{osc}_j(f)\\
&\le b_* \sum_{j=1}^{n} \operatorname{osc}_j(f) + \sum_{j=n+1}^{\infty} \operatorname{osc}_j(f).
\end{aligned}$$

The third inequality used $b_* \le 1$.

To finish the proof we can assume that $D(f) < \infty$. Then the last series in (7.20) vanishes in the $n \to \infty$ limit. To dominate $D(Af)$ with the help of (7.20) note that uniform convergence $f_n \to f$ implies $\operatorname{osc}_j(f_n) \to \operatorname{osc}_j(f)$ $\forall j \in \mathbb{I}$. Apply Fatou's lemma to the series:

$$\begin{aligned}
D(Af) &= \sum_j \operatorname{osc}_j(Af) = \sum_j \lim_{n\to\infty} \operatorname{osc}_j(\rho_1\rho_2\cdots\rho_n f)\\
&\le \varliminf_{n\to\infty} \sum_j \operatorname{osc}_j(\rho_1\rho_2\cdots\rho_n f) = \varliminf_{n\to\infty} D(\rho_1\rho_2\cdots\rho_n f)\\
&\le b_* \sum_{j=1}^{\infty} \operatorname{osc}_j(f) = b_* D(f).
\end{aligned}$$

□

The qualitative significance of Lemma 7.32 is clear: if $b_* < 1$, Af oscillates less than f. Hence if we iterate A, we should arrive at a constant in the limit.

Lemma 7.33. *Let $f \in \mathscr{C}(\Omega)$ satisfy $D(f) < \infty$. Suppose $b_* < 1$. Then $A^n f$ converges to a constant in the sup norm.*

Proof. First observe that sequences $\{\sup A^n f\}$ and $\{\inf A^n f\}$ are both monotone. Namely, $\sup \rho_i f \le \sup f$ and consequently $\sup Af \le \sup f$, and similarly $\inf Af \ge \inf f$. Hence there are constants $a \le b$ such that $\inf A^n f \nearrow a$ and $\sup A^n f \searrow b$. By Lemmas 7.29 and 7.32

$$\sup A^n f - \inf A^n f \le D(A^n f) \le b_*^n D(f) \to 0.$$

Consequently $a = b$, and this suffices for $\|A^n f - a\|_\infty \to 0$. □

We are ready for the proof of the main theorem.

Proof of Dobrushin's uniqueness theorem. Suppose $\nu \in \mathcal{M}_1(\Omega)$ satisfies $\nu\rho_i = \nu$ for all $i \in \mathbb{I}$. Let $f \in \mathscr{C}(\Omega)$ satisfy $D(f) < \infty$. Then, as $n \to \infty$,

$$E^\nu(f) = E^\nu(\rho_1\rho_2\cdots\rho_n f) \to E^\nu(Af).$$

Continuing,

$$E^\nu(f) = E^\nu(Af) = \cdots = E^\nu(A^n f) \to a(f)$$

where $a(f)$ is the constant limit of $A^n f$ from Lemma 7.33. If $\mu \in \mathcal{M}_1(\Omega)$ also satisfies $\mu\rho_i = \mu$ for all $i \in \mathbb{I}$, we conclude that $E^\mu(f) = a(f) = E^\nu(f)$ for all $f \in \mathscr{C}(\Omega)$ such that $D(f) < \infty$. This is enough for $\mu = \nu$. □

***Exercise 7.34.** (a) Show that, under the assumption of continuity of $\alpha_i(\omega)$ as a function of ω, $b_* = 0$ implies the existence of probability measures α_i such that $\alpha_i(\omega, \cdot\,) = \alpha_i \ \forall \omega \in \Omega$.

(b) Show that part (a) can fail if $\alpha_i(\omega)$ is only assumed measurable as a function of ω.

Hint: An example for (b) can be found along these lines: on $\Omega = \{0,1\}^{\mathbb{N}}$, let $\alpha_i(\omega)$ depend on whether ω has finitely or infinitely many zeroes.

***Exercise 7.35.** Provide the details of the proof that the limit in (7.18) exists in sup-norm.

Hint: By compactness, a function $f \in \mathscr{C}(\Omega)$ is uniformly continuous. Fix a point $z \in \mathcal{S}$ and use the product metric (A.3) to show that the local functions $f_k(\omega) = f(\omega_1, \ldots, \omega_k, z, z, z, \ldots)$ converge to f uniformly. Then (7.19) shows that Af_k is a Cauchy sequence in the sup-norm of $\mathscr{C}(\Omega)$. By completeness of $\mathscr{C}(\Omega)$, Af_k has a sup-norm limit Af in $\mathscr{C}(\Omega)$. Show that Af is well-defined in the sense that the same limit Af arises from Ag_k for any sequence of local functions g_k that converge to f in $\mathscr{C}(\Omega)$. To control $\|Af - \rho_1\cdots\rho_n f\|_\infty$, replace f with f_k and use the triangle inequality.

Literature notes. This chapter draws from a number of sources, such as Ellis [**32**], Föllmer [**35**], Georgii [**41**], Israel [**50**], and Simon [**77**]. The reader can turn to these for further development of the subject. Equivalence of ensembles is discussed in the appendix to Section III.4 in [**77**] and article [**19**] provides a large deviations approach to this question.

Chapter 8

Large deviations and equilibrium statistical mechanics

Large deviation theory is concerned with fluctuations of empirical averages of microscopic quantities. Perhaps not surprisingly it is related to equilibrium statistical mechanics. In this chapter we extend specific entropy and i.i.d. process level large deviations to Gibbs measures. As a corollary we obtain a variational characterization of shift-invariant Gibbs measures, an infinite-volume analogue of (7.3).

8.1. Thermodynamic limit of the pressure

As in the previous chapter, $\Omega = \mathcal{S}^{\mathbb{Z}^d}$ for a Polish spin space $\mathcal{S}$. Let Φ be an absolutely summable shift-invariant continuous interaction potential. The inverse temperature is fixed at $\beta = 1$. To have β explicitly in the results, simply replace Φ by $\beta\Phi$.

Recall that λ is a fixed i.i.d. product measure on Ω. We defined the volumes $V_n = \{i \in \mathbb{Z}^d : -n < i_1, \dots, i_d < n\}$ and we let $\tau \in \Omega \cup \{\text{free}\}$ be a given boundary condition. The partition function was defined by

$$Z_\Lambda^\tau = \int e^{-\mathcal{H}_\Lambda^\tau(\sigma_\Lambda)} \lambda(d\sigma_\Lambda).$$

A quantity that encodes the thermodynamic properties of the system is the *pressure*, defined as the limit (if it exists)

$$P(\Phi) = \lim_{n\to\infty} \frac{1}{|V_n|} \log Z^{\tau}_{V_n}$$

of the finite-volume pressure $P^{\tau}_{\Lambda}(\Phi) = \frac{1}{|\Lambda|}\log Z^{\tau}_{\Lambda}$.

The term "pressure" makes sense in the case of the lattice gas (Example 7.3). For ferromagnetic models it is the same as "minus the free energy per site".

The energy contribution of the origin is represented by the microscopic observable

$$f_\Phi = \sum_{A:0\in A} \frac{\Phi_A}{|A|} \in \mathscr{C}_b(\Omega).$$

Lemma 8.2 below allows us to replace the Hamiltonian with a sum of $f_\Phi \circ \theta_i$. Then the existence of $P(\Phi)$ independent of boundary conditions follows from the large deviation principle for empirical fields (Theorem 6.13) and Varadhan's theorem (page 38). Large deviation theory also gives a variational representation of $P(\Phi)$. Recall the definition of $p(f)$ in (4.2) and Proposition 6.14.

Theorem 8.1. *There exists an infinite-volume pressure $P(\Phi) \in \mathbb{R}$ that satisfies the following statements:*

(a) $\lim_{n\to\infty} \sup_{\tau\in\Omega\cup\{\text{free}\}} |P^{\tau}_{V_n}(\Phi) - P(\Phi)| = 0.$

(b) $P(\Phi) = p(-f_\Phi) = \lim_{n\to\infty} \frac{1}{|V_n|} \log E^{\lambda}[e^{-\sum_{i\in V_n} f_\Phi\circ\theta_i}].$

(c) $-P(\Phi) = \inf_{\nu\in\mathcal{M}_\theta(\Omega)} \{E^{\nu}[f_\Phi] + h(\nu\,|\,\lambda)\}.$

Before proving the theorem we develop two estimates. For finite $\Delta \subset \Lambda \subset \mathbb{Z}^d$ define $\Lambda(\Delta) = \{i \in \Lambda : i + \Delta \subset \Lambda\}$ and

$$b_{\Delta,\Lambda} = |\Lambda| \sum_{A:0\in A\not\subset\Delta} \|\Phi_A\|_\infty + \|\Phi\|\,|\Lambda \smallsetminus \Lambda(\Delta)|.$$

Note that for any finite $\Delta \subset \mathbb{Z}^d$,

$$\lim_{\Lambda\nearrow\mathbb{Z}^d} \frac{b_{\Delta,\Lambda}}{|\Lambda|} = \sum_{A:0\in A\not\subset\Delta} \|\Phi_A\|_\infty\,,$$

which can be made arbitrarily small by taking Δ large.

Lemma 8.2. *For finite $\Delta \subset \Lambda \subset \mathbb{Z}^d$,*

$$\sup_{\sigma\in\Omega} \Big| \mathcal{H}^{\text{free}}_{\Lambda}(\sigma_\Lambda) - \sum_{i\in\Lambda} f_\Phi(\theta_i\sigma) \Big| \le b_{\Delta,\Lambda} \tag{8.1}$$

and

$$\sup_{\sigma\in\Omega}\Big|\mathcal{H}_\Lambda^{\text{free}}(\sigma_\Lambda)-\mathcal{H}_\Lambda^{\sigma_{\Lambda^c}}(\sigma_\Lambda)\Big|\le b_{\Delta,\Lambda}. \tag{8.2}$$

Proof. Write

$$\begin{aligned}\mathcal{H}_\Lambda^{\text{free}}&=\sum_{A\subset\Lambda}\Phi_A=\sum_{A\subset\Lambda}\sum_{i\in A}\frac{\Phi_A}{|A|}=\sum_{i\in\Lambda}\sum_{A:i\in A\subset\Lambda}\frac{\Phi_A}{|A|}\\&=\sum_{i\in\Lambda}\sum_{A:i\in A}\frac{\Phi_A}{|A|}-\sum_{i\in\Lambda}\sum_{A:i\in A\not\subset\Lambda}\frac{\Phi_A}{|A|}.\end{aligned}$$

But

$$\sum_{A:i\in A}\frac{\Phi_A}{|A|}=\sum_{B:0\in B}\frac{\Phi_{i+B}}{|B|}=f_\Phi\circ\theta_i.$$

Thus,

$$\begin{aligned}&\Big|\mathcal{H}_\Lambda^{\text{free}}(\sigma_\Lambda)-\sum_{i\in\Lambda}f_\Phi(\theta_i\sigma)\Big|\\&\le\sum_{i\in\Lambda(\Delta)}\sum_{A:i\in A\not\subset\Lambda}\frac{\|\Phi_A\|_\infty}{|A|}+\sum_{i\in\Lambda\smallsetminus\Lambda(\Delta)}\sum_{A:i\in A\not\subset\Lambda}\frac{\|\Phi_A\|_\infty}{|A|}\\&\le\sum_{i\in\Lambda(\Delta)}\sum_{\substack{B:0\in B\\ i+B\not\subset\Lambda}}\frac{\|\Phi_B\|_\infty}{|B|}+\sum_{i\in\Lambda\smallsetminus\Lambda(\Delta)}\sum_{A:i\in A}\frac{\|\Phi_A\|_\infty}{|A|}\\&\le\sum_{i\in\Lambda(\Delta)}\sum_{B:0\in B\not\subset\Delta}\frac{\|\Phi_B\|_\infty}{|B|}+\sum_{i\in\Lambda\smallsetminus\Lambda(\Delta)}\sum_{B:0\in B}\frac{\|\Phi_B\|_\infty}{|B|}\\&\le b_{\Delta,\Lambda},\end{aligned}$$

and (8.1) is proved. Next write

$$\begin{aligned}&\Big|\mathcal{H}_\Lambda^{\text{free}}(\sigma_\Lambda)-\mathcal{H}_\Lambda^{\sigma_{\Lambda^c}}(\sigma_\Lambda)\Big|\le\sum_{\substack{A\cap\Lambda\neq\varnothing\\ A\cap\Lambda^c\neq\varnothing}}\|\Phi_A\|_\infty\\&\le\sum_{i\in\Lambda(\Delta)}\sum_{A:i\in A\not\subset\Lambda}\|\Phi_A\|_\infty+\sum_{i\in\Lambda\smallsetminus\Lambda(\Delta)}\sum_{A:i\in A}\|\Phi_A\|_\infty\\&\le|\Lambda(\Delta)|\sum_{B:0\in B\not\subset\Delta}\|\Phi_B\|_\infty+|\Lambda\smallsetminus\Lambda(\Delta)|\sum_{B:0\in B}\|\Phi_B\|_\infty\\&\le b_{\Delta,\Lambda},\end{aligned}$$

proving (8.2). ☐

Proof of Theorem 8.1. Theorem 6.13 gives us a large deviation principle for the distributions of the empirical fields under the i.i.d. measure λ. Then, taking $\mathcal{E}=\mathcal{M}_1(\Omega)$, $\mathcal{X}=\mathcal{M}(\Omega)$, and $\mathcal{Y}=\mathscr{C}_b(\Omega)$, Theorem 4.25 implies the existence of the limit defining $p(-f_\Phi)$ in part (b), as well as the variational formula in part (c), provided we use (b) as the definition of $P(\Phi)$. Part (a) simply follows from (8.1) and (8.2). Indeed, take $\Delta = V_m$ and $\Lambda = V_n$, for $n > m$. Then

$$\sup_{\tau\in\Omega\cup\{\text{free}\}}\ \sup_{\sigma\in\Omega}\Big| -\mathcal{H}^\tau_{V_n}(\sigma_{V_n}) + \sum_{i\in V_n} f_\Phi(\theta_i\sigma)\Big| \le 2b_{V_m,V_n}.$$

In general

$$\log E^\lambda[e^f] = \log E^\lambda[e^g\, e^{f-g}] \le \log E^\lambda[e^g]e^{\|f-g\|_\infty} = \log E^\lambda[e^g] + \|f-g\|_\infty.$$

Thus,

$$\sup_{\tau\in\Omega\cup\{\text{free}\}}\Big|\log Z^\tau_{V_n} - \log E^\lambda[e^{-\sum_{i\in V_n} f_\Phi\circ\theta_i}]\Big| \le 2b_{V_m,V_n}.$$

Now divide by $|V_n|$ and take $n\to\infty$ and then $m\to\infty$. □

8.2. Entropy and large deviations under Gibbs measures

Define the finite-volume Gibbs measures with free boundary conditions by

$$\pi^{\text{free}}_\Lambda(d\sigma_\Lambda) = \frac{e^{-\mathcal{H}^{\text{free}}_\Lambda(\sigma_\Lambda)}}{Z^{\text{free}}_\Lambda}\lambda_\Lambda(d\sigma_\Lambda).$$

The expectation relative to $\pi^{\text{free}}_\Lambda$ will be denoted by $\pi^{\text{free}}_\Lambda(f) = E^{\pi^{\text{free}}_\Lambda}[f]$.

Theorem 8.3. *For $\nu\in\mathcal{M}_\theta(\Omega)$, the limit*

$$h(\nu\,|\,\Phi) = \lim_{n\to\infty}\frac{1}{|V_n|}H(\nu_{V_n}\,|\,\pi^{\text{free}}_{V_n})$$

exists and satisfies

$$h(\nu\,|\,\Phi) = E^\nu[f_\Phi] + h(\nu\,|\,\lambda) + P(\Phi). \tag{8.3}$$

Furthermore, for any Gibbs measure $\gamma\in\mathscr{G}^\Phi$,

$$h(\nu\,|\,\gamma) = \lim_{n\to\infty}\frac{1}{|V_n|}H(\nu_{V_n}\,|\,\gamma_{V_n})$$

exists and $h(\nu\,|\,\gamma) = h(\nu\,|\,\Phi)$.

Remark 8.4. Theorem 8.1 says that the free energy $-P(\Phi)$ is the smallest value of the average energy contribution per site less the physical entropy, where the minimization takes place over all shift-invariant probability measures. Theorem 8.3 implies that this minimum is attained at any shift-invariant Gibbs measure γ because $h(\gamma\,|\,\Phi)=0$. This is analogous to (7.8) in the Curie-Weiss model and to Exercise 5.19.

***Exercise 8.5.** Use (8.3) to prove that $h(\cdot\,|\,\Phi)$ is affine, lower semicontinuous, and has compact sublevel sets.

Proof of Theorem 8.3. Recall that when the index is the volume V_n, we abbreviate things by simply using index n. Let $g \in b\mathscr{F}_n$. Then

$$\begin{aligned} E^{\nu_n}[g] - \log \pi_n^{\text{free}}(e^g) &= E^{\nu_n}[g] - \log E^{\lambda_n}[e^{g-\mathcal{H}_n^{\text{free}}}] + \log Z_n^{\text{free}} \\ &= E^{\nu_n}[g - \mathcal{H}_n^{\text{free}}] - \log E^{\lambda_n}[e^{g-\mathcal{H}_n^{\text{free}}}] + E^{\nu_n}[\mathcal{H}_n^{\text{free}}] + \log Z_n^{\text{free}}. \end{aligned}$$

As g varies over $b\mathscr{F}_n$, so does $g - \mathcal{H}_n^{\text{free}}$, and we conclude that

$$\frac{1}{|V_n|} H(\nu_n \,|\, \pi_n^{\text{free}}) = \frac{1}{|V_n|} H(\nu_n \,|\, \lambda_n) + \frac{1}{|V_n|} E^{\nu_n}[\mathcal{H}_n^{\text{free}}] + P_{V_n}^{\text{free}}.$$

The first term on the right-hand side converges to $h(\nu \,|\, \lambda)$ by Theorem 6.7. The third term converges to $P(\Phi)$ by Theorem 8.1. By (8.1) the second term $|V_n|^{-1} E^{\nu}[\mathcal{H}_n^{\text{free}}]$ has the same limit as $|V_n|^{-1} \sum_{i \in V_n} E^{\nu}[f_\Phi \circ \theta_i] = E^{\nu}[f_\Phi]$. This gives the existence of $h(\nu \,|\, \Phi)$ as well as (8.3).

Next, observe that if $\gamma \in \mathscr{G}^{\Phi}$, $g \in b\mathscr{F}_n$, and π_n is the Gibbs specification corresponding to Φ and volume V_n, then

$$\begin{aligned} \log E^{\gamma_n}[e^g] &= \log E^{\gamma}[\pi_n(e^g)] \\ &= \log \int \frac{E^{\lambda_n}[e^{g-\mathcal{H}_n^{\tau}}]}{E^{\lambda_n}[e^{-\mathcal{H}_n^{\tau}}]} \gamma(d\tau) \\ &= \log \int \frac{E^{\lambda_n}[e^{g-\mathcal{H}_n^{\text{free}}} e^{\mathcal{H}_n^{\text{free}} - \mathcal{H}_n^{\tau}}]}{E^{\lambda_n}[e^{-\mathcal{H}_n^{\text{free}}} e^{\mathcal{H}_n^{\text{free}} - \mathcal{H}_n^{\tau}}]} \gamma(d\tau) \\ &\le \log \frac{E^{\lambda_n}[e^{g-\mathcal{H}_n^{\text{free}}}]}{E^{\lambda_n}[e^{-\mathcal{H}_n^{\text{free}}}]} + 2 \sup_{\tau} \|\mathcal{H}_n^{\text{free}} - \mathcal{H}_n^{\tau}\|_\infty \qquad (8.4) \\ &\le \log \pi_n^{\text{free}}(e^g) + 2b_{V_m,V_n}, \end{aligned}$$

where we used (8.2) on the last line, with $m < n$. Consequently,

$$E^{\nu_n}[g] - \log \pi_n^{\text{free}}(e^g) \le E^{\nu_n}[g] - \log E^{\gamma_n}[e^g] + 2b_{V_m,V_n}$$

and by taking supremum over $g \in b\mathscr{F}_n$,

$$H(\nu_n \,|\, \pi_n^{\text{free}}) \le H(\nu_n \,|\, \gamma_n) + 2b_{V_m,V_n}.$$

The opposite inequality comes by reversing the inequality on line (8.4), and then we have

$$|H(\nu_n \,|\, \gamma_n) - H(\nu_n \,|\, \pi_n^{\text{free}})| \le 2b_{V_m,V_n}.$$

Thus, $h(\nu \,|\, \gamma)$ exists and equals $h(\nu \,|\, \Phi)$. □

Using Varadhan's theorem and (8.2), one can deduce a uniform large deviation principle for the empirical fields under Gibbs kernels and thus also under Gibbs measures. Recall the periodized empirical fields $\widetilde{R}_n$ defined in (6.3).

Theorem 8.6. *For any Borel set $A \subset \mathcal{M}_\theta(\Omega)$,*

$$
\begin{aligned}
-\inf_{\nu \in A^\circ} h(\nu \,|\, \Phi) &\le \varliminf_{n\to\infty} \inf_{\tau \in \Omega \cup \{\text{free}\}} \frac{1}{|V_n|} \log \pi^\tau_{V_n}\{\widetilde{R}_n \in A\} \\
&\le \varlimsup_{n\to\infty} \sup_{\tau \in \Omega \cup \{\text{free}\}} \frac{1}{|V_n|} \log \pi^\tau_{V_n}\{\widetilde{R}_n \in A\} \\
&\le -\sup_{\nu \in \overline{A}} h(\nu \,|\, \Phi).
\end{aligned}
\tag{8.5}
$$

Also, if $\gamma \in \mathscr{G}^\Phi$, then the distributions $\gamma\{R_n \in \cdot\}$ on $\mathcal{M}_1(\Omega)$ satisfy a large deviation principle with rate function $h(\cdot \,|\, \gamma)$.

Proof. Note that

$$\gamma\{\widetilde{R}_n \in A\} = \int \pi_{V_n}(\tau, \{\widetilde{R}_n \in A\}) \gamma(d\tau).$$

So if we replace R_n by $\widetilde{R}_n$, the last statement follows from (8.5) and $h(\cdot \,|\, \gamma) = h(\cdot \,|\, \Phi)$. It then follows for R_n from Exercises 6.5 and 2.16 and the tightness of the rate function $h(\cdot \,|\, \gamma)$. Let us thus prove (8.5).

Define the probability measures ρ_n on $\mathcal{M}_\theta(\Omega)$ by

$$\rho_n(A) = \frac{1}{c_n} \int \mathbb{1}\{\widetilde{R}_n \in A\} e^{-|V_n| \widetilde{R}_n(f_\Phi)} d\lambda,$$

where $c_n = E^\lambda\big[e^{-|V_n|\widetilde{R}_n(f_\Phi)}\big]$. Since Theorem 6.13 gives a large deviation principle for the distributions of $\widetilde{R}_n$ under the i.i.d. measure λ with rate function $h(\cdot \,|\, \lambda)$, Exercise 3.9 implies that the large deviation principle holds for ρ_n with normalization $|V_n|$ and rate function

$$
\begin{aligned}
I(\nu) &= E^\nu[f_\Phi] + h(\nu \,|\, \lambda) - \inf_{\mu \in \mathcal{M}_\theta(\Omega)} \{E^\mu[f_\Phi] + h(\mu \,|\, \lambda)\} \\
&= E^\nu[f_\Phi] + h(\nu \,|\, \lambda) + P(\Phi) = h(\nu \,|\, \Phi).
\end{aligned}
$$

Now, the same approximation step as before allows us to switch from ρ_n to the Gibbs kernels. To see this, write for $\tau \in \Omega \cup \{\text{free}\}$,

$$
\begin{aligned}
&\log \rho_n(A) = \log \int \mathbb{1}_A(\widetilde{R}_n) e^{-|V_n|\widetilde{R}_n(f_\Phi)} d\lambda - \log \int e^{-|V_n|\widetilde{R}_n(f_\Phi)} d\lambda \\
&\quad \le \log \int \mathbb{1}_A(\widetilde{R}_n) e^{-\mathcal{H}^\tau_n} d\lambda - \log \int e^{-\mathcal{H}^\tau_n} d\lambda + 2\big\|\mathcal{H}^\tau_n - |V_n|\,\widetilde{R}_n(f_\Phi)\big\|_\infty \\
&\quad \le \log \pi^\tau_n\{\widetilde{R}_n \in A\} + 2\,\|\mathcal{H}^\tau_n - |V_n|\,R_n(f_\Phi)\|_\infty + 2\,|V_n|\,\big\|\widetilde{R}_n(f_\Phi) - R_n(f_\Phi)\big\|_\infty.
\end{aligned}
$$

Abbreviate $f_\Phi^{(r)} = \sum_{0\in A, |A|\le r} \Phi_A/|A|$. By Exercise 6.5

$$\begin{aligned}\lim_{n\to\infty} \|\widetilde{R}_n(f_\Phi) - R_n(f_\Phi)\| &\le 2 \sum_{0\in A, |A|>r} \|\Phi_A\|_\infty + \lim_{n\to\infty} \|\widetilde{R}_n(f_\Phi^{(r)}) - R_n(f_\Phi^{(r)})\| \\ &= 2 \sum_{0\in A, |A|>r} \|\Phi_A\|_\infty \underset{r\to\infty}{\longrightarrow} 0.\end{aligned}$$

Then, for m fixed,

$$\begin{aligned}\varliminf_{n\to\infty} \inf_{\tau\in\Omega\cup\{\text{free}\}} \frac{1}{|V_n|} \log \pi_n^\tau\{\widetilde{R}_n \in A\} &\ge \varliminf_{n\to\infty} \frac{1}{|V_n|} \log \rho_n(A) - 4 \varlimsup_{n\to\infty} \frac{b_{V_m,V_n}}{|V_n|} \\ &\ge -\inf_{\nu\in A^\circ} h(\nu\,|\,\Phi) - 4 \varlimsup_{n\to\infty} \frac{b_{V_m,V_n}}{|V_n|}.\end{aligned}$$

Taking $m \to \infty$ kills the right-most term and proves the lower large deviation bound. The upper bound is similar. □

8.3. Dobrushin-Lanford-Ruelle (DLR) variational principle

We have so far seen that shift-invariant Gibbs measures are minimizers in the variational formula for the pressure $P(\Phi)$ (Remark 8.4). The next question is whether all the minimizers are shift-invariant Gibbs measures. This theorem completes the picture.

Dobrushin-Lanford-Ruelle variational principle. *Fix a shift-invariant absolutely summable continuous interaction potential Φ. Let $\gamma \in \mathcal{M}_\theta(\Omega)$. The following three statements are equivalent.*

(a) $\gamma \in \mathscr{G}^\Phi$.

(b) $h(\gamma\,|\,\Phi) = 0$.

(c) $E^\gamma[f_\Phi] + h(\gamma\,|\,\lambda) = -P(\Phi) = \inf\{E^\nu[f_\Phi] + h(\nu\,|\,\lambda) : \nu \in \mathcal{M}_\theta(\Omega)\}$.

The equivalence of (a) and (b) implies that the large deviation rate function under Gibbs measures has multiple zeroes if there is more than one shift-invariant Gibbs measure. The next chapter establishes this for the multidimensional Ising model at low temperature.

Proof of the DLR variational principle. (a) implies (b) due to Theorem 8.3. Also, (b) is equivalent to (c) due to (8.3). The delicate part is to show that (b) implies (a). Suppose (a) does not hold. Then there exists an integer ℓ such that $\gamma \ne \gamma\pi_{V_\ell}$ and hence $H(\gamma\,|\,\gamma\pi_{V_\ell}) > 0$. Fix $\eta > 0$ such that $H(\gamma\,|\,\gamma\pi_{V_\ell}) \ge 2\eta$. By Exercise 6.11, $H(\gamma_{V_k}\,|\,(\gamma\pi_{V_\ell})_{V_k}) \ge \eta$ for k large.

For a probability measure $\mu \in \mathcal{M}_1(\Omega)$ and two sets $A, B \subset \mathbb{Z}^d$, let $\mu_B^A = \mu_B(\cdot\,|\,\mathscr{F}_A)$ be the conditional distribution of $\sigma_B = (\sigma_i)_{i\in B}$ given $\sigma_A = (\sigma_i)_{i\in A}$. It is a probability measure on Ω_B that depends functionally on σ_A. (The existence of such conditional distributions is stated in Theorem B.5.)

Lemma 8.7. *Let $\Delta \subset \Lambda \subset \Gamma \subset \mathbb{Z}^d$ be finite. Then*

$$\Big| \int_{\Omega_{\Lambda\smallsetminus\Delta}} H(\gamma_\Lambda^{\Lambda\smallsetminus\Delta} \,|\, (\gamma\pi_\Delta)_\Lambda^{\Lambda\smallsetminus\Delta})\, d\gamma \;-\; \int_{\Omega_{\Lambda\smallsetminus\Delta}} H(\gamma_\Lambda^{\Lambda\smallsetminus\Delta} \,|\, (\pi_\Gamma^{\text{free}})_\Lambda^{\Lambda\smallsetminus\Delta})\, d\gamma \Big| \le 4d_{\Delta,\Lambda}\,,$$

where

$$d_{\Delta,\Lambda} = \sum_{\substack{A:A\cap\Delta\neq\varnothing \\ A\not\subset\Lambda}} \|\Phi_A\|_\infty\,.$$

Quantity $d_{\Delta,\Lambda}$ bounds the energy contribution from outside Λ to the volume Δ. For fixed Δ it decays to 0 as Λ grows to $\mathbb{Z}^d$.

Proof of Lemma 8.7. Define temporarily the Δ-volume Hamiltonian of interactions confined to Λ,

$$\mathcal{H}_{\Delta,\Lambda} = \sum_{\substack{A:A\cap\Delta\neq\varnothing \\ A\subset\Lambda}} \Phi_A.$$

Let φ denote an arbitrary bounded $\mathscr{F}_\Lambda$-measurable function. Define measure $\pi_{\Delta,\Lambda}(\sigma_{\Lambda\smallsetminus\Delta}) \in \mathcal{M}_1(\Omega_\Lambda)$ by

$$E^{\pi_{\Delta,\Lambda}(\sigma_{\Lambda\smallsetminus\Delta})}[\varphi] = \frac{\int \varphi(\omega_\Delta,\sigma_{\Lambda\smallsetminus\Delta}) e^{-\mathcal{H}_{\Delta,\Lambda}(\omega_\Delta,\sigma_{\Lambda\smallsetminus\Delta})}\lambda_\Delta(d\omega_\Delta)}{\int e^{-\mathcal{H}_{\Delta,\Lambda}(\omega_\Delta,\sigma_{\Lambda\smallsetminus\Delta})}\lambda_\Delta(d\omega_\Delta)}\,.$$

As usual, we will abbreviate $E^{\pi_{\Delta,\Lambda}(\sigma_{\Lambda\smallsetminus\Delta})}[\varphi] = \pi_{\Delta,\Lambda}(\sigma_{\Lambda\smallsetminus\Delta},\varphi)$.

In the calculations that follow equalities hold almost surely because they involve conditional expectations. But all the finite-volume measures involved are mutually absolutely continuous relative to λ, restricted to the volume in question. Hence the statements hold almost surely relative to any of these finite-volume measures. Now write:

$$\begin{aligned}
&E^{\pi_\Gamma^{\text{free}}}[\varphi \,|\, \sigma_{\Lambda\smallsetminus\Delta}] \\
&= \frac{\int_{\Omega_{\Gamma\smallsetminus\Lambda}}\int_{\Omega_\Delta} \varphi(\omega_\Delta,\sigma_{\Lambda\smallsetminus\Delta}) e^{-\mathcal{H}_\Gamma^{\text{free}}(\omega_\Delta,\sigma_{\Lambda\smallsetminus\Delta},\tau_{\Gamma\smallsetminus\Lambda})}\,\lambda_\Delta(d\omega_\Delta)\,\lambda_{\Gamma\smallsetminus\Lambda}(d\tau_{\Gamma\smallsetminus\Lambda})}{\int_{\Omega_{\Gamma\smallsetminus\Lambda}}\int_{\Omega_\Delta} e^{-\mathcal{H}_\Gamma^{\text{free}}(\omega_\Delta,\sigma_{\Lambda\smallsetminus\Delta},\tau_{\Gamma\smallsetminus\Lambda})}\,\lambda_\Delta(d\omega_\Delta)\,\lambda_{\Gamma\smallsetminus\Lambda}(d\tau_{\Gamma\smallsetminus\Lambda})} \\
&= \frac{\int_{\Omega_{\Gamma\smallsetminus\Lambda}}\int_{\Omega_\Delta} \varphi(\omega_\Delta,\sigma_{\Lambda\smallsetminus\Delta}) e^{-\mathcal{H}_{\Delta,\Lambda}(\omega_\Delta,\sigma_{\Lambda\smallsetminus\Delta})-\Sigma'-\Sigma''}\,\lambda_\Delta(d\omega_\Delta)\,\lambda_{\Gamma\smallsetminus\Lambda}(d\tau_{\Gamma\smallsetminus\Lambda})}{\int_{\Omega_{\Gamma\smallsetminus\Lambda}}\int_{\Omega_\Delta} e^{-\mathcal{H}_{\Delta,\Lambda}(\omega_\Delta,\sigma_{\Lambda\smallsetminus\Delta})-\Sigma'-\Sigma''}\,\lambda_\Delta(d\omega_\Delta)\,\lambda_{\Gamma\smallsetminus\Lambda}(d\tau_{\Gamma\smallsetminus\Lambda})},
\end{aligned}$$

where

$$\Sigma' = \sum_{\substack{A:A\cap\Delta\neq\varnothing \\ A\not\subset\Lambda,A\subset\Gamma}} \Phi_A \quad \text{and} \quad \Sigma'' = \sum_{\substack{A:A\cap\Delta=\varnothing \\ A\subset\Gamma}} \Phi_A.$$

In the next calculation we bound the (small) sum Σ' uniformly in ω_Δ to decouple $\mathcal{H}_{\Delta,\Lambda}$ and Σ''. Then the integrals of the (large) sum Σ'' can be cancelled. Define the function $C_1(\sigma_{\Lambda\smallsetminus\Delta})$ by the first equality of the display below. Abbreviate temporarily $A = \Lambda \smallsetminus \Delta$. We have

$$\begin{aligned}
&E^{\pi_\Gamma^{\text{free}}}[\varphi \,|\, \sigma_{\Lambda\smallsetminus\Delta}] \\
&= e^{C_1(\sigma_A)} \frac{\int_{\Omega_{\Gamma\smallsetminus\Lambda}}\int_{\Omega_\Delta} \varphi(\omega_\Delta, \sigma_A) e^{-\mathcal{H}_{\Delta,\Lambda}(\omega_\Delta,\sigma_A) - \Sigma''} \lambda_\Delta(d\omega_\Delta)\lambda_{\Gamma\smallsetminus\Lambda}(d\tau_{\Gamma\smallsetminus\Lambda})}{\int_{\Omega_{\Gamma\smallsetminus\Lambda}}\int_{\Omega_\Delta} e^{-\mathcal{H}_{\Delta,\Lambda}(\omega_\Delta,\sigma_A) - \Sigma''} \lambda_\Delta(d\omega_\Delta)\,\lambda_{\Gamma\smallsetminus\Lambda}(d\tau_{\Gamma\smallsetminus\Lambda})} \\
&= e^{C_1(\sigma_A)} \frac{\int_{\Omega_\Delta} \varphi(\omega_\Delta, \sigma_A) e^{-\mathcal{H}_{\Delta,\Lambda}(\omega_\Delta,\sigma_A)} \lambda_\Delta(d\omega_\Delta) \int_{\Omega_{\Gamma\smallsetminus\Lambda}} e^{-\Sigma''} \lambda_{\Gamma\smallsetminus\Lambda}(d\tau_{\Gamma\smallsetminus\Lambda})}{\int_{\Omega_\Delta} e^{-\mathcal{H}_{\Delta,\Lambda}(\omega_\Delta,\sigma_A)} \lambda_\Delta(d\omega_\Delta) \int_{\Omega_{\Gamma\smallsetminus\Lambda}} e^{-\Sigma''} \lambda_{\Gamma\smallsetminus\Lambda}(d\tau_{\Gamma\smallsetminus\Lambda})} \\
&= e^{C_1(\sigma_A)} \frac{\int_{\Omega_\Delta} \varphi(\omega_\Delta, \sigma_A) e^{-\mathcal{H}_{\Delta,\Lambda}(\omega_\Delta,\sigma_A)} \lambda_\Delta(d\omega_\Delta)}{\int_{\Omega_\Delta} e^{-\mathcal{H}_{\Delta,\Lambda}(\omega_\Delta,\sigma_A)} \lambda_\Delta(d\omega_\Delta)} \\
&= e^{C_1(\sigma_{\Lambda\smallsetminus\Delta})} \pi_{\Delta,\Lambda}(\sigma_{\Lambda\smallsetminus\Delta}, \varphi).
\end{aligned}$$

From $\sup|\Sigma'| \le d_{\Delta,\Lambda}$ it follows that $\sup|C_1| \le 2d_{\Delta,\Lambda}$.

***Exercise 8.8.** Show that, similarly to the computation above,

$$\pi_\Delta\varphi(\sigma_{\Lambda\smallsetminus\Delta}, \tau_{\Lambda^c}) = e^{C_2(\sigma_{\Lambda\smallsetminus\Delta}, \tau_{\Lambda^c})} \pi_{\Delta,\Lambda}(\sigma_{\Lambda\smallsetminus\Delta}, \varphi)$$

with $\sup|C_2| \le 2d_{\Delta,\Lambda}$.

Next we show that

$$E^{\gamma\pi_\Delta}[\varphi \,|\, \sigma_{\Lambda\smallsetminus\Delta}] = \int \pi_\Delta\varphi(\sigma_{\Lambda\smallsetminus\Delta}, \tau_{\Lambda^c})\, \gamma(d\tau_{\Lambda^c} \,|\, \sigma_{\Lambda\smallsetminus\Delta}).$$

Indeed, if ψ is bounded and $\mathscr{F}_{\Lambda\smallsetminus\Delta}$-measurable, then

$$\begin{aligned}
E^{\gamma\pi_\Delta}[\varphi\psi] &= \int \pi_\Delta(\varphi\psi)\, d\gamma = \int \psi\, \pi_\Delta\varphi\, d\gamma = \int E^\gamma[\psi\, \pi_\Delta\varphi \,|\, \mathscr{F}_{\Lambda\smallsetminus\Delta}]\, d\gamma \\
&= \int \psi\, E^\gamma[\pi_\Delta\varphi \,|\, \mathscr{F}_{\Lambda\smallsetminus\Delta}]\, d\gamma = \int \psi\, E^\gamma[\pi_\Delta\varphi \,|\, \mathscr{F}_{\Lambda\smallsetminus\Delta}]\, d(\gamma\pi_\Delta).
\end{aligned}$$

The last equality used the fact that γ and $\gamma\pi_\Delta$ agree on $\mathscr{F}_{\Lambda\smallsetminus\Delta}$.

Combine the steps above and define yet another function $C_3(\sigma_{\Lambda\smallsetminus\Delta})$ by the third equality below:

$$\begin{aligned}
E^{\gamma\pi_\Delta}[\varphi \,|\, \sigma_{\Lambda\smallsetminus\Delta}] &= \int \pi_\Delta\varphi(\sigma_{\Lambda\smallsetminus\Delta}, \tau_{\Lambda^c})\, \gamma(d\tau_{\Lambda^c} \,|\, \sigma_{\Lambda\smallsetminus\Delta}) \\
&= \int e^{C_2(\sigma_{\Lambda\smallsetminus\Delta}, \tau_{\Lambda^c})} \pi_{\Delta,\Lambda}(\sigma_{\Lambda\smallsetminus\Delta}, \varphi)\, \gamma(d\tau_{\Lambda^c} \,|\, \sigma_{\Lambda\smallsetminus\Delta}) \\
&= e^{C_3(\sigma_{\Lambda\smallsetminus\Delta})} \pi_{\Delta,\Lambda}(\sigma_{\Lambda\smallsetminus\Delta}, \varphi) \\
&= e^{-C_1(\sigma_{\Lambda\smallsetminus\Delta}) + C_3(\sigma_{\Lambda\smallsetminus\Delta})} E^{\pi_\Gamma^{\text{free}}}[\varphi \,|\, \sigma_{\Lambda\smallsetminus\Delta}],
\end{aligned}$$

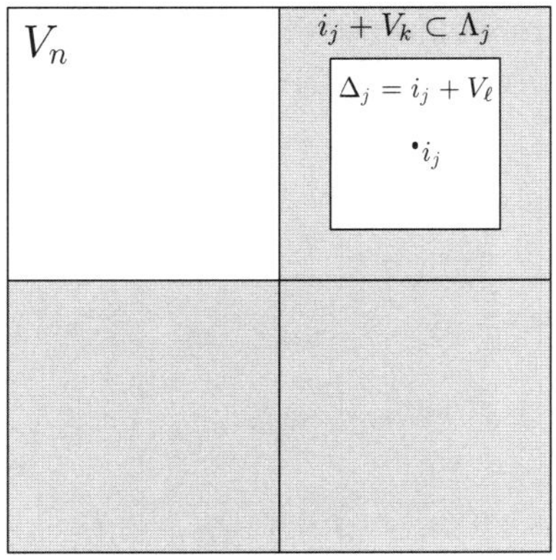

Figure 8.1. The shaded region is $\Lambda_j \smallsetminus \Delta_j$.

with $\sup|C_3| \le 2d_{\Delta,\Lambda}$. This implies that for $g \in b\mathscr{F}_\Lambda$,

$$\sup\Big|\log E^{(\gamma\pi_\Delta)_\Lambda^{\Lambda\smallsetminus\Delta}}[e^g] - \log E^{(\pi_\Gamma^{\text{free}})_\Lambda^{\Lambda\smallsetminus\Delta}}[e^g]\Big| \le 4d_{\Delta,\Lambda},$$

which implies the claim of the lemma by the variational formula for relative entropy (Theorem 5.4). □

Returning to the proof of the theorem, take k large enough such that $d_{V_\ell,V_k} < \eta/8$. Fix an odd $m \in \mathbb{N}$ and pick $n = n(m)$ such that V_n is a disjoint union of m^d translates of V_k: $V_n = \bigcup_{s=1}^{m^d}(i_s + V_k)$. Abbreviate $\Delta_j = i_j + V_\ell$ and $\Lambda_j = \bigcup_{s=1}^{j}(i_s + V_k)$, for $j = 1,\dots,m^d$. See Figure 8.1. Abbreviate $\pi = \pi_{V_n}^{\text{free}}$. Recall Exercise 6.11 and the conditional entropy formula from Exercise 5.13. Compute for $j \ge 2$:

$$\begin{aligned}
&H(\gamma_{\Lambda_j}\,|\,\pi_{\Lambda_j}) - H(\gamma_{\Lambda_{j-1}}\,|\,\pi_{\Lambda_{j-1}})\\
&\qquad\ge H(\gamma_{\Lambda_j}\,|\,\pi_{\Lambda_j}) - H(\gamma_{\Lambda_j\smallsetminus\Delta_j}\,|\,\pi_{\Lambda_j\smallsetminus\Delta_j})\\
&\qquad= \int H(\gamma_{\Lambda_j}^{\Lambda_j\smallsetminus\Delta_j}\,|\,\pi_{\Lambda_j}^{\Lambda_j\smallsetminus\Delta_j})\,d\gamma_{\Lambda_j}\\
&\qquad\ge \int H(\gamma_{\Lambda_j}^{\Lambda_j\smallsetminus\Delta_j}\,|\,(\gamma\pi_{\Delta_j})_{\Lambda_j}^{\Lambda_j\smallsetminus\Delta_j})\,d\gamma_{\Lambda_j} - 4d_{V_\ell,V_k}\\
&\qquad= H(\gamma_{\Lambda_j}\,|\,(\gamma\pi_{\Delta_j})_{\Lambda_j}) - 4d_{V_\ell,V_k}\\
&\qquad\ge H(\gamma_{i_j+V_k}\,|\,(\gamma\pi_{i_j+V_\ell})_{i_j+V_k}) - 4d_{V_\ell,V_k}\\
&\qquad\ge \eta - \frac{4\eta}{8} = \eta/2.
\end{aligned}$$

The second equality used $(\gamma\pi_{\Delta_j})_{\Lambda_j\smallsetminus\Delta_j} = \gamma_{\Lambda_j\smallsetminus\Delta_j}$. The third inequality restricted entropy to $i_j + V_k$. The last inequality used the shift invariance of γ.

The computation above holds also for $j = 1$, with the first line replaced by $H(\gamma_{\Lambda_1} \mid \pi_{\Lambda_1})$. Adding these inequalities over j gives

$$H(\gamma_{V_n} \mid \pi_{V_n}^{\text{free}}) \geq m^d \eta/2.$$

Consequently for $n = n(m)$,

$$\frac{1}{|V_n|} H(\gamma_{V_n} \mid \pi_{V_n}^{\text{free}}) \geq \frac{\eta}{2|V_k|}.$$

Letting $m \nearrow \infty$ leads to $h(\gamma \mid \Phi) \geq \frac{1}{2}\eta\,|V_k|^{-1} > 0$, contradicting (b). The proof of the DLR variational principle is complete. □

A corollary of the variational principle is the existence of shift-invariant Gibbs measures.

Corollary 8.9. *Let $\Omega = \mathcal{S}^{\mathbb{Z}^d}$ for a Polish spin space $\mathcal{S}$ and let Φ be an absolutely summable shift-invariant continuous interaction potential. Then $\mathscr{G}^\Phi \cap \mathcal{M}_\theta(\Omega)$ is nonempty, convex, and compact.*

Proof. We already know that $\mathscr{G}^\Phi \cap \mathcal{M}_\theta(\Omega)$ is convex. We have shown that

$$\begin{aligned} \mathscr{G}^\Phi \cap \mathcal{M}_\theta(\Omega) &= \{\gamma \in \mathcal{M}_\theta(\Omega) : h(\gamma \mid \Phi) = 0\} \\ &= \bigcap_{n \geq 1} \{\gamma \in \mathcal{M}_\theta(\Omega) : h(\gamma \mid \Phi) \leq 1/n\}. \end{aligned} \tag{8.6}$$

By Theorem 8.6, $h(\cdot \mid \Phi)$ is a rate function for an LDP on the space $\mathcal{M}_\theta(\Omega)$. The infimum of $h(\cdot \mid \Phi)$ over $\mathcal{M}_\theta(\Omega)$ must then be 0, and consequently the nested sets $\{\gamma \in \mathcal{M}_\theta(\Omega) : h(\gamma \mid \Phi) \leq 1/n\}$ in the intersection are nonempty. By Exercise 8.5 these are compact sets. The intersection in (8.6) is nonempty and compact by the finite intersection property of nested compact sets (Theorem A.5(i) in Appendix A or Proposition 4.21 of [**34**]). □

Using the notion of subdifferential, we can give a geometric interpretation of a Gibbs state. Let $\mathbb{B}$ be the Banach space of absolutely summable, shift-invariant continuous interaction potentials. Set $\mathbb{B}$ and $\mathcal{M}_\theta(\Omega)$ in duality by $\langle \Phi, \nu \rangle = -E^\nu[f_\Phi]$. By part (c) of Theorem 8.1, $P(\Phi)$ is a supremum over affine continuous functions and is hence convex and lower semicontinuous on $\mathbb{B}$.

Corollary 8.10. *Suppose $\gamma \in \mathcal{M}_\theta(\Omega)$. Then $\gamma \in \mathscr{G}^\Phi$ if and only if $\gamma \in \partial P(\Phi)$, that is, γ is a tangent to P at Φ.*

Exercise 8.11. Show that the duality above between $\mathbb{B}$ and $\mathcal{M}_\theta(\Omega)$ satisfies Assumption 4.1 so that the induced weak topologies are Hausdorff.

Literature note. The connections between statistical mechanics, thermodynamics and large deviation theory are explored further in Ellis's monograph [**32**].

Chapter 9

Phase transition in the Ising model

This chapter marks a departure from large deviations. We study the phase transition of the Ising model. Even though in the previous chapter we reached a complete characterization of shift-invariant Gibbs measures in terms of entropy and the variational principle of the pressure, those general results play no role in this chapter. Perhaps large deviation results are too crude and general to penetrate the question of phase transition.

We saw in Section 3.4 that the Curie-Weiss model of ferromagnetism captures the phase transition phenomenon known as spontaneous magnetization. A serious shortcoming of the Curie-Weiss model is the absence of spatial structure: every pair of spins interacts with the same strength.

A more realistic model of ferromagnetism is the Ising model where the spins are placed on the sites of an integer lattice and only neighboring spins interact. This is still a rough approximation because in reality spins are carried by electrons, but the Ising model is a worthwhile test case for ideas of statistical mechanics. It has also been a rich source of deep mathematics, as evidenced by Stanislav Smirnov's 2010 Fields Medal awarded in part for work on the planar Ising model. The Ising model was introduced by Wilhelm Lenz in 1920. The one-dimensional Ising model was studied by his student Ernst Ising.

Fix a dimension $d \in \mathbb{N}$. The (single spin) state space is $\mathcal{S} = \{-1, +1\}$ and $\Omega = \mathcal{S}^{\mathbb{Z}^d}$. The choice of $\mathcal{S}$ represents the simplifying assumption that spins can take only two values. The Ising model on $\mathbb{Z}^d$ has a nearest-neighbor potential described by three parameters: the strength of the nearest-neighbor coupling ($J > 0$), the strength of the external magnetic field ($h \in \mathbb{R}$), and the inverse temperature ($\beta > 0$). The potential takes the form

$$\Phi_A(\sigma) = \begin{cases} -J\sigma_i\sigma_j, & A = \{i,j\} \text{ with } |i-j|_1 = 1, \\ -h\sigma_i, & A = \{i\}, \\ 0, & \text{otherwise.} \end{cases}$$

Here $|i|_1$ is the ℓ^1 norm and $|i-j|_1 = 1$ means that i and j are nearest neighbors, also expressed by $i \sim j$. Consequently, the Hamiltonian in volume Λ with boundary condition ω is

$$\mathcal{H}^\omega_\Lambda(\sigma_\Lambda) = -J \sum_{\substack{\{i,j\}\subset\Lambda \\ |i-j|_1=1}} \sigma_i\sigma_j - J \sum_{\substack{i\in\Lambda,\, j\in\Lambda^c \\ |i-j|_1=1}} \sigma_i\omega_j - h\sum_{i\in\Lambda}\sigma_i\,.$$

Note that each nearest-neighbor pair appears in the sums exactly once. The natural "fair coin-tossing" reference measure is $\lambda = (\frac{1}{2}\delta_{-1} + \frac{1}{2}\delta_1)^{\otimes\mathbb{Z}^d}$. From these ingredients we define the Feller-continuous specification $\Pi_{\beta,h,J} = \{\pi_\Lambda : \Lambda \subset \mathbb{Z}^d \text{ finite}\}$ by

$$\pi_\Lambda(\omega, B) = \frac{1}{Z^\omega_{\Lambda,\beta,h,J}} \int_{\Omega_\Lambda} \mathbb{1}_B(\sigma_\Lambda, \omega_{\Lambda^c})\, e^{-\beta\mathcal{H}^\omega_\Lambda(\sigma_\Lambda)}\, \lambda_\Lambda(d\sigma_\Lambda), \quad B \in \mathscr{F},$$

with partition function

$$Z^\omega_{\Lambda,\beta,h,J} = \int_{\Omega_\Lambda} e^{-\beta\mathcal{H}^\omega_\Lambda(\sigma_\Lambda)}\, \lambda_\Lambda(d\sigma_\Lambda).$$

We use the notation $\mu^\omega_{\Lambda,\beta,h,J} = \pi_\Lambda(\omega,\cdot)$ for finite-volume Ising Gibbs measures. These can be regarded as probability measures on Ω_Λ or on the full configuration space Ω, whichever is convenient. Due to the Feller-continuity, limit points of the measures $\mu^\omega_{\Lambda,\beta,h,J}$ as $\Lambda \nearrow \mathbb{Z}^d$ are infinite-volume Gibbs measures (Theorem 7.17).

The free Hamiltonian and the finite-volume Gibbs measures with free boundary conditions are defined as above, but without the boundary terms ω_j. This is equivalent to setting $\omega_j = 0$ and thus we will use the notation $\mathcal{H}^0_\Lambda(\sigma_\Lambda)$ and $\mu^0_{\Lambda,\beta,h,J}$. We will also use the notation $\mu^+_{\Lambda,\beta,h,J}$ for the case of $+$ boundary condition, that is, $\omega_j = 1$ for all j, and similarly $\mu^-_{\Lambda,\beta,h,J}$ for the Gibbs measure with boundary condition $\omega = -1$.

The mapping $T\sigma = -\sigma$ is called the *global spin flip*.

***Exercise 9.1.** Check the following symmetry in the absence of an external magnetic field: $\mu^+_{\Lambda,\beta,0,J} = \mu^-_{\Lambda,\beta,0,J} \circ T$.

We shall only consider the ferromagnetic case $J > 0$. When $h = 0$, $\mathcal{H}^0_\Lambda$ is minimized by two ground states: all spins $+1$ and all spins -1. At positive temperature and without external field, $\mu^+_{\Lambda,\beta,0,J}$ prefers $+$ spins while $\mu^-_{\Lambda,\beta,0,J}$ prefers $-$ spins. Based on what happens in the Curie-Weiss model, we might expect that at low temperature the infinite-volume limits of the two measures give distinct Gibbs states and thereby phase transition occurs. This would happen if boundary effects can reach the origin regardless of how large the volume Λ is. On the other hand, at high temperature boundary effects should be overwhelmed by stochastic noise and a unique Gibbs state is expected.

The free Gibbs measure $\mu^0_{\Lambda\,\beta,h,J}$ prefers $+$ or $-$ spins depending on the sign of h. When an external magnetic field and opposite boundary conditions are present, they are in competition. However, since the magnetic field acts on the whole volume one would expect it to dominate boundary effects. So there should not be a phase transition when $h \neq 0$.

Here is the main theorem that validates these informal ideas.

Theorem 9.2. *The weak limits*

$$\mu^+_{\beta,h,J} = \lim_{\Lambda \nearrow \mathbb{Z}^d} \mu^+_{\Lambda,\beta,h,J} \quad \textit{and} \quad \mu^-_{\beta,h,J} = \lim_{\Lambda \nearrow \mathbb{Z}^d} \mu^-_{\Lambda,\beta,h,J}$$

exist in $\mathcal{M}_1(\Omega)$ *for all* $d \in \mathbb{N}$, $h \in \mathbb{R}$, $J > 0$, *and* $\beta > 0$. *The measures* $\mu^\pm_{\beta,h,J}$ *are shift invariant.*

(a) *When* $d = 1$ *there is a unique Gibbs measure for all* $h \in \mathbb{R}$, $J > 0$, *and* $\beta > 0$ *and in particular* $\mu^+_{\beta,h,J} = \mu^-_{\beta,h,J}$.

(b) *When* $d \geq 2$ *there exists a finite, positive critical inverse temperature* $\beta_c = \beta_c(J,d)$ *such that this holds: if* $h \neq 0$ *or* $\beta < \beta_c$, *then* $\mu^+_{\beta,h,J} = \mu^-_{\beta,h,J}$ *and the Gibbs measure is unique, while if* $h = 0$ *and* $\beta > \beta_c$, *then* $\mu^+_{\beta,h,J} \neq \mu^-_{\beta,h,J}$ *and we have a phase transition.*

We complement the theorem with some additional facts about the Ising model. Part (b) left out the case $h = 0$, $\beta = \beta_c$. However, it is known that the Gibbs measure is unique for these parameter values in all dimensions $d \geq 2$. The last remaining case $d = 3$ was treated in article [**3**].

Onsager's famous 1944 result identified the critical temperature in $d = 2$ as $\beta_c = \frac{1}{2J}\log(1+\sqrt{2})$ [**61**]. In $d = 2$, $\mu^+_{\beta,h,J}$ and $\mu^-_{\beta,h,J}$ are the only extreme Gibbs measures (Aizenman [1], [**2**] and Higuchi [**46**]). Consequently all Gibbs measures are shift invariant in dimension two. By contrast, Dobrushin [**24**] showed that when $d \geq 3$, $h = 0$, and β is large, finite-volume Gibbs measures with $+$ boundary conditions on one side of a half-plane and $-$ on the other side converge weakly to extreme Gibbs measures. Far enough from the interface these measures are close to either $\mu^+_{\beta,0,J}$ or $\mu^-_{\beta,0,J}$, depending on the side. Hence they are not shift invariant. The critical temperature at which these measures appear is the *roughening temperature*. Whether it is equal to the critical temperature of phase transition is a major open question.

The rest of this chapter is devoted to the proof of Theorem 9.2. It is worthwhile at this point to take a look at page 108 where the familiar liquid-gas phase transition was discussed.

9.1. One-dimensional Ising model

In this section we prove part (a) of Theorem 9.2. The situation is related to the one in Example 7.21 but this time we are not given a transition probability. Instead of appealing to the Markov chain convergence theorem, we diagonalize a transfer matrix.

Compactness of Ω implies that the set of Gibbs measures $\mathscr{G}$ is nonempty for all values of the parameters $\beta > 0$, $J > 0$, and $h \in \mathbb{R}$ (Corollary 7.18). Next, fix $\gamma \in \mathscr{G}$, $m \in \mathbb{N}$, and $b_{-m}, \dots, b_m \in \{-1, +1\}$. Consider the cylinder set $A = \{\sigma \in \Omega : \sigma_i = b_i \text{ for } |i| \le m\}$. The backward-martingale convergence theorem (Theorem 5.6.3 of [**27**] or page 155 of [**54**]) implies the existence of the γ-a.s. limit of

$$Y_n = E^\gamma[\mathbb{1}_A | \mathscr{F}_{[-n,n]^c}] \quad \text{as } n \to \infty.$$

The goal will be to prove that this limit is a constant $c(A)$ that does not depend on γ. Then by the dominated convergence theorem

$$\begin{aligned}\gamma(A) &= \lim_{n\to\infty} E^\gamma\big[E^\gamma(\mathbb{1}_A | \mathscr{F}_{[-n,n]^c})\big] = E^\gamma\big[\lim_{n\to\infty} E^\gamma(\mathbb{1}_A | \mathscr{F}_{[-n,n]^c})\big] \\ &= c(A).\end{aligned} \tag{9.1}$$

This implies that there can be only one $\gamma \in \mathscr{G}$.

Without loss of generality, we can absorb β into J and h and hence set $\beta = 1$. Fix a boundary condition $\tau \in \Omega$ and $n > m$. Write

$$Y_n(\tau) = E^\gamma[\mathbb{1}_A | \mathscr{F}_{[-n,n]^c}](\tau) = \mu^\tau_{[-n,n],1,h,J}\{\sigma_i = b_i \text{ for } |i| \le m\}$$

$$= \frac{\displaystyle\sum_{\substack{\sigma_{-n},\dots,\sigma_{-m-1}\\ \sigma_{m+1},\dots,\sigma_n}} \exp\big\{-\mathcal{H}^\tau_{[-n,n]}(\sigma_{-n},\dots,\sigma_{-m-1}, b_{-m},\dots,b_m, \sigma_{m+1},\dots,\sigma_n)\big\}}{\displaystyle\sum_{\omega_{-n},\dots,\omega_n} \exp\big\{-\mathcal{H}^\tau_{[-n,n]}(\omega_{-n},\dots,\omega_n)\big\}}.$$

Define, for $x, y \in \{-1, +1\}$, $A(x,y) = e^{Jxy+h(x+y)/2}$. Then an exponential term in the numerator above can be rewritten as

$$A(\tau_{-n-1}, \sigma_{-n}) \left[\prod_{i=-n}^{-m-2} A(\sigma_i, \sigma_{i+1}) \right] A(\sigma_{-m-1}, b_{-m}) \left[\prod_{i=-m}^{m-1} A(b_i, b_{i+1}) \right]$$
$$\times A(b_m, \sigma_{m+1}) \left[\prod_{i=m+1}^{n-1} A(\sigma_i, \sigma_{i+1}) \right] A(\sigma_n, \tau_{n+1}) e^{h(\tau_{-n-1}+\tau_{n+1})/2}.$$

If we think of A as a 2-by-2 matrix (called the *transfer matrix*)

$$A = \begin{bmatrix} A(-1,-1) & A(-1,1) \\ A(1,-1) & A(1,1) \end{bmatrix} = \begin{bmatrix} e^{J-h} & e^{-J} \\ e^{-J} & e^{J+h} \end{bmatrix}$$

and define

$$a_0 = \prod_{i=-m}^{m-1} A(b_i, b_{i+1}),$$

then

$$Y_n(\tau) = \frac{A^{n-m+1}(\tau_{-n-1}, b_{-m})\, a_0\, A^{n-m+1}(b_m, \tau_{n+1})}{A^{2n+2}(\tau_{-n-1}, \tau_{n+1})}.$$

The matrix A is real and symmetric. It has two real eigenvalues, say λ and μ, with orthonormal eigenvectors u and v. Then

$$A = \begin{bmatrix} u & v \end{bmatrix} \begin{bmatrix} \lambda & 0 \\ 0 & \mu \end{bmatrix} \begin{bmatrix} u & v \end{bmatrix}^T.$$

Observe that $\det A = e^{2J} - e^{-2J} > 0$ and $\operatorname{trace} A = e^{J-h} + e^{J+h} > 0$, so the two eigenvalues are positive. They cannot be equal since otherwise A would be a multiple of the identity matrix. We can, therefore, assume $\lambda > \mu > 0$. Since $(A/\lambda)^k = uu^T + (\mu/\lambda)^k vv^T$, we see that the largest eigenvalue and its eigenvector determine the asymptotics of A. We have

$$\begin{aligned} Y_n(\tau) &= a_0 \lambda^{-2m} \frac{(A/\lambda)^{n-m+1}(\tau_{-n-1}, b_{-m})(A/\lambda)^{n-m+1}(b_m, \tau_{n+1})}{(A/\lambda)^{2n+2}(\tau_{-n-1}, \tau_{n+1})} \\ &= a_0 \lambda^{-2m} \frac{\big(u(\tau_{-n-1})u(b_{-m}) + o(1)\big)\big(u(b_m)u(\tau_{n+1}) + o(1)\big)}{u(\tau_{-n-1})u(\tau_{n+1}) + o(1)} \\ &= a_0 \lambda^{-2m} \frac{u(\tau_{-n-1})u(b_{-m})u(b_m)u(\tau_{n+1}) + o(1)}{u(\tau_{-n-1})u(\tau_{n+1}) + o(1)}. \end{aligned}$$

Direct inspection shows that if u has a zero entry, then it cannot be an eigenvector of A. Thus, $Y_n(\tau)$ converges to $a_0\lambda^{-2m}u(b_{-m})u(b_m)$, which is independent of τ. Hence, (9.1) implies that $\gamma(A) = a_0\lambda^{-2m}u(b_{-m})u(b_m)$. This characterizes $\gamma \in \mathscr{G}$ and part (a) of Theorem 9.2 is proved.

We proved a little bit more than the absence of phase transition in the one-dimensional Ising model.

Exercise 9.3. Define the vector $\rho(x) = u(x)^2$ and the matrix

$$P(x,y) = \frac{A(x,y)u(y)}{\lambda u(x)}, \quad x,y \in \{-1,+1\}.$$

Check that P is a Markov transition matrix and that $\rho P = P$, that is, ρ is an invariant measure for P. Check also that γ is precisely the distribution of the Markov chain with marginal ρ and transition matrix P. In other words, this stationary Markov chain is the unique Gibbs measure for the one-dimensional Ising model.

9.2. Phase transition at low temperature

The absence of phase transition in the one-dimensional case led initially to the dismissal of the Ising model as a model of ferromagnetism. In 1936 Peierls discovered that phase transition does happen in two dimensions. The argument is based on quantifying the energy needed to create a contour that separates a droplet of one spin surrounded by a sea of the opposite. In order for the boundary conditions to not influence the origin, such contours must be possible with nontrivial probability.

The next theorem is the first step towards proving part (b) of Theorem 9.2. The rigorous proof is due to Dobrushin [**21**] and Griffiths [**43**].

Theorem 9.4. *Let $d = 2$ and set $h = 0$. Then for all large enough β the following holds: as $\Lambda \nearrow \mathbb{Z}^d$, every weak limit point of the sequence $\{\mu^+_{\Lambda,\beta,0,J} : \Lambda \subset \mathbb{Z}^d$ finite$\}$ is distinct from every limit point of $\{\mu^-_{\Lambda,\beta,0,J} : \Lambda \subset \mathbb{Z}^d$ finite$\}$. Consequently there exists more than one Gibbs measure and phase transition happens at $h = 0$ and large β.*

Proof. Let us temporarily drop $\beta > 0$, $h = 0$, and $J > 0$ from the notation and write $\mu^\pm_\Lambda = \mu^\pm_{\Lambda,\beta,0,J}$. We expect the $+$ boundary condition to make it more likely for the spin at 0 to be $+$.

Claim 9.5. *There exist $\beta_0 < \infty$ and $\delta > 0$ such that, if $\beta \geq \beta_0$, then $\mu^+_\Lambda\{\sigma_0 = -1\} \leq \frac{1}{2} - \delta$ for all finite $\Lambda \subset \mathbb{Z}^2$.*

Let us assume this claim for the moment and finish the proof of the theorem. By the spin-flip symmetry (Exercise 9.1)

$$\mu^-_\Lambda\{\sigma_0 = -1\} = 1 - \mu^-_\Lambda\{\sigma_0 = 1\} = 1 - \mu^+_\Lambda\{\sigma_0 = -1\} \geq \tfrac{1}{2} + \delta$$

for all finite $\Lambda \subset \mathbb{Z}^2$. Let μ^+ be an arbitrary limit point of $\{\mu^+_\Lambda\}$. Pick a sequence $\Lambda_n \nearrow \mathbb{Z}^2$ such that $\mu^+_{\Lambda_n} \to \mu^+$ weakly. Then since $\{\sigma_0 = -1\}$ is both open and closed

$$\mu^+\{\sigma_0 = -1\} = \lim_{n\to\infty} \mu^+_{\Lambda_n}\{\sigma_0 = -1\} \leq \tfrac{1}{2} - \delta.$$

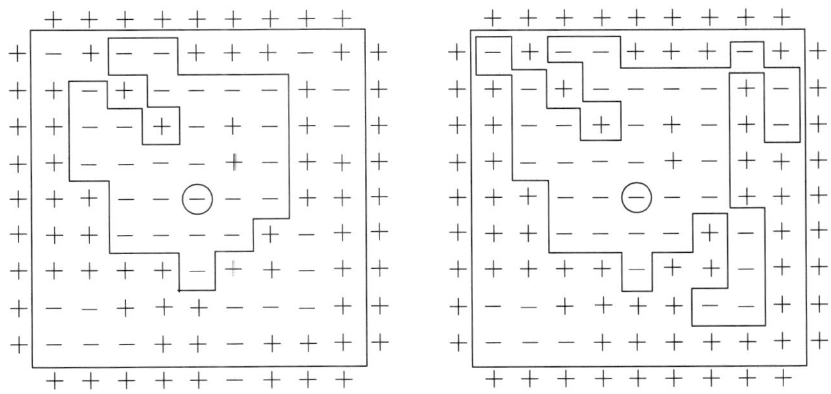

Figure 9.1. Peierls's contours. The outer square is Λ with $+$ boundary conditions. The circle marks the minus spin at the origin and the inside path marks two contours surrounding it.

By the same reasoning an arbitrary limit point μ^- of $\{\mu_\Lambda^-\}$ satisfies $\mu^-\{\sigma_0 = -1\} \geq \frac{1}{2} + \delta$. This implies that $\mu^- \neq \mu^+$ and the theorem is proved. It remains to prove Claim 9.5.

Proof of Claim 9.5. First, define a *circuit* to be a closed path formed by edges between nearest-neighbor points of the dual lattice $(1/2, 1/2)+\mathbb{Z}^2$ that does not repeat edges. Formally, $\gamma = \{(u_i, u_{i+1}) : 0 \leq i \leq k-1\}$ is a circuit of length k if $u_i \in (1/2, 1/2)+\mathbb{Z}^2$ and $|u_{i+1} - u_i|_1 = 1$ for all i, $u_0 = u_k$, and $(u_i, u_{i+1}) \notin \{(u_j, u_{j+1}), (u_{j+1}, u_j)\}$ for all $i \neq j$. Circuit γ is said to surround 0 in Λ if any nearest-neighbor path from the origin to a point outside Λ must cross γ. That is, if $\{x_i\}_{i=0}^m$ is a nearest-neighbor path such that $x_0 = 0$ and $x_m \in \Lambda^c$, then $\exists i, j \geq 0$ such that $(x_i + x_{i+1})/2 = (u_j + u_{j+1})/2$.

Given a configuration $\sigma_\Lambda \in \Omega_\Lambda$ with $+$ boundary condition, a circuit γ is a *contour* if each edge of γ separates a $\{+,-\}$-spin pair. As Figure 9.1 illustrates, there may be many ways to enclose the origin in a contour. Due to the $+$ boundary condition, if $\sigma_0 = -1$, there is at least one contour surrounding 0, namely, the boundary of the *connected minus cluster of the origin.*

For a circuit γ surrounding the origin, let $\Omega_{\Lambda,\gamma}$ denote the event that $\sigma_0 = -1$ and γ is a contour. Let Γ_k be the set of circuits of length k surrounding the origin. Then

$$\mu_\Lambda^+\{\sigma_0 = -1\} \leq \sum_{k=1}^{\infty} \sum_{\gamma \in \Gamma_k} \mu_\Lambda^+(\Omega_{\Lambda,\gamma}).$$

Lemma 9.6. *The cardinality* $|\Gamma_k| \leq k3^{k-1}$ *for all* $k \geq 1$. *If* $\gamma \in \Gamma_k$, *then* $\mu_\Lambda^+(\Omega_{\Lambda,\gamma}) \leq e^{-2k\beta J}$.

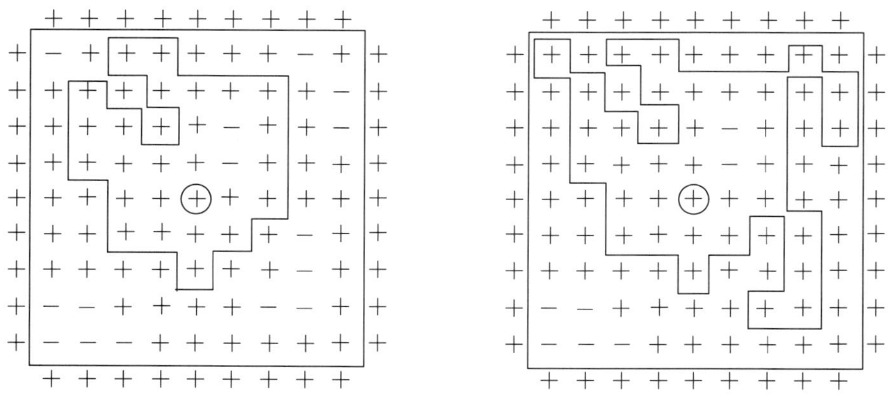

Figure 9.2. The configurations resulting from applying (9.2) to the contours in Figure 9.1.

Proof. In order to surround the origin, a circuit $\gamma \in \Gamma_k$ must contain one of the vertical line segments $[(i+1/2,-1/2),(i+1/2,1/2)]$, $0 \le i < k$. Choose one of these k edges. Starting from the top endpoint of this edge, build the circuit by choosing the remaining edges in order. At each of the remaining $k-1$ steps there are at most 3 available edges.

Suppose that under configuration σ_Λ there is a minus cluster at the origin bounded by contour $\gamma \in \Gamma_k$. Flip the spins inside γ to get the new configuration

$$\bar{\sigma}_j = \begin{cases} -\sigma_j & \text{if } j \text{ is inside } \gamma, \\ \sigma_j & \text{if } j \text{ is outside } \gamma. \end{cases} \tag{9.2}$$

(The inside of γ is defined, for example, as the sites to the left of γ as it is traversed counterclockwise. See Figure 9.2.)

This mapping aligns every nearest-neighbor pair of spins separated by γ, but the alignment or disagreement of a spin pair entirely inside or outside γ is not affected. Consequently

$$J\bar{\sigma}_i\bar{\sigma}_j = \begin{cases} J\sigma_i\sigma_j & \text{if } \gamma \text{ does not separate } i,j, \\ J\sigma_i\sigma_j + 2J & \text{if } \gamma \text{ separates } i,j. \end{cases}$$

The same formula applies also to pairs $\{i,j\}$ where one site lies outside Λ and so its spin is part of the $+$ boundary condition. For the total change of the Hamiltonian we get

$$\begin{aligned}\mathcal{H}_\Lambda^+(\bar\sigma_\Lambda) &= -J \sum_{\substack{\{i,j\}\subset\Lambda\\ |i-j|_1=1}} \bar\sigma_i\bar\sigma_j - J \sum_{\substack{i\in\Lambda, j\in\Lambda^c\\ |i-j|_1=1}} \bar\sigma_i\\ &= -J \sum_{\substack{\{i,j\}\subset\Lambda\\ |i-j|_1=1}} \sigma_i\sigma_j - J \sum_{\substack{i\in\Lambda, j\in\Lambda^c\\ |i-j|_1=1}} \sigma_i - 2kJ\\ &= \mathcal{H}_\Lambda^+(\sigma_\Lambda) - 2kJ.\end{aligned}$$

$2kJ$ is then the energy cost of the circuit. We have

$$\mu_\Lambda^+(\Omega_{\Lambda,\gamma}) = \frac{\sum_{\sigma_\Lambda\in\Omega_{\Lambda,\gamma}} e^{-\beta\mathcal{H}_\Lambda^+(\sigma_\Lambda)}}{\sum_{\sigma_\Lambda\in\Omega_\Lambda} e^{-\beta\mathcal{H}_\Lambda^+(\sigma_\Lambda)}} = e^{-2k\beta J}\frac{\sum_{\sigma_\Lambda\in\Omega_{\Lambda,\gamma}} e^{-\beta\mathcal{H}_\Lambda^+(\bar\sigma_\Lambda)}}{\sum_{\sigma_\Lambda\in\Omega_\Lambda} e^{-\beta\mathcal{H}_\Lambda^+(\sigma_\Lambda)}}.$$

For a given γ the mapping (9.2) is one-to-one. Hence the sum in the numerator of the right-most fraction contains a subset of the terms in the denominator. Thus, the ratio is less than one and the lemma is proved. □

To complete the proof of Claim 9.5, and thus of the theorem, write

$$\mu_\Lambda^+\{\sigma_0=-1\} \le \sum_{k=1}^\infty k3^{k-1}e^{-2k\beta J} \le \sum_{k=1}^\infty e^{-2(\beta J-\log 3)k} \le \frac{1}{2}-\delta,$$

for all β large enough. □

Note that this argument fails in one dimension because the cost to flip the spins in any interval is $e^{-4\beta J}$, while there are n^2 possible intervals surrounding 0 in $(-n,n)$.

9.3. Case of no external field

The second step in the proof of Theorem 9.2(b) is the high temperature counterpart of Theorem 9.4. Namely, for small β the interaction is weak and we can apply Dobrushin's uniqueness condition from Theorem 7.27.

Theorem 9.7. *Let $d\in\mathbb{N}$, $J>0$, and $h=0$. Then, if $\beta>0$ is small enough, $\mathcal{G}$ is a singleton.*

Proof. The interaction with the spin at the origin is given by $\beta\Phi_{\{0,\pm e_i\}}(\sigma) = -\beta J\sigma_0\sigma_{\pm e_i}$ for $1\le i\le d$. Condition (7.13) for uniqueness is $2d\beta J<1$. □

The next task is to complete Theorems 9.4 and 9.7 into a proof of part (b) of Theorem 9.2 for the case $h=0$. The following informal observations about the effect of β and d guide our way.

Increasing the temperature increases the effect of noise. Hence, if at a given temperature there is no phase transition, the same should hold for higher temperatures. In other words, if there is no phase transition at $\beta = \beta_0$, then there should be no phase transition at $\beta \le \beta_0$.

Increasing dimension increases the boundary relative to the volume, thereby enhancing the influence of boundary conditions. Thus we might expect that if there is phase transition in dimension d at a particular β, then there will be phase transition in higher dimensions at this same β.

To make use of the natural monotonicity $-1 \le 1$ of the spin values, we develop a partial order on probability measures. In general, a *partial order* on an arbitrary set is a binary relation $\le$ that satisfies (i) $x \le x$, (ii) $x \le y$ and $y \le z$ imply $x \le z$, and (iii) $x \le y$ and $y \le x$ imply $x = y$. (That is, a partial order is reflexive, transitive, and antisymmetric.) A *total order* satisfies the additional requirement that each pair of elements x, y satisfies either $x \le y$ or $y \le x$.

A partial order on Ω is defined by $\omega \le \sigma$ if $\omega_i \le \sigma_i$ for all $i \in \mathbb{Z}^d$. A function $f : \Omega \to \mathbb{R}$ is *increasing* if $f(\omega) \le f(\sigma)$ whenever $\omega \le \sigma$. We call an event A increasing if the indicator function $\mathbb{1}_A$ is increasing. We can then define a partial order on $\mathcal{M}_1(\Omega)$ by saying that $\mu \le \nu$ (ν *stochastically dominates* μ) if $E^\mu[f] \le E^\nu[f]$ for all increasing functions $f \in \mathscr{C}_b(\Omega)$.

***Exercise 9.8.** Check that this relation does define a partial order on $\mathcal{M}_1(\Omega)$. In particular, show that if $\mu \le \nu$ and $\nu \le \mu$, then $\mu = \nu$.

Hint: Events of the type $\{\sigma : \sigma_i = 1 \ \forall i \in D\}$ for $D \subset \mathbb{Z}^d$ are increasing. Show that the indicator function of a cylinder event $\{\sigma : \sigma_i = 1 \ \forall i \in A, \ \sigma_j = -1 \ \forall j \in B\}$, for disjoint finite $A, B \subset \mathbb{Z}^d$, can be expressed in terms of sums and differences of indicators of events of the type $\{\sigma_D = 1\}$. Induction on the size of B is straightforward. Alternatively, expand the product in

$$\begin{aligned} &\mathbb{1}\{\sigma = 1 \text{ on } A \text{ and } \sigma = -1 \text{ on } B\} \\ &\quad = \prod_{i \in B} \big(\mathbb{1}\{\sigma = 1 \text{ on } A\} - \mathbb{1}\{\sigma = 1 \text{ on } A \cup \{i\}\}\big). \end{aligned}$$

The function

$$\rho(\Lambda) = \mu\{\sigma_i = 1 \ \forall i \in \Lambda\}, \qquad \Lambda \subset \mathbb{Z}^d \text{ finite},$$

is sometimes called the *correlation function* of a probability measure μ. The exercise above implies that the values $\{\rho(\Lambda)\}$ determine the measure μ.

The proof of the next lemma can be found in Section B.7. It uses the idea of the previous exercise.

Strassen's lemma. *Recall that $\Omega = \{-1, 1\}^{\mathbb{Z}^d}$. Let $\mu, \nu \in \mathcal{M}_1(\Omega)$ be such that $\mu \le \nu$. Assume that $\mu\{\sigma_i = 1\} = \nu\{\sigma_i = 1\}$ for all $i \in \mathbb{Z}^d$. Then $\mu = \nu$.*

The next theorem expresses the intuition that flipping some $-$ spins to $+$ spins at the boundary is favorable for $+$ spins inside the volume.

Theorem 9.9. *Fix $d \in \mathbb{N}$, $\beta > 0$, $J > 0$, and $h \in \mathbb{R}$. Also fix a finite $\Lambda \subset \mathbb{Z}^d$. If $\omega \le \sigma$, then $\mu^{\omega}_{\Lambda,\beta,h,J} \le \mu^{\sigma}_{\Lambda,\beta,h,J}$.*

Proof. The proof uses Holley's theorem, a proof of which is given in Section B.7. Define pointwise minima and maxima by $(\eta \wedge \xi)_i = \eta_i \wedge \xi_i$ and $(\eta \vee \xi)_i = \eta_i \vee \xi_i$.

Holley's theorem. *Let $\Omega_\Lambda = \{-1, 1\}^\Lambda$ for some finite $\Lambda \subset \mathbb{Z}^d$. Let μ and ν be strictly positive probability measures on Ω_Λ. If*

$$\mu(\eta \wedge \xi)\nu(\eta \vee \xi) \ge \mu(\eta)\nu(\xi)$$

for all η and ξ in Ω_Λ, then $\mu \le \nu$.

Let us drop the indices β, h, and J. For Theorem 9.9 it is enough to prove that

$$\mu^{\omega}_{\Lambda}(\eta \wedge \xi)\mu^{\sigma}_{\Lambda}(\eta \vee \xi) \ge \mu^{\omega}_{\Lambda}(\eta)\mu^{\sigma}_{\Lambda}(\xi).$$

This translates into

$$\mathcal{H}^{\omega}_{\Lambda}(\eta \wedge \xi) + \mathcal{H}^{\sigma}_{\Lambda}(\eta \vee \xi) \le \mathcal{H}^{\omega}_{\Lambda}(\eta) + \mathcal{H}^{\sigma}_{\Lambda}(\xi).$$

It is enough to show that the inequality above holds term by term in the sums defining the Hamiltonians. That is,

$$\begin{aligned}
(\eta_i \wedge \xi_i)(\eta_j \wedge \xi_j) + (\eta_i \vee \xi_i)(\eta_j \vee \xi_j) \;&\ge\; \eta_i\eta_j + \xi_i\xi_j \quad \forall i, j \in \Lambda,\\
(\eta_i \wedge \xi_i)\omega_j + (\eta_i \vee \xi_i)\sigma_j \;&\ge\; \eta_i\omega_j + \xi_i\sigma_j \quad \forall i \in \Lambda, \forall j \notin \Lambda,\\
h(\eta_i \wedge \xi_i + \eta_i \vee \xi_i) \;&\le\; h(\eta_i + \xi_i) \quad \forall i \in \Lambda.
\end{aligned}$$

The third line is true for an h of either sign because it is an equality. Next, observe that if $\eta_i = -1$ or $\xi_i = 1$, then $\eta_i \wedge \xi_i = \eta_i$ and $\eta_i \vee \xi_i = \xi_i$. Thus, in this case there is equality in the second line. On the other hand, when $\eta_i = 1$ and $\xi_i = -1$ the second line reads $\sigma_j - \omega_j \ge \omega_j - \sigma_j$, which is true since $\omega \le \sigma$.

As to the first line, it is again an equality unless $\eta_i = \xi_j = 1$ and $\xi_i = \eta_j = -1$ or $\eta_j = \xi_i = 1$ and $\xi_j = \eta_i = -1$, in which case it reads $2 \ge -2$. □

As a consequence we see that $-$ and $+$ boundary conditions are extreme in a sense, and everything else is sandwiched between them.

Theorem 9.10. *Fix $\beta > 0$, $J > 0$, $h \in \mathbb{R}$, and $d \in \mathbb{N}$.*

(a) $\mu^{-}_{\Lambda,\beta,h,J} \le \mu^{\omega}_{\Lambda,\beta,h,J} \le \mu^{+}_{\Lambda,\beta,h,J}$ *for all $\Lambda \subset \mathbb{Z}^d$.*

(b) $\mu^{+}_{\Lambda,\beta,h,J} \le \mu^{+}_{\Delta,\beta,h,J}$ *and* $\mu^{-}_{\Lambda,\beta,h,J} \ge \mu^{-}_{\Delta,\beta,h,J}$ *for all finite $\Delta \subset \Lambda \subset \mathbb{Z}^d$.*

(c) *As Λ increases to $\mathbb{Z}^d$, measures $\mu^+_{\Lambda,\beta,h,J}$ (respectively, $\mu^-_{\Lambda,\beta,h,J}$) converge weakly to a limit $\mu^+_{\beta,h,J}$ (respectively, $\mu^-_{\beta,h,J}$).*

(d) *$\mu^+_{\beta,h,J}$ and $\mu^-_{\beta,h,J}$ are shift-invariant Gibbs measures.*

(e) *$\mu^-_{\beta,h,J} \le \mu \le \mu^+_{\beta,h,J}$ for all Gibbs measures μ.*

Proof. Part (a) is an immediate consequence of the previous theorem. Part (b) follows from part (a) and the definition of a specification. Indeed, let $f \in \mathscr{C}_b(\Omega)$ be increasing. Then, part (c) of Definition 7.8 implies

$$E^{\mu^+_{\Lambda,\beta,h,J}}[f] = \int E^{\mu^{\omega}_{\Delta,\beta,h,J}}(f)\,\mu^+_{\Lambda,\beta,h,J}(d\omega) \le E^{\mu^+_{\Delta,\beta,h,J}}[f].$$

For part (c) observe first that by part (b), for any finite volume $V \subset \mathbb{Z}^d$, the sequence $\mu^+_{\Lambda,\beta,h,J}\{\sigma_i = 1\,\forall i \in V\}$ is nonincreasing as Λ increases. Hence the limit

$$\rho(V) = \lim_{\Lambda \nearrow \mathbb{Z}^d} \mu^+_{\Lambda,\beta,h,J}\{\sigma_i = 1\,\forall i \in V\}$$

exists as Λ exhausts $\mathbb{Z}^d$. By this we mean that $\forall \varepsilon > 0\ \exists$ finite $A \subset \mathbb{Z}^d$ such that

$$|\,\rho(V) - \mu^+_{\Lambda,\beta,h,J}\{\sigma_i = 1\,\forall i \in V\}\,| \le \varepsilon \quad \text{for all finite } \Lambda \supset A.$$

Equivalently, the limit is the same for all sequences $\Lambda_k \nearrow \mathbb{Z}^d$ that eventually contain any fixed finite volume.

Inclusion-exclusion shows that for any event $A \in \mathscr{F}_V$, $\mu^+_{\Lambda,\beta,h,J}(A)$ converges to a quantity $\alpha_V(A)$, determined by $\{\rho(\Delta) : \Delta \subset V\}$ (Exercise 9.11 below). Measures $\alpha_V \in \mathcal{M}_1(\Omega_V)$ are consistent and Kolmogorov's extension theorem produces a unique probability measure $\mu^+_{\beta,h,J}$ that has these marginals. This is the weak limit of the measures $\mu^+_{\Lambda,\beta,h,J}$ on the full space Ω because weak convergence follows from convergence of finite-dimensional marginals (Exercise 6.1). The same argument works for $\mu^-_{\beta,h,J}$.

These limits are Gibbs measures due to Theorem 7.17. Since $V_{n-|i|_1} \subset V_n - i \subset V_{n+|i|_1}$ for $n > |i|_1$, part (b) combined with Exercise 7.4 gives

$$\mu^+_{V_{n-|i|_1},\beta,h,J} \ge \mu^+_{V_n - i,\beta,h,J} = \mu^+_{V_n,\beta,h,J} \circ \theta_{-i} \ge \mu^+_{V_{n+|i|_1},\beta,h,J}.$$

Take expectations of an increasing $f \in \mathscr{C}_b(\Omega)$ and let $n \to \infty$. In the limit we get $\mu^+_{\beta,h,J} \ge \mu^+_{\beta,h,J} \circ \theta_{-i} \ge \mu^+_{\beta,h,J}$, which implies $\mu^+_{\beta,h,J} = \mu^+_{\beta,h,J} \circ \theta_{-i}$. Shift invariance has been checked.

To prove part (e) let μ be a Gibbs measure. Let $f \in \mathscr{C}_b(\Omega)$ be a bounded increasing function. By the existence of the limit from part (c),

$$E^\mu[f] = \lim_{n\to\infty} E^\mu\big[E^\mu[f \,|\, \mathscr{F}_{V_n^c}]\big] = \lim_{n\to\infty} \int E^{\mu^\omega_{V_n,\beta,h,J}}(f)\,\mu(d\omega)$$
$$\le \lim_{n\to\infty} E^{\mu^+_{V_n,\beta,h,J}}[f] = E^{\mu^+_{\beta,h,J}}[f].$$

A similar argument holds for the other inequality. □

***Exercise 9.11.** Let $\{\mu_k\}$ be a sequence of probability measures on $\Omega = \{\pm 1\}^{\mathbb{Z}^d}$. Show that convergence of probabilities $\mu_k(\sigma_A = 1)$ $\forall$ finite $A \subset \mathbb{Z}^d$ implies the convergence of all cylinder probabilities $\mu_k(\sigma_A = 1, \sigma_B = -1)$ for disjoint finite $A, B \subset \mathbb{Z}^d$.

Hint: See the hints for Exercise 9.8.

Proof of part (b) of Theorem 9.2 when $h = 0$. Due to part (e) in the theorem above, phase transition at $(\beta, 0, J)$ is equivalent to $\mu^-_{\beta,0,J} \neq \mu^+_{\beta,0,J}$, which, by shift invariance and Strassen's lemma, is equivalent to $\mu^-_{\beta,0,J}\{\sigma_0 = 1\} \neq \mu^+_{\beta,0,J}\{\sigma_0 = 1\}$. By (e), Exercise 9.1, and (c) this in turn is equivalent to having $\mu^+_{\beta,0,J}\{\sigma_0 = 1\} > 1/2$.

We invoke another useful inequality. The proof can be found in Section C.1.

Griffiths's inequality. *Fix a finite volume $\Lambda \subset \mathbb{Z}^d$. Let $\mathcal{E} = \{\{i,j\} : |i-j|_1 = 1, i \in \Lambda\}$ be the set of nearest-neighbor edges with at least one endpoint in Λ. Define the function $F : [0,\infty)^{\mathcal{E}} \to [-1,1]$ by*

$$F(\mathbf{J}) = \frac{\sum_{\sigma_\Lambda} \sigma_0 \exp\big(\sum_{\{i,j\}\in\mathcal{E}} J_{i,j}\,\sigma_i\sigma_j\big)}{\sum_{\sigma_\Lambda} \exp\big(\sum_{\{i,j\}\in\mathcal{E}} J_{i,j}\,\sigma_i\sigma_j\big)},$$

where $\mathbf{J} = (J_{i,j})_{\{i,j\}\in\mathcal{E}}$, the sums run over $\sigma_\Lambda \in \{-1,1\}^\Lambda$, and for $j \notin \Lambda$ we set $\sigma_j = 1$ (+ boundary condition). Then

$$\frac{\partial F}{\partial J_{i,j}}(\mathbf{J}) \ge 0 \ \forall \{i,j\} \in \mathcal{E} \text{ and } \forall \mathbf{J} \in [0,\infty)^{\mathcal{E}}.$$

***Exercise 9.12.** Fix $J > 0$ and $h = 0$. Use Griffiths's inequality to show that $E^{\mu^+_{\Lambda,\beta,0,J}}[\sigma_0]$ increases with β. Also, show that if $\Lambda_n^{(d)} = \{i \in \mathbb{Z}^d : |i|_1 \le n\}$, then

$$\int \sigma_0 \, d\mu^+_{\Lambda_n^{(d)},\beta,0,J} \le \int \sigma_0 \, d\mu^+_{\Lambda_n^{(d+1)},\beta,0,J}.$$

Conclude that $\mu^+_{\beta,0,J}\{\sigma_0 = 1\}$ increases with β and with d.

Theorem 9.4 showed that for $J > 0$, $d = 2$, and large $\beta > 0$ phase transition happens and thereby $\mu^+_{\beta,0,J}\{\sigma_0 = 1\} > 1/2$. By the exercise above, phase transition happens in any $d \ge 2$, provided β is large enough.

Now, fix $J > 0$ and $d \geq 2$. Let $\beta_c = \sup\{\beta : |\mathscr{G}^{\Pi_{\beta,0,J}}| = 1\}$. By the previous argument, $\beta_c < \infty$. By Theorem 9.7, $\beta_c > 0$.

There is phase transition at all $\beta > \beta_c$ by the definition of β_c. There cannot be phase transition at any $\beta' < \beta_c$ because then Griffiths's inequality would imply a phase transition at all $\beta \in (\beta', \infty)$, contradicting the choice of β_c. Part (b) of Theorem 9.2 is thus proved in the case $h = 0$. □

9.4. Case of nonzero external field

This section finishes the proof of part (b) of Theorem 9.2. Parameters $J > 0$ and $\beta > 0$ are fixed. We drop the dependence on (J, β) from the notation and write $P(h) = P(\beta, h, J) = P(\Phi)$ for the infinite-volume pressure from Theorem 8.1. The argument goes by linking the phase transition at h to the differentiability of P at h. Recall that the subscript V_n is abbreviated as n: e.g. $\mathcal{H}_n^+ = \mathcal{H}_{V_n}^+$. We first record the derivative in finite volume.

***Exercise 9.13.** The finite-volume pressure functions with $+$ and $-$ boundary conditions are $P_n^\pm(h) = |V_n|^{-1} \log E^\lambda[e^{-\beta \mathcal{H}_n^\pm}]$. Check that

$$\frac{\partial P_n^\pm}{\partial h} = \beta E^{\mu_{n,h}^\pm}\Big[\frac{1}{|V_n|}\sum_{i\in V_n} \sigma_i\Big].$$

Lemma 9.14. *For both + and − boundary conditions,*

$$\lim_{n\to\infty} \frac{\partial P_n^\pm}{\partial h} = \beta E^{\mu_h^\pm}[\sigma_0].$$

Proof. We treat the $+$ boundary condition, the other being similar. Fix an integer m and use the limit $\frac{|V_n \smallsetminus V_{n-m}|}{|V_n|} \to 0$ as $n \to \infty$ to write

$$\varlimsup_{n\to\infty} \frac{\partial P_n^+}{\partial h} = \varlimsup_{n\to\infty} \beta E^{\mu_{n,h}^+}\Big[\frac{1}{|V_n|}\sum_{i\in V_n}\sigma_i\Big] = \varlimsup_{n\to\infty} \beta E^{\mu_{n,h}^+}\Big[\frac{1}{|V_{n-m}|}\sum_{i\in V_{n-m}}\sigma_i\Big],$$

with the same series of equalities for the liminf. If $i \in V_{n-m}$, then $i + V_m \subset V_n$. By (b) and (d) of Theorem 9.10,

$$E^{\mu_h^+}[\sigma_0] = E^{\mu_h^+}[\sigma_i] \leq E^{\mu_{n,h}^+}[\sigma_i] \leq E^{\mu_{i+V_m,h}^+}[\sigma_i] = E^{\mu_{m,h}^+}[\sigma_0].$$

Average the inequalities above over $i \in V_{n-m}$ and let $n \to \infty$:

$$\beta E^{\mu_h^+}[\sigma_0] \leq \varliminf_{n\to\infty} \frac{\partial P_n^+}{\partial h} \leq \varlimsup_{n\to\infty} \frac{\partial P_n^+}{\partial h} \leq \beta E^{\mu_{m,h}^+}[\sigma_0].$$

Let $m \to \infty$ so that $E^{\mu_{m,h}^+}[\sigma_0] \to E^{\mu_h^+}[\sigma_0]$. □

Next we give a property of convex functions.

***Exercise 9.15.** Let f_n be a sequence of convex functions on an interval (a, b). Assume $f_n(y) \to f(y)$ for all $y \in (a, b)$ and that $f_n'(x)$ and $f'(x)$ exist for some $x \in (a, b)$. Prove that $f_n'(x) \to f'(x)$.

Hint: For $y < x < z$,

$$\frac{f_n(x) - f_n(y)}{x - y} \leq f_n'(x) \leq \frac{f_n(z) - f_n(x)}{z - x}.$$

***Exercise 9.16.** Use Hölder's inequality to check that $P_n^+(h)$ is a convex function of h.

Theorem 9.17. *Consider the Ising model in dimension $d \geq 1$ and fix $\beta > 0$ and $J > 0$. If the infinite-volume pressure $P(\beta, h, J)$ is differentiable with respect to h at $h = h_0$, then $|\mathscr{G}^{\mathbb{I}_{\beta,h_0,J}}| = 1$.*

Proof. By Exercise 9.13 and Theorem 8.1, P_n^+ is a sequence of convex differentiable functions of h that converge pointwise to $P(h)$. By Exercise 9.15, if $P'(h_0)$ exists, then $(P_n^+)'(h_0) \to P'(h_0)$. Then Lemma 9.14 implies that $P'(h_0) = \beta E^{\mu_{h_0}^+}[\sigma_0]$. The same argument implies that $P'(h_0) = \beta E^{\mu_{h_0}^-}[\sigma_0]$. Thus, $\mu_{h_0}^+\{\sigma_0 = 1\} = \mu_{h_0}^-\{\sigma_0 = 1\}$ and Strassen's lemma along with shift invariance imply that $\mu_{h_0}^+ = \mu_{h_0}^-$. By part (e) of Theorem 9.10 this implies the uniqueness of the Gibbs measure. $\square$

The final step is that $P(h)$ is differentiable at all $h \neq 0$.

Theorem 9.18. *Consider the Ising model in dimension $d \geq 1$. Then, for all $\beta > 0$ and $J > 0$, $P(\beta, h, J)$ is differentiable in h at $h \neq 0$.*

Proof. Fix $\beta > 0$ and $J > 0$. We start with the case $h > 0$. Let

$$M_n(h) = E^{\mu_{n,h}^+}\Big[\frac{1}{|V_n|}\sum_{i \in V_n} \sigma_i\Big] \quad \text{and} \quad M(h) = E^{\mu_h^+}[\sigma_0].$$

(M is for magnetization.) Lemma 9.14 gives the pointwise limit $M_n(h) \to M(h)$. Exercise 9.13 gives

$$P_n^+(h) - P_n^+(0) = \beta \int_0^h M_n(s)\, ds.$$

Since $|M_n(h)| \leq 1$, dominated convergence applies and in the limit we have

$$P(h) - P(0) = \beta \int_0^h M(s)\, ds.$$

Differentiability of P at $h > 0$ would follow if we knew that M is continuous on $(0, \infty)$. Concavity of M would be sufficient for continuity (Exercise 4.8). Concavity of M follows from concavity of each M_n on $(0, \infty)$. This last point is a consequence of the next inequality, whose proof is given in Section C.2.

Griffiths-Hurst-Sherman inequality. *Consider the Ising model in* $d \geq 1$. *Fix* $\beta > 0$, $J > 0$, *and the volume* V_n. *Then* $\frac{\partial^2 M_n}{\partial h^2} \leq 0$ *for* $h > 0$.

The theorem is now proved for $h > 0$. When $h < 0$ symmetry yields

$$E^{\mu^-_{n,h}}\Big[\frac{1}{|V_n|}\sum_{i\in V_n}\sigma_i\Big] = -E^{\mu^+_{n,-h}}\Big[\frac{1}{|V_n|}\sum_{i\in V_n}\sigma_i\Big] = -M_n(-h)$$

and then by Exercise 9.13

$$P^-_n(0) - P^-_n(h) = -\beta\int_h^0 M_n(-s)\,ds.$$

After $n \to \infty$, continuity of $M(-h)$ gives differentiability of P at $h < 0$. □

Theorems 9.17 and 9.18 combined show that there is no phase transition when $h \neq 0$. All the cases of Theorem 9.2 have now been proved.

Exercise 9.19. Fix $h \in \mathbb{R}$, $0 < J, \beta < \infty$, and $d \in \mathbb{N}$. Let $S_n = \sum_{i\in V_n}\sigma_i$. Show that there exists a rate function I on $\mathbb{R}$ such that, for any infinite-volume Ising Gibbs measure γ, the distributions $\gamma(|V_n|^{-1} S_n \in \cdot\,)$ satisfy an LDP with normalization $|V_n|$ and rate I. Show that I is symmetric and that if $d \geq 2$, $\beta > \beta_c$, and $h = 0$, the zero set $\{I = 0\}$ is a nontrivial interval.

Literature notes. For further information about the Ising model we refer to Georgii [**41**], including the bibliographical notes of its Section 6.2. The phase transition of long-range Ising models in both dimension one and higher are discussed in Ellis [**32**].

Chapter 10

Percolation approach to phase transition

In the previous chapter we understood how phase transition in the Ising model depends on whether the influence of the boundary reaches all the way back to the origin, as the volume grows. In this chapter we give an alternative proof of the Ising model phase transition that makes this percolation idea explicit. The next section introduces percolation and the Fortuin-Kasteleyn random cluster model, which is then used to re-prove the Ising phase transition.

10.1. Bernoulli bond percolation and random cluster measures

Let $\mathbb{E}^d = \{e = \{i,j\} : i,j \in \mathbb{Z}^d, i \sim j\}$ denote the set of nearest-neighbor edges or *bonds* of $\mathbb{Z}^d$. (It is convenient to use the words edge and bond interchangeably.) A configuration of open and closed bonds is denoted by $\eta = (\eta(e) : e \in \mathbb{E}^d) \in \{0,1\}^{\mathbb{E}^d}$. The value $\eta(e) = 1$ signifies that bond e is open or present, while $\eta(e) = 0$ means that bond e is closed, or not available. Equivalently we can think of η as the set of open bonds $\{e : \eta(e) = 1\}$.

For $0 < p < 1$ let P_p denote the product probability measure on $\Omega = \{0,1\}^{\mathbb{E}^d}$ under which the bond variables are i.i.d Bernoulli with marginal distributions $P_p(\eta(e) = 1) = p = 1 - P_p(\eta(e) = 0)$. This gives us a random graph $(\mathbb{Z}^d, \eta)$ with vertex set $\mathbb{Z}^d$ and edges specified by η. The subject of *percolation* studies the large scale properties of this random graph and other similar models.

The first question is whether an infinite connected component exists. By shift invariance the probability that a particular vertex lies in an infinite connected component is the same for all vertices, so we might as well focus on the origin. Here are some notational conventions. For $x \in \mathbb{Z}^d$ and $B \subset \mathbb{Z}^d$, $\{x \leftrightarrow B\}$ is the event that there is a path of open edges from x to some vertex in the set B. To be precise about it,

$$\{x \leftrightarrow B\} = \{\, \exists x_0, x_1, \ldots, x_m \in \mathbb{Z}^d \text{ such that } x_0 = x, x_m \in B, \text{ and } \forall i,\ x_i \sim x_{i+1} \text{ and } \eta(\{x_i, x_{i+1}\}) = 1\}.$$

The connected component or cluster at the origin is $C = \{x : 0 \leftrightarrow x\}$. The event that the cluster at the origin is infinite is expressed in two ways: $\{|C| = \infty\} = \{0 \leftrightarrow \infty\}$ where the second event reads "the origin is connected to infinity". The notation for the probability is $\theta(p) = P_p(0 \leftrightarrow \infty)$.

Lemma 10.1. *As a function of* $0 \le p \le 1$, $\theta(p)$ *is nondecreasing.*

Proof. This intuitively obvious fact is proved elegantly with a simple coupling. Let $\{U_e : e \in \mathbb{E}^d\}$ be i.i.d. uniform $[0,1]$-valued random variables. For each $p \in [0,1]$ define $\eta_p(e) = \mathbb{1}\{U_e \le p\}$. Then $\{\eta_p(e) : e \in \mathbb{E}^d\}$ is P_p-distributed. We have simultaneously constructed percolation processes for all edge probabilities p. The construction has this monotonicity: $p_1 < p_2$ implies $\eta_{p_1}(e) \le \eta_{p_2}(e)$ and hence $\{0 \leftrightarrow \infty \text{ along } \eta_{p_1}\text{-bonds}\} \subset \{0 \leftrightarrow \infty \text{ along } \eta_{p_2}\text{-bonds}\}$. □

The fundamental theorem of the subject is that there is a nontrivial phase transition where the critical probability is strictly between 0 and 1.

Theorem 10.2. *For each dimension* $d \ge 2$ *there is a critical probability* $0 < p_c(d) < 1$ *such that for Bernoulli bond percolation on* $\mathbb{Z}^d$, $\theta(p) = 0$ *for* $p < p_c(d)$ *and* $\theta(p) > 0$ *for* $p > p_c(d)$.

Proof. A path is self-avoiding if it does not intersect itself. The number of self-avoiding paths of n edges that start from the origin is bounded by $(2d)^n$. Any particular self-avoiding path is open with probability p^n because it uses n distinct edges. If the origin lies in an infinite cluster, there must be an open self-avoiding path from the origin of any length n. Thus if we choose $0 < p < (2d)^{-1}$,

$$\theta(p) \le (2dp)^n \longrightarrow 0 \quad \text{as } n \to \infty.$$

This proves that $p_c(d) > 0$ for all $d \ge 2$.

To prove that $p_c(d) < 1$ we first reduce the task to the planar case $d = 2$. It is intuitively clear that it is easier to percolate in higher dimensions and we can make this precise (again) by a coupling. Consider $(\mathbb{Z}^d, \mathbb{E}^d)$ as a subgraph of $(\mathbb{Z}^{d+1}, \mathbb{E}^{d+1})$ by letting $(\mathbb{Z}^d, \mathbb{E}^d)$ be represented by the subset of

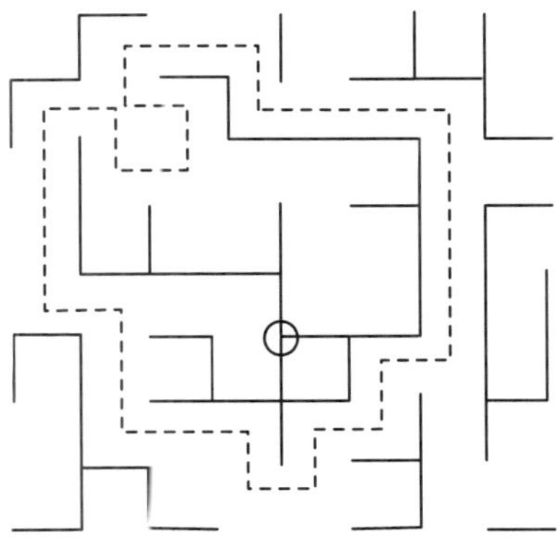

Figure 10.1. Peierls's contour. The circle marks the origin. The lines mark open bonds. The dashed path marks a circuit of open dual bonds surrounding the origin.

vertices $V = \{(i, 0) : i \in \mathbb{Z}^d\} \subset \mathbb{Z}^{d+1}$ and the edges between them. Then the percolation event $\{0 \leftrightarrow \infty$ along edges of $V\}$ implies that $0 \leftrightarrow \infty$ also in $\mathbb{Z}^{d+1}$. Consequently $p_c(d+1) \le p_c(d)$.

To show $p_c(2) < 1$ we use the already familiar Peierls argument. The dual lattice of $(\mathbb{Z}^2, \mathbb{E}^2)$ is the graph with vertices $(\frac{1}{2}, \frac{1}{2}) + \mathbb{Z}^2 = \{(i+\frac{1}{2}, j+\frac{1}{2}) : (i,j) \in \mathbb{Z}^2\}$ and edges between nearest-neighbor pairs of points $\{x, y\}$ that satisfy $|x - y|_1 = 1$. Each original edge $e \in \mathbb{E}^d$ is crossed by exactly one dual edge e^*, and vice versa. Given a bond configuration η, define its dual by $\eta^*(e^*) = 1 - \eta(e)$. In other words, a bond is open if and only if its dual is closed. Thus the dual configuration obeys the probability measure P_{1-p}.

The connected cluster at the origin is finite if and only if the origin is surrounded by a circuit of open dual bonds. This statement is intuitively fairly obvious (see Figure 10.1), so we do not endeavor to justify it in any technical manner. In Lemma 9.6 we deduced that the number of circuits of length k is at most $k3^{k-1}$. A dual circuit of k edges is open with probability $(1-p)^k$. We get the bound

$$1 - \theta(p) = P_p(|C| < \infty) \le \sum_{k=1}^{\infty} k3^{k-1}(1-p)^k.$$

This series is finite and can be made strictly less than 1 if p is close enough to 1. Consequently for $d = 2$, $\theta(p) > 0$ for p close enough to 1, which says that $p_c(2) < 1$. ☐

Random cluster measures are generalizations of Bernoulli percolation that allow correlated bonds. These measures are parametrized by an edge probability $0 < p < 1$ and a second parameter $0 < q < \infty$. For the purpose of this chapter we only need them on finite graphs. Let $G = (V, E)$ be a

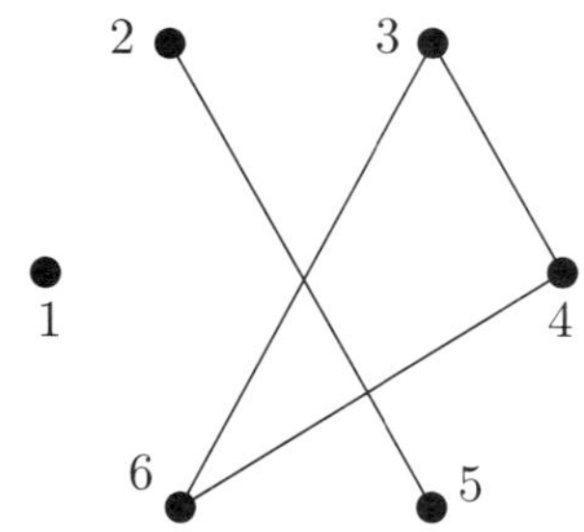

Figure 10.2. The vertex set is $V = \{1,2,3,4,5,6\}$. The figure illustrates the configuration η such that $\eta(e) = 1$ for $e = \{2,5\}, \{3,4\}, \{3,6\}$, and $\{4,6\}$. The connected components are $\{1\}$, $\{2,5\}$, and $\{3,4,6\}$. Hence $k(\eta) = 3$. The original edge set E is not displayed. E could be any set of edges that contains the edges in the figure.

finite graph with vertex set V and edge set E. Write $i \sim j$ if vertices i, j are linked by an edge, in other words, if $\{i, j\} \in E$. The random cluster measure is a probability distribution on the space $\Omega = \{0,1\}^E$ of edge configurations $\eta = (\eta(e))_{e \in E}$. Let $k(\eta)$ denote the number of connected components in the graph (V, η). An isolated vertex, that is, a vertex with no edges incident to it, is a connected component of size 1. See Figure 10.2.

The random cluster measure is defined by

$$\rho_{p,q}(\eta) = \frac{1}{Z_{p,q}} \Big\{ \prod_{e \in E} p^{\eta(e)} (1-p)^{1-\eta(e)} \Big\} q^{k(\eta)}.$$

$Z_{p,q}$ is the normalization that makes $\rho_{p,q}$ a probability measure on Ω. The case $q = 1$ is Bernoulli percolation with independent bonds. The next lemma shows that these measures are stochastically increasing in p when $q \geq 1$.

We generalize the model slightly for the next section. Let E_b be an additional set of edges among the vertices V, disjoint from the original edge set E. The edges in E_b are permanently in place and not random, and we regard E_b as a *boundary condition* for the random cluster measure. The measure with boundary condition E_b is defined by

$$\rho^{E_b}_{p,q}(\eta) = \frac{1}{Z^{E_b}_{p,q}} \Big\{ \prod_{e \in E} p^{\eta(e)} (1-p)^{1-\eta(e)} \Big\} q^{k(\eta, E_b)}$$

where $k(\eta, E_b)$ is the number of connected components in the graph $(V, \eta \cup E_b)$.

Lemma 10.3. *For $q \geq 1$ and $p \leq \tilde{p}$, $\rho^{E_b}_{p,q} \leq \rho^{E_b}_{\tilde{p},q}$.*

Proof. Let $\eta, \zeta \in \Omega$. By Holley's theorem (Section B.7), it suffices to show

$$\rho^{E_b}_{p,q}(\eta \wedge \zeta)\, \rho^{E_b}_{\tilde{p},q}(\eta \vee \zeta) \;\geq\; \rho^{E_b}_{p,q}(\eta)\, \rho^{E_b}_{\tilde{p},q}(\zeta),$$

which rearranges to

$$\left\{\prod_{e\in E}\frac{\tilde{p}^{\eta(e)\vee\zeta(e)-\zeta(e)}(1-p)^{\eta(e)-\eta(e)\wedge\zeta(e)}}{p^{\eta(e)-\eta(e)\wedge\zeta(e)}(1-\tilde{p})^{\eta(e)\vee\zeta(e)-\zeta(e)}}\right\}q^{k(\eta\wedge\zeta,E_b)-k(\zeta,E_b)}$$
$$\geq q^{k(\eta,E_b)-k(\eta\vee\zeta,E_b)}.$$

The factor in braces is ≥ 1 by the assumption and because $a\vee b-b=a-a\wedge b$. It remains to show

$$k(\eta\wedge\zeta,E_b)-k(\zeta,E_b)\geq k(\eta,E_b)-k(\eta\vee\zeta,E_b). \tag{10.1}$$

In general, suppose $\gamma\leq\tilde{\gamma}$ are bond configurations on E and e is an edge in E that is not present in either γ or $\tilde{\gamma}$, that is $\gamma(e)=\tilde{\gamma}(e)=0$. Let $\gamma\cup e$ and $\tilde{\gamma}\cup e$ represent the configurations obtained by adding e to γ and $\tilde{\gamma}$. Then

$$k(\gamma,E_b)-k(\gamma\cup e,E_b)\geq k(\tilde{\gamma},E_b)-k(\tilde{\gamma}\cup e,E_b). \tag{10.2}$$

Adding the edge $e=\{i,j\}$ either reduces the number of components by one or does not. The left side of (10.2) is 0 if and only if i and j were already connected by the edges in $\gamma\cup E_b$. But then i and j were already connected also by the edges in $\tilde{\gamma}\cup E_b$. This verifies (10.2).

This idea proves (10.1). Let $e_1,\dots,e_m$ be the edges in $\zeta\smallsetminus\eta$. Adding these edges one by one to $\eta\wedge\zeta$ creates ζ, and adding these edges one by one to η creates $\eta\vee\zeta$. Denote the intermediate configurations by

$$\gamma_0=\eta\wedge\zeta,\ \gamma_1=\gamma_0\cup e_1,\ \dots,\ \gamma_i=\gamma_{i-1}\cup e_i,\dots,\ \gamma_m=\zeta$$

and

$$\tilde{\gamma}_0=\eta,\ \tilde{\gamma}_1=\tilde{\gamma}_0\cup e_1,\ \dots,\ \tilde{\gamma}_i=\tilde{\gamma}_{i-1}\cup e_i,\dots,\ \tilde{\gamma}_m=\eta\vee\zeta$$

at each stage $\gamma_i\leq\tilde{\gamma}_i$. On the left of (10.1)

$$k(\eta\wedge\zeta,E_b)-k(\zeta,E_b)=\sum_{i=1}^{m}\big(k(\gamma_{i-1},E_b)-k(\gamma_i,E_b)\big)$$

with the analogous identity on the right. Apply (10.2) to each term. ☐

Exercise 10.4. Extend Lemma 10.3 to the statement $\rho^{E_b}_{p,q}\leq\rho^{E_b}_{\tilde{p},\tilde{q}}$ under some hypotheses on the pairs (p,q) and $(\tilde{p},\tilde{q})$.

10.2. Ising phase transition revisited

In this section we give a second proof of the phase transition of the Ising model in dimensions $d\geq 2$ and zero external field $h=0$. Fix the coupling constant $J=1$. Let $\mu^{\pm}=\mu^{\pm}_{\beta,0,1}$ denote the infinite-volume Gibbs measures with $+$ and $-$ boundary conditions. We take for granted two facts from the previous chapter: that phase transition is equivalent to $\mu^{+}(\sigma_0=1)>1/2$

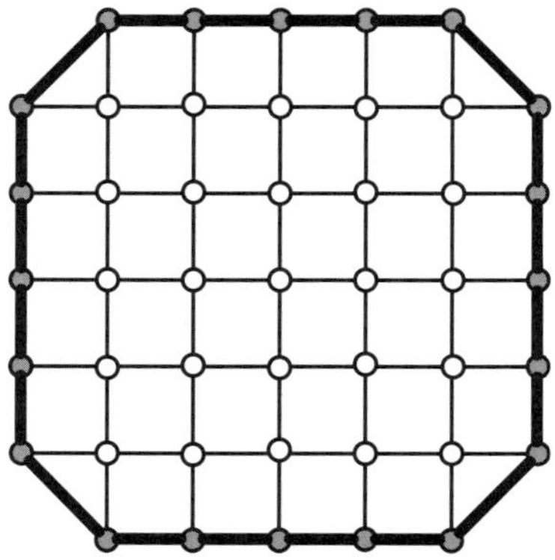

Figure 10.3. Graph $G_n = (\Gamma_n, E_n)$ with boundary condition $E_{n,b}$ in dimension $d = 2$ with $n = 2$. Vertices of Λ_n are denoted by open circles, vertices of $\partial\Lambda_n$ by solid circles, edges of E_n by thin lines, and edges of $E_{n,b}$ by thick lines.

(see the argument on page 145) and the limit $\mu^+_{\Lambda_n}(\sigma_0 = 1) \to \mu^+(\sigma_0 = 1)$ as $\Lambda_n \nearrow \mathbb{Z}^d$ (Theorem 9.10(c)).

Theorem 10.5. *Let $d \geq 2$, $J = 1$, and $h = 0$. There exists $0 < \beta_c < \infty$ such that $\mu^- = \mu^+$ for $\beta < \beta_c$ and $\mu^- \neq \mu^+$ for $\beta > \beta_c$.*

The strategy is to construct a coupling of the Ising model and a random cluster measure. The edge probability $0 < p < 1$ and the inverse temperature $0 < \beta < \infty$ are connected by

$$p = 1 - e^{-2\beta} \tag{10.3}$$

and $q = 2$ because the Ising spin has two values ± 1. The q-state Potts model (Example 7.3) has an analogous connection with the random cluster measure with parameter q.

The measures to be constructed live on graphs $G_n = (\Gamma_n, E_n)$. The vertex set $\Gamma_n = \Lambda_n \cup \partial\Lambda_n$ where

$$\Lambda_n = \{i \in \mathbb{Z}^d : |i|_\infty \leq n\}$$

and

$$\partial\Lambda_n = \{i \in \mathbb{Z}^d \smallsetminus \Lambda_n : \exists j \in \Lambda_n \text{ such that } i \sim j\}.$$

$\partial\Lambda_n$ is the outer boundary of the cube Λ_n. The edge set

$$E_n = \{\{i,j\} : i, j \in \Lambda_n,\ i \sim j\} \cup \{(i,j) : i \in \Lambda_n,\ j \in \partial\Lambda_n,\ i \sim j\}.$$

In other words, E_n has all the nearest-neighbor edges in Λ_n and all the nearest-neighbor edges between Λ_n and $\partial\Lambda_n$. See Figure 10.3.

We define a joint distribution Q on the spin variables $\sigma = (\sigma_i)_{i \in \Lambda_n}$ and bond variables $\eta = (\eta(e))_{e \in E_n}$. The measure Q has plus boundary conditions ω on the spins and wired boundary conditions $E_{n,b}$ on the edges, defined as follows: $\omega_j = 1$ for $j \in \partial\Lambda_n$ and $E_{n,b}$ is a set of edges that makes $\partial\Lambda_n$ into a single connected component. In the formulas below $\sigma_j = \omega_j$ whenever $j \in \partial\Lambda_n$. The measure Q is defined by

$$Q(\sigma,\eta) = \frac{1}{Z} \prod_{e=\{i,j\}\in E_n} p^{\eta(e)}(1-p)^{1-\eta(e)}\big(\mathbb{1}_{\sigma_i=\sigma_j} + (1-\eta(e))\mathbb{1}_{\sigma_i\neq\sigma_j}\big).$$

Z is the normalization that makes Q a probability measure.

Let us derive the marginals of Q. Here is the marginal on a spin configuration $\sigma \in \{\pm 1\}^{\Lambda_n}$:

$$\begin{aligned}
&\sum_{\eta\in\{0,1\}^{E_n}} Q(\sigma,\eta)\\
&= \frac{1}{Z} \sum_{\eta\in\{0,1\}^{E_n}} \prod_{e=\{i,j\}\in E_n} \Big[p^{\eta(e)}(1-p)^{1-\eta(e)}\mathbb{1}_{\sigma_i=\sigma_j} + (1-p)(1-\eta(e))\mathbb{1}_{\sigma_i\neq\sigma_j}\Big]\\
&= \frac{1}{Z} \prod_{e=\{i,j\}\in E_n} \sum_{b\in\{0,1\}} \Big[p^{b}(1-p)^{1-b}\mathbb{1}_{\sigma_i=\sigma_j} + (1-p)(1-b)\mathbb{1}_{\sigma_i\neq\sigma_j}\Big]\\
&= \frac{1}{Z} \prod_{e=\{i,j\}\in E_n} \big[\mathbb{1}_{\sigma_i=\sigma_j} + (1-p)\mathbb{1}_{\sigma_i\neq\sigma_j}\big]\\
&= \frac{1}{Z}(1-p)^{\sum_{\{i,j\}\in E_n} \mathbb{1}\{\sigma_i\neq\sigma_j\}} = \frac{1}{Z} e^{-2\beta\sum_{\{i,j\}\in E_n} \mathbb{1}\{\sigma_i\neq\sigma_j\}}\\
&= \frac{1}{Z^+_{\Lambda_n}} \exp\Big\{\beta \sum_{i,j\in\Lambda_n:\, i\sim j} \sigma_i\sigma_j + \beta \sum_{i\in\Lambda_n,\, j\in\partial\Lambda_n:\, i\sim j} \sigma_i\Big\} = \mu^+_{\Lambda_n}(\sigma).
\end{aligned}$$

The next to last equality came from $-2\mathbb{1}_{\sigma_i\neq\sigma_j} = \sigma_i\sigma_j - 1$. The -1's were subsumed into the normalizing constant. Boundary condition $\omega = 1$ on $\partial\Lambda_n$ is the same as $+$ boundary conditions outside Λ_n on the whole lattice $\mathbb{Z}^d$. Thus in the end we get the finite-volume Ising measure $\mu^+_{\Lambda_n}(\sigma)$. Note that we need not check that the resulting normalization is the correct Ising partition function $Z^+_{\Lambda_n}$. This follows simply by summing over σ.

Here is the marginal on a bond configuration $\eta \in \{0,1\}^{E_n}$:

$$\begin{aligned}
&\sum_{\sigma\in\{\pm1\}^{\Lambda_n}} Q(\sigma,\eta)\\
&= \frac{1}{Z} \sum_{\sigma\in\{\pm1\}^{\Lambda_n}} \prod_{e=\{i,j\}\in E_n} p^{\eta(e)}(1-p)^{1-\eta(e)}\big(\mathbb{1}_{\sigma_i=\sigma_j} + (1-\eta(e))\mathbb{1}_{\sigma_i\neq\sigma_j}\big)\\
&= \frac{1}{Z} \prod_{e=\{i,j\}\in E_n} p^{\eta(e)}(1-p)^{1-\eta(e)}\\
&\qquad \times \sum_{\sigma\in\{\pm1\}^{\Lambda_n}} \prod_{e=\{i,j\}\in E_n} \big(\mathbb{1}_{\sigma_i=\sigma_j} + (1-\eta(e))\mathbb{1}_{\sigma_i\neq\sigma_j}\big).
\end{aligned}$$

The last product is 1 if and only if $\sigma_i = 1$ on every vertex $i \in \Lambda_n$ connected to $\partial\Lambda_n$ and σ is constant on other connected components of $(\Gamma_n, \eta \cup E_{n,b})$. Thus for a given $\eta \in \{0,1\}^{E_n}$,

$$\sum_{\sigma\in\{\pm 1\}^{\Lambda_n}} \prod_{e=\{i,j\}\in E_n} \big(\mathbb{1}_{\sigma_i=\sigma_j} + (1-\eta(e))\mathbb{1}_{\sigma_i\neq\sigma_j}\big) = 2^{k(\eta,E_{n,b})-1}$$

because we are free to assign ± 1 to each connected component of $(\Gamma_n, \eta \cup E_{n,b})$ except the one that contains $\partial\Lambda_n$, where we assigned $+$ boundary conditions. The factor 2^{-1} can be subsumed into the normalizing constant, and so

$$\begin{aligned}\sum_{\sigma\in\{\pm 1\}^{\Lambda_n}} Q(\sigma,\eta) &= \frac{1}{2Z}\Big\{\prod_{e=\{i,j\}\in E_n} p^{\eta(e)}(1-p)^{1-\eta(e)}\Big\}2^{k(\eta,E_{n,b})}\\ &= \rho^{E_{n,b}}_{p,2,G_n}(\eta),\end{aligned}$$

the random cluster measure on G_n with boundary condition $E_{n,b}$ and parameters $(p,q) = (p,2)$.

By rewriting the joint distribution we can understand the construction of a Q-distributed pair (σ,η):

$$Q(\sigma,\eta) = \rho^{E_{n,b}}_{p,2,G_n}(\eta)\, 2^{1-k(\eta,E_{n,b})} \prod_{e=\{i,j\}\in E_n} \big(\mathbb{1}_{\sigma_i=\sigma_j} + (1-\eta(e))\mathbb{1}_{\sigma_i\neq\sigma_j}\big).$$

The recipe reads as follows. Pick η with probability $\rho^{E_{n,b}}_{p,2,G_n}(\eta)$. Set $\sigma_i = 1$ on all vertices i connected to $\partial\Lambda_n$. For each of the remaining $k(\eta, E_{n,b}) - 1$ components flip a fair coin to determine whether the spins in that component are all $+$ or all $-$.

Part (ii) of the next lemma captures the percolation idea of phase transition.

Lemma 10.6. (i) *The limit* $\lim_{n\to\infty} \rho^{E_{n,b}}_{p,2,G_n}(0 \leftrightarrow \partial\Lambda_n)$ *exists.*

(ii) *The Ising model on* $\mathbb{Z}^d$ *has a unique Gibbs measure if and only if* $\lim_{n\to\infty} \rho^{E_{n,b}}_{p,2,G_n}(0 \leftrightarrow \partial\Lambda_n) = 0$.

Proof. Use the coupling Q. Given η, $\sigma_0 = 1$ is certain if $0 \leftrightarrow \partial\Lambda_n$ and otherwise it happens with even chances. Thus

$$\begin{aligned}\mu^+_{\Lambda_n}(\sigma_0 = 1) &= \sum_{\eta\in\{0,1\}^{E_n}} \rho^{E_{n,b}}_{p,2,G_n}(\eta)\, Q(\sigma_0 = 1\,|\,\eta)\\ &= \sum_{\eta\in\{0,1\}^{E_n}} \rho^{E_{n,b}}_{p,2,G_n}(\eta)\big(\tfrac{1}{2}\mathbb{1}\{0 \not\leftrightarrow \partial\Lambda_n\} + \mathbb{1}\{0 \leftrightarrow \partial\Lambda_n\}\big)\\ &= \tfrac{1}{2} + \tfrac{1}{2}\rho^{E_{n,b}}_{p,2,G_n}(0 \leftrightarrow \partial\Lambda_n).\end{aligned}$$

On the left $\mu^+_{\Lambda_n}(\sigma_0 = 1) \to \mu^+(\sigma_0 = 1)$ (Theorem 9.10(c)). Hence the limit exists also on the right. The criterion in (ii) follows because uniqueness of the Gibbs measure in the Ising model is equivalent to $\mu^+(\sigma_0 = 1) = 1/2$. □

By Lemma 10.3 the measures $\rho^{E_{n,b}}_{p,2,G_n}$ are nondecreasing in p, in the sense of stochastic ordering. Consequently there exists a unique $\bar{p}_c \in [0,1]$ such that

$$(10.4) \qquad \lim_{n\to\infty} \rho^{E_{n,b}}_{p,2,G_n}(0 \leftrightarrow \partial\Lambda_n) \begin{cases} = 0, & p < \bar{p}_c, \\ > 0, & p > \bar{p}_c. \end{cases}$$

From this monotonicity, (10.3), and Lemma 10.6, we deduce that the Ising phase transition is monotone; that is, there is a unique $\beta_c \in [0,\infty]$ such that $\mu^- = \mu^+$ for $\beta < \beta_c$ and $\mu^- \neq \mu^+$ for $\beta > \beta_c$. The missing piece is the nontriviality of the critical point: that $0 < \bar{p}_c < 1$ or equivalently $0 < \beta_c < \infty$. This comes from comparison with Bernoulli percolation.

Lemma 10.7. *For measures on the graph G_n,*

$$\rho^{E_{n,b}}_{\frac{p}{2-p},1,G_n}(0 \leftrightarrow \partial\Lambda_n) \le \rho^{E_{n,b}}_{p,2,G_n}(0 \leftrightarrow \partial\Lambda_n) \le \rho^{E_{n,b}}_{p,1,G_n}(0 \leftrightarrow \partial\Lambda_n).$$

Proof. The event $0 \leftrightarrow \partial\Lambda_n$ is increasing (adding an edge cannot switch it from true to false). Hence it suffices to check this stochastic ordering of measures on the edges in E_n:

$$(10.5) \qquad \rho^{E_{n,b}}_{\frac{p}{2-p},1,G_n} \le \rho^{E_{n,b}}_{p,2,G_n} \le \rho^{E_{n,b}}_{p,1,G_n}.$$

We apply Holley's theorem (Section B.7). Let $\eta, \zeta \in \{0,1\}^{E_n}$.

For ease of notation abbreviate $s = p/(2-p)$. The first domination in (10.5) follows from

$$\rho^{E_{n,b}}_{s,1,G_n}(\eta \wedge \zeta)\, \rho^{E_{n,b}}_{p,2,G_n}(\eta \vee \zeta) \;\ge\; \rho^{E_{n,b}}_{s,1,G_n}(\eta)\, \rho^{E_{n,b}}_{p,2,G_n}(\zeta)$$

$$\iff$$

$$2^{k(\eta\vee\zeta,E_{n,b})} \prod_e s^{\eta(e)\wedge\zeta(e)}(1-s)^{1-\eta(e)\wedge\zeta(e)} p^{\eta(e)\vee\zeta(e)}(1-p)^{1-\eta(e)\vee\zeta(e)}$$
$$\ge\; 2^{k(\zeta,E_{n,b})} \prod_e s^{\eta(e)}(1-s)^{1-\eta(e)} p^{\zeta(e)}(1-p)^{1-\zeta(e)}$$

$$\iff \quad \prod_e 2^{\eta(e)-\eta(e)\wedge\zeta(e)} \ge 2^{k(\zeta,E_{n,b})-k(\eta\vee\zeta,E_{n,b})}.$$

To see the truth of the last inequality, imagine going from $\eta \wedge \zeta$ to η by adding in one by one the edges of $\eta \smallsetminus \zeta$. Adding these same edges to ζ produces $\eta \vee \zeta$. Each edge reduces the number of connected components by at most one. The reduction in components is counted on the right.

The second domination in (10.5) follows similarly and is simpler. □

Completion of the proof of Theorem 10.5. Let $p_c = p_c(d)$ be the critical probability of Bernoulli percolation in $\mathbb{Z}^d$ from Theorem 10.2. Since the edges in E_n are i.i.d. Bernoulli under $\rho^{E_{n,b}}_{p,1,G_n}$, we have

$$\rho^{E_{n,b}}_{p,1,G_n}(0 \leftrightarrow \partial\Lambda_n) = P_p(0 \leftrightarrow \partial\Lambda_n)$$

and consequently, by continuity of probability along a monotone sequence of events,

$$\theta(p) = P_p(0 \leftrightarrow \infty) = \lim_{n\to\infty} \rho^{E_{n,b}}_{p,1,G_n}(0 \leftrightarrow \partial\Lambda_n).$$

Combining with Lemma 10.7,

$$\theta\big(\tfrac{p}{2-p}\big) \;\le\; \lim_{n\to\infty} \rho^{E_{n,b}}_{p,2,G_n}(0 \leftrightarrow \partial\Lambda_n) \;\le\; \theta(p).$$

This and Theorem 10.2 imply that $\bar{p}_c$ in (10.4) satisfies

$$0 < p_c \le \bar{p}_c \le \frac{2p_c}{1+p_c} < 1. \qquad \square$$

Exercise 10.8. Check the second domination in (10.5).

Literature notes. The phase transition of percolation was first proved in [**12**]. For further reading on percolation the reader should begin with Grimmett's monograph [**44**]. Random cluster measures were introduced by Fortuin and Kasteleyn [**36**]. The random cluster measure proof of the Ising phase transition given above follows mainly the presentation in [**45**].

Part III

Additional large deviation topics

Chapter 11

Further asymptotics for i.i.d. random variables

This chapter is somewhat separate from the main themes of the book, but it addresses questions that are a natural complement to the large deviation theory of i.i.d. variables. The first question concerns finer asymptotics beyond the exponential scale of the LDP. The second issue is that of moderate deviations intermediate between the central limit theorem and the LDP, briefly alluded to in Section 2.5.

11.1. Refinement of Cramér's theorem

We work with familiar conventions. $X, X_1, X_2, \ldots$ are i.i.d. real-valued random variables with nondegenerate distribution μ and $S_n = X_1 + \cdots + X_n$. $M(\theta) = E[e^{\theta X}]$ and $I(x) = \sup_{\theta \in \mathbb{R}}\{\theta x - \log M(\theta)\}$ for $x \in \mathbb{R}$. Let

$$\theta^+ = \sup\{\theta : M(\theta) < \infty\} \quad \text{and} \quad \theta^- = \inf\{\theta : M(\theta) < \infty\}. \tag{11.1}$$

Suppose $\theta^- < 0 < \theta^+$. Then $m = \int x\,\mu(dx)$ exists as a finite number. Suppose the point $a > m$ is in the interior of the interval $\{I < \infty\}$. Then Cramér's theorem (page 26) gives

$$\begin{aligned} -I(a) &\le \varliminf_{n\to\infty} \frac{1}{n} \log P\{S_n/n > a\} \\ &\le \varlimsup_{n\to\infty} \frac{1}{n} \log P\{S_n/n \ge a\} \le -I(a). \end{aligned}$$

In other words,

$$P\{S_n/n > a\} = e^{-nI(a)+o(n)}.$$

In the next theorem we identify the term $o(n)$ in the exponent under some hypotheses.

Theorem 11.1. *Suppose $\theta^- < 0 < \theta^+$ and $m < a < c^+ = \lim_{\theta \nearrow \theta^+} \frac{M'(\theta)}{M(\theta)}$. Then there exist real numbers $0 < c_1(a) < c_2(a) < \infty$ such that $\forall n \in \mathbb{N}$*

$$c_1(a) n^{-1/2} e^{-nI(a)} \le P\{S_n/n > a\} \le c_2(a) n^{-1/2} e^{-nI(a)}. \tag{11.2}$$

Similarly, if $\theta^- < 0 < \theta^+$ and $c^- = \lim_{\theta \searrow \theta^-} \frac{M'(\theta)}{M(\theta)} < a < m$, then there exist real numbers $0 < c_3(a) < c_4(a) < \infty$ such that $\forall n \in \mathbb{N}$

$$c_3(a) n^{-1/2} e^{-nI(a)} \le P\{S_n/n < a\} \le c_4(a) n^{-1/2} e^{-nI(a)}. \tag{11.3}$$

An explicit upper bound on $\sqrt{n}\, e^{nI(a)} P\{S_n/n > a\}$ is given in the proof. See (11.8) below. The same results are true for $P\{S_n/n \ge a\}$ and $P\{S_n/n \le a\}$. In the proof one then needs to consider the possibility that $\widetilde{S}_n$ has an atom at zero, but that requires only minor adjustments.

Proof. We verify (11.2). The other statement then follows by replacing X with $-X$. Let $m < a < c^+$. Then

$$P\{S_n/n > a\} = e^{-n(\theta a - \log M(\theta))} J_n(\theta) \tag{11.4}$$

where

$$\begin{aligned} J_n(\theta) = \int \cdots \int e^{-\theta \sum_{j=1}^n (x_j - a)} \mathbb{1}\Big\{ \sum_{j=1}^n (x_j - a) > 0 \Big\} \\ \times \frac{e^{\theta(x_1 + \cdots + x_n)}}{M(\theta)^n} \mu(dx_1) \cdots \mu(dx_n). \end{aligned} \tag{11.5}$$

Since μ is nondegenerate, direct differentiation shows that $\frac{M'(\theta)}{M(\theta)}$ is strictly increasing and continuous on $(0, \theta^+)$. Thus the limit defining c^+ exists as an extended real. Since $\frac{M'(0)}{M(0)} = m$ we have $m < c^+$, and $m < a < c^+$ implies that there exists a unique $\theta_a \in (0, \theta^+)$ such that $\frac{M'(\theta_a)}{M(\theta_a)} = a$. By Exercise 2.25

$$I(a) = \sup_{\theta \ge 0} \{\theta a - \log M(\theta)\} = \theta_a a - \log M(\theta_a).$$

Taking $\theta = \theta_a$ in (11.4) gives

$$P\{S_n/n > a\} = e^{-nI(a)} J_n(\theta_a). \tag{11.6}$$

Now

$$J_n(\theta_a) = E[e^{-\theta_a \widetilde{S}_n} \mathbb{1}\{\widetilde{S}_n > 0\}], \tag{11.7}$$

where

$$\widetilde{S}_n = \sum_{j=1}^{n} (Z_j - a)$$

and $Z, Z_1, Z_2, \dots$ are i.i.d. with distribution ν_{θ_a} defined by the Radon-Nikodym derivative

$$\frac{d\nu_{\theta_a}}{d\mu}(x) = \frac{e^{\theta_a x}}{M(\theta_a)}.$$

Letting $F_n(u) = P\{\widetilde{S}_n \le u\sqrt{n}\}$, we see

$$\begin{aligned} J_n(\theta_a) &= \int_{(0,\infty)} e^{-\sqrt{n}\,\theta_a u} dF_n(u) \\ &= \int_{(0,\infty)} \int_u^\infty \sqrt{n}\,\theta_a e^{-\sqrt{n}\,\theta_a x} dx\, dF_n(u) \\ &= \int_0^\infty (F_n(x) - F_n(0))\sqrt{n}\,\theta_a e^{-\sqrt{n}\,\theta_a x} dx. \end{aligned}$$

Define

$$\rho = \int |x-a|^3 \frac{e^{\theta_a x}}{M(\theta_a)}\,\mu(dx) \quad \text{and} \quad \sigma^2 = \int (x-a)^2 \frac{e^{\theta_a x}}{M(\theta_a)}\,\mu(dx).$$

By the Berry-Esseen theorem [**27**, Theorem 3.4.9]

$$\sup_x |\Phi_\sigma(x) - F_n(x)| \le 3\frac{E[|Z-a|^3]}{\sigma^3\sqrt{n}} = 3\rho/(\sigma^3\sqrt{n}),$$

where $\Phi_\sigma(x)$ is the cumulative distribution function of a mean zero Gaussian random variable G_σ with variance σ^2. Hence

$$J_n(\theta_a) \le \int_0^\infty (\Phi_\sigma(x) - \Phi_\sigma(0))\sqrt{n}\,\theta_a e^{-\sqrt{n}\,\theta_a x} dx + \frac{6\rho}{\sigma^3\sqrt{n}} \int_0^\infty \sqrt{n}\,\theta_a e^{-\sqrt{n}\,\theta_a x} dx.$$

Reversing the calculation above with Φ_σ replacing F_n gives

$$\begin{aligned} J_n(\theta_a) &\le \int_{(0,\infty)} e^{-\sqrt{n}\,\theta_a x} d\Phi_\sigma(x) + \frac{6\rho}{\sigma^3\sqrt{n}} \\ &= \int_0^\infty e^{-\sqrt{n}\,\theta_a x - x^2/(2\sigma^2)} \frac{dx}{\sqrt{2\pi\sigma^2}} + \frac{6\rho}{\sigma^3\sqrt{n}}. \end{aligned}$$

Drop $-x^2/(2\sigma^2)$ from the exponent and integrate to get

$$\sqrt{n}\, J_n(\theta_a) \le (2\pi\sigma^2\theta_a^2)^{-\frac{1}{2}} + \frac{6\rho}{\sigma^3}. \tag{11.8}$$

Combining (11.6) with the above gives the upper bound in (11.2).

It remains to check the lower bound. Now

$$\begin{aligned}
J_n(\theta_a) &\geq e^{-2A\theta_a} P\{n^{-1/2}\widetilde{S}_n \in (n^{-1/2}A, 2n^{-1/2}A)\} \\
&\geq e^{-2A\theta_a}\Big(P\{G_\sigma \in (n^{-1/2}A, 2n^{-1/2}A)\} - 6\rho/(\sigma^3\sqrt{n})\Big) \\
&\sim e^{-2A\theta_a}(A/\sqrt{2\pi\sigma^2 n} - 6\rho/(\sigma^3\sqrt{n}))
\end{aligned}$$

as $n \to \infty$. Thus taking A sufficiently large we see

$$\varliminf_{n\to\infty} \sqrt{n}\, J_n(\theta_a) > 0.$$

The lower bound in (11.2) then follows from observing that for each $n \geq 1$ the quantity in question is positive. Otherwise, we would have $P(X > a)^n \leq P(S_n > na) = 0$ and $X \leq a$ P-a.s. But then $M'(\theta) \leq aM(\theta)$, for all $\theta \in (\theta^-, \theta^+)$, and $c_+ \leq a$, which contradicts $a < c^+$. The theorem is proved. □

Literature notes. The classic reference for refined estimates such as those in Theorem 11.1 is the paper [**5**]. See also Section 3.7 of [**15**]. Analogous results are known for random variables with values in $\mathbb{R}^d$ and even in separable Banach spaces, when the intervals (a, ∞) and $(-\infty, a)$ are replaced with convex open sets D. One of the first difficulties in establishing such formulas is to find a suitable replacement for the endpoint of the interval when the set D is of higher dimension. The replacement is known as the *dominating point* of the set D. Their existence and uniqueness is now reasonably well understood. See article [**29**] and its references.

11.2. Moderate deviations

Continue with a sum $S_n = X_1 + \cdots + X_n$ of i.i.d. real-valued random variables and let $\theta^\pm$ be as in (11.1). As we saw on page 34, order 1 deviations of S_n/n are handled by Cramér's theorem while order $n^{-1/2}$ deviations are described by the central limit theorem. Here we consider deviations of intermediate order $n^{-1/2+\alpha}$ for $\alpha \in (0, 1/2)$. We state the result for the right tail. The same result for the left tail comes by replacing X_i with $-X_i$.

Theorem 11.2. *Let $\theta^- < 0 < \theta^+$, $E[X] = 0$, and $0 < \sigma^2 = E[X^2] < \infty$. Assume $\{b_n : n \geq 1\}$ is a sequence of positive numbers such that*

$$\lim_{n\to\infty} b_n/n^{\frac{1}{2}} = \infty \quad \textit{and} \quad \lim_{n\to\infty} b_n/n = 0.$$

Then for all $a > 0$

$$\lim_{n\to\infty} \frac{n}{b_n^2} \log P(S_n/b_n \geq a) = -\frac{a^2}{2\sigma^2}. \tag{11.9}$$

Remark 11.3. The proof of the lower bound part of (11.9) relies on the central limit theorem and hence uses only $0 < \sigma^2 < \infty$. The upper bound proof uses the assumption that the moment generating function is finite near zero.

Proof. Step 1. Upper bound. The upper bound is basically Chebyshev's inequality: for $a > 0$ and $s \geq 0$

$$P\{S_n/b_n \geq a\} = P\{b_n S_n/n \geq ab_n^2/n\} \leq E[e^{sb_n S_n/n}]e^{-sab_n^2/n}. \tag{11.10}$$

Below we prove that

$$\lim_{n\to\infty} \frac{n}{b_n^2} \log E[e^{sb_n S_n/n}] = s^2\sigma^2/2. \tag{11.11}$$

Combine this with (11.10) and set $s = a/\sigma^2$ to obtain the upper bound:

$$\overline{\lim_{n\to\infty}} \frac{n}{b_n^2} \log P\{S_n/b_n \geq a\} \leq \frac{s^2\sigma^2}{2} - sa = -\frac{a^2}{2\sigma^2}.$$

To prove (11.11) first observe that by Taylor's formula for all real t there exists $\rho = \rho(t) \in (0,1)$ such that $e^t = 1 + t + \frac{t^2}{2}e^{\rho t}$. Hence we have

$$E[e^{sb_nX/n}] = E\Big[1 + sb_nX/n + \tfrac{1}{2}(sb_nX/n)^2 e^{\rho sb_nX/n}\Big],$$

where $|\rho(\omega)| \leq 1$. Since $E[X] = 0$ we thus have

$$E[e^{sb_nX/n}] = 1 + E\Big[\tfrac{1}{2}(sb_nX/n)^2 e^{\rho sb_nX/n}\Big],$$

and since $E[(tX)^2 e^{t|X|}] < \infty$ for all $t < \min\{|\theta^-|, \theta^+\}$ the dominated convergence theorem implies

$$\lim_{n\to\infty} E[(sX)^2 e^{\rho sb_nX/n}] = s^2 E[X^2] = s^2\sigma^2.$$

Consequently, since $x^{-1}\log(1+x) \to 1$ as $x \to 0$,

$$\begin{aligned}\lim_{n\to\infty} \frac{n}{b_n^2} \log E[e^{sb_nS_n/n}] &= \lim_{n\to\infty} \frac{n^2}{b_n^2} \log E[e^{sb_nX/n}] \\ &= \lim_{n\to\infty} \frac{n^2}{b_n^2} \log\Big(1 + \tfrac{1}{2}E[(sX)^2 e^{\rho sb_nX/n}]b_n^2/n^2\Big) \\ &= \tfrac{1}{2}s^2\sigma^2.\end{aligned}$$

Thus (11.11) holds.

Step 2. Lower bound. Recall that $\lfloor x \rfloor$ denotes the greatest integer less than or equal to x. Let $t > 0$ and set $p_n = \lfloor t^2n^2/b_n^2 \rfloor$, $q_n = \lfloor n/p_n \rfloor$, and $r_n = \frac{b_n}{tq_n}$. Next, fix $a > 0$ and $\varepsilon > 0$ and note that

$$\begin{aligned}P\{S_{p_n}/r_n \geq t(a+\varepsilon)\}^{q_n} &\leq P\{S_{p_nq_n}/r_n \geq tq_n(a+\varepsilon)\} \\ &= P\{S_{p_nq_n}/b_n \geq a+\varepsilon\}.\end{aligned} \tag{11.12}$$

Since $p_n q_n \le n$ and $S_n = S_{p_n q_n} + (S_n - S_{p_n q_n})$,

$$P\{S_n/b_n \ge a\} \ge P\{S_{p_n q_n}/b_n \ge a+\varepsilon, |S_n - S_{p_n q_n}| < \varepsilon b_n\}. \tag{11.13}$$

From the definitions $n - p_n q_n \le p_n \le t^2 n^2/b_n^2$. Then, since $n/b_n^2 \to 0$ by assumption,

$$\begin{aligned} \frac{n}{b_n^2}\log P\{|S_n - S_{p_n q_n}| \ge \varepsilon\sqrt{n}\} &\le \frac{n}{b_n^2}\log \frac{E[(S_n - S_{p_n q_n})^2]}{\varepsilon^2 n} \\ &= \frac{n}{b_n^2}\log\frac{(n-p_n q_n)\sigma^2}{\varepsilon^2 n} \le \frac{n}{b_n^2}\log\frac{nt^2\sigma^2}{b_n^2\varepsilon^2} \to 0 \end{aligned} \tag{11.14}$$

as $n \to \infty$. By the independence of $S_n - S_{p_n q_n}$ and $S_{p_n q_n}$, (11.13) and (11.14) combine to yield

$$\varliminf_{n\to\infty} \frac{n}{b_n^2}\log P\{S_n/b_n \ge a\} \ge \varliminf_{n\to\infty}\frac{n}{b_n^2}\log P\{S_{p_n q_n}/b_n \ge a+\varepsilon\}.$$

Thus (11.12) implies

$$\varliminf_{n\to\infty} \frac{n}{b_n^2}\log P\{S_n/b_n \ge a\} \ge \varliminf_{n\to\infty}\frac{n}{b_n^2} q_n \log P\{S_{p_n}/r_n \ge t(a+\varepsilon)\}.$$

Asymptotically (recall that $a_n \sim b_n$ means $a_n/b_n \to 1$)

$$\frac{n}{b_n^2}q_n = \frac{n}{b_n^2}\lfloor n/p_n\rfloor \sim \frac{n^2}{b_n^2 p_n} \sim t^{-2}$$

and

$$r_n = \frac{b_n}{tq_n} = \frac{b_n}{t\lfloor\frac{n}{p_n}\rfloor} \sim \frac{b_n p_n}{nt} \sim p_n^{\frac{1}{2}}.$$

Hence the central limit theorem implies

$$\varliminf_{n\to\infty} \frac{n}{b_n^2}\log P\{S_n/b_n \ge a\} \ge t^{-2}\log P\{G_\sigma \ge t(a+\varepsilon)\}$$

where G_σ is normal with mean zero and variance σ^2. Letting $\varepsilon \searrow 0$ turns this into

$$\varliminf_{n\to\infty} \frac{n}{b_n^2}\log P\{S_n/b_n \ge a\} \ge t^{-2}\log P\{G_\sigma > ta\}. \tag{11.15}$$

Now for $s > 0$

$$\sqrt{2\pi\sigma^2}P\{G_\sigma > s\} \ge \int_s^{s+1/s} e^{-\frac{u^2}{2\sigma^2}}\,du \ge s^{-1}e^{-\frac{(s+1/s)^2}{2\sigma^2}},$$

and hence

$$\lim_{t\to\infty} t^{-2}\log P\{G_\sigma > ta\} = -\frac{a^2}{2\sigma^2},$$

which when combined with (11.15) implies the lower bound

$$\varliminf_{n\to\infty} \frac{n}{b_n^2}\log P\{S_n/b_n \ge a\} \ge -\frac{a^2}{2\sigma^2}.$$

The theorem is proved. □

Chapter 12

Large deviations through the limiting generating function

In this chapter we prove a general LDP for a sequence $\{\mu_n\}$ of probability distributions on $\mathbb{R}^d$. The result is due to Gärtner [**40**] and Ellis [**31**]. No structural assumptions such as independence are made. Instead the assumptions are on the existence and regularity of the limiting logarithmic moment generating function. Before tackling the LDP, we need to go deeper into convex analysis. An application of the general LDP is given in Section 12.3.

12.1. Essential smoothness and exposed points

We develop further the material from Section 4.1, this time restricted to convex functions $f : \mathbb{R}^d \to (-\infty, \infty]$. The inner product on $\mathbb{R}^d$ is denoted by $\langle x, y\rangle$. The epigraph of a convex function f is the convex set $\operatorname{epi} f = \{(x, s) : x \in \mathbb{R}^d,\ f(x) \le s < \infty\}$ in $\mathbb{R}^{d+1}$. The *effective domain* $\operatorname{dom} f$ is the projection of $\operatorname{epi} f$ onto $\mathbb{R}^d$, namely $\operatorname{dom} f = \{x \in \mathbb{R}^d : f(x) < \infty\}$.

As usual, a function $f : \mathbb{R}^d \to (-\infty, \infty]$ is differentiable at x if f is finite in a neighborhood of x and there exists a gradient vector $\nabla f(x) \in \mathbb{R}^d$ that satisfies

$$\lim_{h\to 0} \frac{f(x + hw) - f(x) - h\langle \nabla f(x), w\rangle}{h} = 0 \qquad \forall w \ne 0.$$

The topological boundary of a set A is $\operatorname{bd} A = \overline{A} \setminus A^\circ$.

Definition 12.1. A convex function $f : \mathbb{R}^d \to (-\infty, \infty]$ is *essentially smooth* if it possesses the following three properties.

(a) The interior $(\operatorname{dom} f)^\circ$ of $\operatorname{dom} f$ is nonempty.

(b) f is differentiable on $(\operatorname{dom} f)^\circ$.

(c) If $\operatorname{dom} f$ is not all of $\mathbb{R}^d$, then $|\nabla f(x_n)| \to \infty$ whenever $(\operatorname{dom} f)^\circ \ni x_n \to x \in \operatorname{bd}(\operatorname{dom} f)$.

Part (c) of the definition is sometimes expressed by saying that f is *steep*. Note that the condition concerns only boundary points $x \in \mathbb{R}^d$. In other words, it is not required that $|\nabla f(x_n)| \to \infty$ if $|x_n| \to \infty$ for a sequence $x_n \in (\operatorname{dom} f)^\circ$.

Exercise 12.2. An essentially smooth function f may fail to be lower semicontinuous because the definition does not constrain f on the boundary of $\operatorname{dom} f$. Consider the example $f(x) = x \log x$ for $x > 0$, $f(x) = \infty$ for $x < 0$, and $0 < f(0) < \infty$.

Recall the subdifferential and the convex conjugate of f:

$$\partial f(x) = \{y \in \mathbb{R}^d : f(w) \ge f(x) + \langle y, w - x\rangle \ \forall w \in \mathbb{R}^d\} \tag{12.1}$$

and

$$f^*(y) = \sup_{x \in \mathbb{R}^d} \{\langle x, y\rangle - f(x)\}, \quad y \in \mathbb{R}^d.$$

Definition 12.3. A point $y_0 \in \mathbb{R}^d$ is an *exposed point* of f^* if $\exists x \in \mathbb{R}^d$ such that

$$f^*(y) > f^*(y_0) + \langle x, y - y_0\rangle \quad \forall y \ne y_0.$$

When this happens x is an *exposing hyperplane* of y_0.

***Exercise 12.4.** Show that y_0 is an exposed point of f^* if and only if $\exists x \in \partial f^*(y_0)$ such that $x \notin \partial f^*(y)$ $\forall y \ne y_0$.

A set $U \subset \mathbb{R}^d$ is *affine* if

$$x, y \in U \text{ implies that } tx + (1-t)y \in U \quad \forall t \in \mathbb{R}.$$

Note the contrast with convexity: for all $t \in \mathbb{R}$. Geometrically the condition is that the entire line through x and y lies in U, for all $x, y \in U$. The familiar notion of (vector) subspace of $\mathbb{R}^d$ can then be defined as an affine set that contains the origin. Every affine set U is a translate $U = a + V$ of a unique subspace V, namely

$$V = U - U = \{x - y : x, y \in U\}. \tag{12.2}$$

(Exercise 12.6 below takes the reader through these claims.) The dimension of U is by definition the dimension of V.

An intersection of affine sets is affine. Hence any set $A \subset \mathbb{R}^d$ has an *affine hull* $\operatorname{aff} A$ defined as the intersection of all affine sets that contain A. Alternatively,

$$\operatorname{aff} A = \Big\{ \sum_{i=1}^m t_i x_i : m \in \mathbb{N},\, x_i \in A,\, t_i \in \mathbb{R},\, \sum_{i=1}^m t_i = 1 \Big\}.$$

The *relative interior* $\operatorname{ri} A$ of A is the interior of A regarded as a subset of $\operatorname{aff} A$. In other words, $x \in \operatorname{ri} A$ if $B(x,r) \cap (\operatorname{aff} A) \subset A$ for some $r > 0$. The relative interior of a nonempty convex set is itself a nonempty convex set (Lemma 12.7 below).

An extreme case is a singleton $A = \{x\}$. Then $\operatorname{aff} A = \operatorname{ri} A = A = \{x\}$. The simplest nontrivial example is a closed line segment $A = \{sx + (1-s)y : 0 \le s \le 1\}$ for $x, y \in \mathbb{R}^d$. Then $\operatorname{aff} A$ is the entire line through x and y while $\operatorname{ri} A = \{sx + (1-s)y : 0 < s < 1\}$.

The main result of this section is the next theorem. It is used in the proof of the LDP in the next section.

Theorem 12.5. *Let $f : \mathbb{R}^d \to (-\infty, \infty]$ be convex, lower semicontinuous, and essentially smooth. Let $y \in \operatorname{ri}(\operatorname{dom} f^*)$. Then y is an exposed point of f^* with an exposing hyperplane $x \in (\operatorname{dom} f)^\circ$.*

By Lemma 4.13, $-\infty < f^* \le \infty$ and $\operatorname{dom} f^*$ is nonempty. So the theorem is never vacuous.

The rest of this section is devoted to the proof of the theorem. The proof involves a variety of valuable ideas from convex analysis. The main step below is the characterization of the subdifferential of an essentially smooth function in Theorem 12.12. The next exercise contains the relationship between affine sets and vector subspaces.

***Exercise 12.6.** Let U be an affine subset of $\mathbb{R}^d$.

(a) Show that for any $y \in U$, $U - y = \{x - y : x \in U\}$ is a vector subspace. In particular, if U contains the origin, then U is a vector subspace.

Hint: Start with scalar multiplication: $t(x-y) = tx + (1-t)y - y$. For addition note that $x_1, x_2 \in U$ implies $(x_1 + x_2)/2 \in U$.

(b) Show that for any $y_1, y_2 \in U$, $U - y_1 = U - y_2$.

Hint: $V_i = U - y_i$ are subspaces that satisfy $U = V_1 + y_1 = V_2 + y_2$. How can two subspaces be translates of each other?

(c) U is a translate of the unique subspace $V = U - U = U - y$ (for any $y \in U$).

Next we establish properties of the relative interior. Parts (a) and (b) together tell us that the relative interior of a nonempty convex set is itself a nonempty convex set. Part (c) below is sometimes used as the definition of the relative interior of a convex set, for example in [**15**].

Lemma 12.7. *Let A be a nonempty convex subset of $\mathbb{R}^d$.*

(a) $\operatorname{ri} A$ *is not empty.*

(b) *If $x \in \operatorname{ri} A$ and $y \in A$, then $sx + (1-s)y \in \operatorname{ri} A$ $\forall s \in (0,1]$. In particular, $\operatorname{ri} A$ is a convex set.*

(c) $\operatorname{ri} A = \{x \in A : \forall y \in A\ \exists \delta > 0$ *such that* $(1+\delta)x - \delta y \in A\}$.

Proof. Part (a). As already observed above, if A is a singleton, then $\operatorname{ri} A = A$. So let us suppose that A is not a singleton. If we translate A, the same translation will be applied to $\operatorname{aff} A$ and $\operatorname{ri} A$. Consequently we can assume that $0 \in A$. Then $V = \operatorname{aff} A$ is a nontrivial vector subspace of $\mathbb{R}^d$. Let $m = \dim V \geq 1$. Let $u_1, \dots, u_k$ be a maximal linearly independent subset of A. Then $A \subset \operatorname{span}\{u_1, \dots, u_k\}$, for otherwise another linearly independent point from A could be added to $u_1, \dots, u_k$. Since $\operatorname{span}\{u_1, \dots, u_k\}$ is affine, it must be that $k = m$ and $V = \operatorname{span}\{u_1, \dots, u_m\}$.

Consider the nonempty subset

$$B = \Big\{ \sum_{i=1}^{m} t_i u_i : t_i > 0, \sum_{i=1}^{m} t_i < 1 \Big\}$$

of V. $B \subset A$ because $0 \in A$ and A is convex. If we can show that B is relatively open in V, then $B \subset \operatorname{ri} A$ and thereby $\operatorname{ri} A \neq \varnothing$.

Extend $u_1, \dots, u_m$ to a basis $u_1, \dots, u_d$ of $\mathbb{R}^d$. Define the (continuous) linear map $T : \mathbb{R}^d \to \mathbb{R}^m$ by

$$Tu_i = \begin{cases} e_i, & 1 \leq i \leq m, \\ 0, & m < i \leq d. \end{cases}$$

Then $B = T^{-1}(B_0) \cap V$ for the open set

$$B_0 = \left\{ (y_1, \dots, y_m) \in \mathbb{R}^m : y_i > 0, \sum_{i=1}^{m} y_i < 1 \right\}.$$

Since $T^{-1}(B_0)$ is open in $\mathbb{R}^d$, B is relatively open in V.

Part (b). Fix $s \in (0,1]$ and set $w = sx + (1-s)y$. Since $x \in \operatorname{ri} A$, $\exists r > 0$ such that $(\operatorname{aff} A) \cap B(x,r) \subset A$. We shall show that $(\operatorname{aff} A) \cap B(w, sr) \subset A$. This gives the desired conclusion $w \in \operatorname{ri} A$.

So let $z \in (\operatorname{aff} A) \cap B(w, sr)$. Then $\tilde{z} = s^{-1}z + (1-s^{-1})y \in \operatorname{aff} A$. Since $x = s^{-1}w + (1-s^{-1})y$, $|\tilde{z} - x| = s^{-1}\,|z - w| < r$, and $\tilde{z} \in (\operatorname{aff} A) \cap B(x,r) \subset A$. By convexity, $z = s\tilde{z} + (1-s)y \in A$.

Part (c). Suppose $x \in \operatorname{ri} A$. Then $\exists r > 0$ such that $B(x, r) \cap (\operatorname{aff} A) \subset A$. If $y \in A$, then $w_\delta = (1+\delta)x - \delta y \in \operatorname{aff} A$ for all $\delta \in \mathbb{R}$. $|x - w_\delta| = \delta\,|x - y| < r$ for small enough δ, and for such δ, $w_\delta \in B(x, r) \cap (\operatorname{aff} A) \subset A$.

Conversely, suppose x satisfies the condition on the right. Since $\operatorname{ri} A \neq \varnothing$, we can fix a point $y \in \operatorname{ri} A$ and then $\delta > 0$ such that $w = (1+\delta)x - \delta y \in A$. Then $x = \frac{\delta}{1+\delta} y + \frac{1}{1+\delta} w \in \operatorname{ri} A$ by part (b). $\square$

We extend the separation theorem to an affine space.

Lemma 12.8. *Let U be an affine subset of $\mathbb{R}^d$, let B be a relatively open, convex subset of U, and let $x_0 \in U \smallsetminus B$. Then $\exists\, z^* \in \mathbb{R}^d$ and $\gamma \in \mathbb{R}$ such that*

$$\langle x_0, z^* \rangle \le \gamma < \langle x, z^* \rangle \quad \forall x \in B.$$

Proof. The quick proof: U is isomorphic and homeomorphic to a translate of a vector space; apply the Hahn-Banach theorem. Here are some details. Let m be the dimension of U.

Let us first take for granted the following basic fact about affine spaces. There exists an $m \times d$ matrix S and $b \in \mathbb{R}^m$ such that the map $Tx = Sx + b$ is a bijection and a homeomorphism (both T and T^{-1} continuous) from U onto $\mathbb{R}^m$. T is an *affine* map.

Assuming the existence of T, the set $C = T(B)$ is a convex open subset of $\mathbb{R}^m$, and $y_0 = Tx_0 \notin C$. By the Hahn-Banach separation theorem $\exists w^* \in \mathbb{R}^m$ and $\beta \in \mathbb{R}$ such that

$$\begin{aligned} &\langle y_0, w^* \rangle \le \beta < \langle y, w^* \rangle \quad \forall y \in C \\ \iff &\langle x_0, S^t w^* \rangle \le \beta - \langle b, w^* \rangle < \langle x, S^t w^* \rangle \quad \forall x \in B. \end{aligned}$$

(S^t is the transpose of S.) This gives the claim.

We sketch the argument for the existence of T. Let $v_1, \dots, v_m$ be a basis for the space V defined in (12.2), and extend it to a basis $v_1, \dots, v_d$ of $\mathbb{R}^d$. Let G be the isomorphism on $\mathbb{R}^d$ that maps $Gv_i = e_i$. Exercise 12.6 showed that $U = a + V$ for some $a \in \mathbb{R}^d$. Define the affine bijection F on $\mathbb{R}^d$ by $Fx = Gx - Ga$. Let P be the $\mathbb{R}^d \to \mathbb{R}^m$ projection $(x_1, \dots, x_d) \overset{P}{\mapsto} (x_1, \dots, x_m)$. Define $T : U \to \mathbb{R}^m$ by $T = P \circ F$. Its inverse $T^{-1} : \mathbb{R}^m \to U$ is given by $T^{-1}y = G^{-1}(y, 0, \dots, 0) + a$ where $y \in \mathbb{R}^m$ is first padded with $d - m$ zeroes to create $(y, 0, \dots, 0) \in \mathbb{R}^d$. $\square$

The next intermediate goal (Theorem 12.10) is to show that $\partial g(x)$ is not empty for relative interior points x of the effective domain of a proper convex function g. As preparation, we describe the relative interior of the epigraph.

Lemma 12.9. *Let $g : \mathbb{R}^d \to (-\infty, \infty]$ be convex. Then* $\mathrm{ri}(\mathrm{epi}\, g) = \{(x, s) : x \in \mathrm{ri}(\mathrm{dom}\, g),\ g(x) < s < \infty\}$.

Proof. Let B denote the set on the right in the statement. To show that $B \subset \mathrm{ri}(\mathrm{epi}\, g)$, let $(x, s) \in B$ and $(w, t) \in \mathrm{epi}\, g$. To satisfy the criterion in Lemma 12.7 we need to show that for some $\delta > 0$,

$$\begin{aligned} &(1+\delta)(x,s) - \delta(w,t) \in \mathrm{epi}\, g \\ \iff &(1+\delta)s - \delta t \geq g((1+\delta)x - \delta w). \end{aligned} \tag{12.3}$$

Let $\varphi(\delta) = g((1+\delta)x - \delta w)$, a finite, convex function on an interval $(-\varepsilon, \varepsilon)$ for some $\varepsilon > 0$ because $x \in \mathrm{ri}(\mathrm{dom}\, g)$ and $w \in \mathrm{dom}\, g$. Hence φ is continuous (Exercise 4.8). Since $s > g(x) = \varphi(0)$, the inequality on the right above is true for small enough $\delta > 0$. Since $(w, t) \in \mathrm{epi}\, g$ was arbitrary, we have verified that $(x, s) \in \mathrm{ri}(\mathrm{epi}\, g)$.

Conversely, suppose $(x, s) \in \mathrm{ri}(\mathrm{epi}\, g)$. Then, for any $(w, t) \in \mathrm{epi}\, g$, $\exists \delta > 0$ such that (12.3) holds. We can take w to be an arbitrary point of $\mathrm{dom}\, g$. This implies that $x \in \mathrm{ri}(\mathrm{dom}\, g)$. Next, take $w = x$ and $t = g(x) + 1$. Then (12.3) says that $s \geq g(x) + \frac{\delta}{1+\delta}$ for some $\delta > 0$. We have checked that $(x, s) \in B$. □

Theorem 12.10. *Let $g : \mathbb{R}^d \to (-\infty, \infty]$ be convex and let $x \in \mathrm{ri}(\mathrm{dom}\, g)$. Then $\partial g(x) \neq \varnothing$.*

Proof. Apply Lemma 12.8 to $U = \mathrm{aff}(\mathrm{epi}\, g)$ in $\mathbb{R}^{d+1}$, the relatively open convex set $A = \mathrm{ri}(\mathrm{epi}\, g)$ in U, and the point $(x, g(x)) \in U \setminus A$. Then we have $y^* \in \mathbb{R}^d$ and $\gamma, \tau \in \mathbb{R}$ such that

$$\langle x, y^* \rangle + \tau g(x) \leq \gamma < \langle w, y^* \rangle + \tau t \qquad \forall (w, t) \in \mathrm{ri}(\mathrm{epi}\, g).$$

Since we can take $w = x \in \mathrm{ri}(\mathrm{dom}\, g)$, $\tau = 0$ is impossible. Taking $w = x$ and $t \nearrow \infty$ rules out $\tau < 0$. With $y = y^*/\tau$ we have

$$\langle x, y \rangle + g(x) < \langle w, y \rangle + t \qquad \forall (w, t) \in \mathrm{ri}(\mathrm{epi}\, g).$$

Let $(z, g(z)) \in \mathrm{epi}\, g$ and $(w, t) \in \mathrm{ri}(\mathrm{epi}\, g)$. By Lemma 12.7(b), $(1-\theta)(w, t) + \theta(z, g(z)) \in \mathrm{ri}(\mathrm{epi}\, g)$ for all $\theta \in [0, 1)$. Substitute this point on the right-hand side and let $\theta \nearrow 1$ to get

$$\langle x, y \rangle + g(x) \leq \langle z, y \rangle + g(z).$$

Since $z \in \mathrm{dom}\, g$ is arbitrary, this says that $-y \in \partial g(x)$. □

Lemma 12.11. *Let $f : \mathbb{R}^d \to (-\infty, \infty]$ be a convex function. Suppose f is differentiable at x_0 (which implies that f is finite in a neighborhood of x_0). Then $\partial f(x_0) = \{\nabla f(x_0)\}$. Assume additionally that f is lower semicontinuous. Then $y_0 = \nabla f(x_0)$ is an exposed point of f^* with exposing hyperplane x_0.*

Proof. To show that $\nabla f(x_0) \in \partial f(x_0)$, we need to check that

$$f(x) \geq f(x_0) + \langle \nabla f(x_0), x - x_0 \rangle \quad \forall x \in \mathbb{R}^d.$$

Write the inner product as a directional derivative and use convexity:

$$\langle \nabla f(x_0), x - x_0 \rangle = \lim_{\varepsilon \searrow 0} \frac{f(x_0 + \varepsilon(x - x_0)) - f(x_0)}{\varepsilon} \leq f(x) - f(x_0).$$

Conversely, if $y \in \partial f(x_0)$, then for $\varepsilon > 0$

$$\frac{f(x_0 + \varepsilon(x - x_0)) - f(x_0)}{\varepsilon} \geq \langle y, x - x_0 \rangle$$

from which, after $\varepsilon \searrow 0$, $\langle \nabla f(x_0), x - x_0 \rangle \geq \langle y, x - x_0 \rangle$ $\forall x \in \mathbb{R}^d$. With $x = x_0 + y - \nabla f(x_0)$ this gives $|y - \nabla f(x_0)| \leq 0$. We have verified $\partial f(x_0) = \{\nabla f(x_0)\}$.

By Theorem 4.21, under the lower semicontinuity assumption, $\partial f(x_0) = \{y_0\}$ is equivalent to $x_0 \in \partial f^*(y_0)$ and $x_0 \notin \partial f^*(y)$ $\forall y \neq y_0$. This is the same as saying that y_0 is an exposed point of f^* with exposing hyperplane x_0 (Exercise 12.4). ∎

The next theorem is pretty technical. But it is easy to verify for a one-dimensional example such as that in Exercise 12.2.

Theorem 12.12. *Let $f : \mathbb{R}^d \to (-\infty, \infty]$ be convex and essentially smooth. Then*

$$\partial f(x) = \begin{cases} \{\nabla f(x)\}, & x \in (\operatorname{dom} f)^\circ, \\ \varnothing, & x \notin (\operatorname{dom} f)^\circ. \end{cases} \tag{12.4}$$

Proof. The first part of Lemma 12.11 gives the case $x \in (\operatorname{dom} f)^\circ$. If $x \notin \operatorname{dom} f$, then $f(x) = \infty$ and the inequality in (12.1) cannot be satisfied for any $w \in \operatorname{dom} f$. Hence $\partial f(x) = \varnothing$.

It remains to prove that if $y_0 \in \partial f(x_0)$, then $x_0 \in (\operatorname{dom} f)^\circ$. Since $x_0 \in \operatorname{dom} f$, $f(x_0)$ is finite. By Theorem 4.21, $f^*(y_0) = \langle x_0, y_0 \rangle - f(x_0)$.

Since $(\operatorname{dom} f)^\circ \neq \varnothing$, we can fix a closed ball $B_1 = \overline{B(x_1, r)}$ with radius $r > 0$ that lies inside $(\operatorname{dom} f)^\circ$. For $t \in [0, 1]$ let $x_t = (1 - t)x_0 + tx_1$, and define the closed balls $B_t = (1-t)x_0 + tB_1 = \overline{B(x_t, tr)}$. See Figure 12.1. By Lemma 12.7(b), for $0 < t \leq 1$ ball B_t lies inside $(\operatorname{dom} f)^\circ$.

We develop an estimate for the oscillation of f inside B_t. This will show that $\nabla f(x_t)$ stays bounded as $x_t \to x_0$. Since f is steep, this tells us that x_0 cannot be a boundary point of $\operatorname{dom} f$, and thereby $x_0 \in (\operatorname{dom} f)^\circ$.

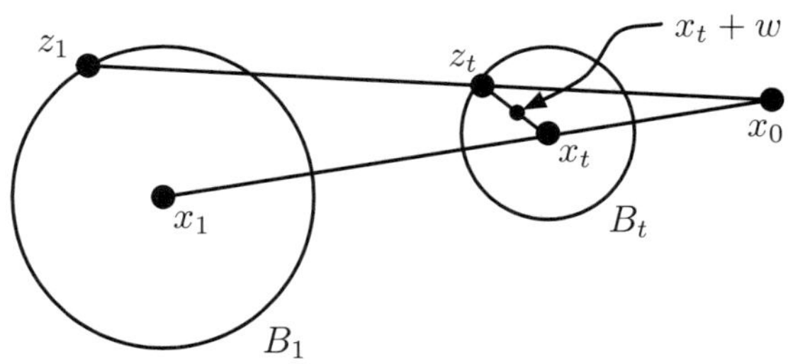

Figure 12.1. Balls B_t.

Let first $x \in B_1 - x_0$ and $s \in [0,1]$. By convexity

$$\begin{aligned} f(x_0 + sx) - f(x_0) &= f((1-s)x_0 + s(x_0 + x)) - f(x_0) \\ &\le s\big(f(x_0+x) - f(x_0)\big) \le s\big(\sup_{B_1} |f| + |f(x_0)|\big). \end{aligned}$$

From

$$\langle x_0, y_0 \rangle - f(x_0) = f^*(y_0) \ge \langle x_0 + sx, y_0 \rangle - f(x_0 + sx)$$

comes

$$f(x_0 + sx) - f(x_0) \ge s\langle x, y_0 \rangle \ge -s(|x_0| + |x_1| + r)\,|y_0|\,.$$

Thus for a constant C,

$$|f(x_0 + sx) - f(x_0)| \le Cs \quad \text{for } x \in B_1 - x_0 \text{ and } s \in [0,1]. \tag{12.5}$$

Now we control oscillation in B_t. Let $0 < t \le 1$. Consider a point $x_t + w \in B_t$ with $0 < |w| \le tr$. Let $z_t = x_t + (tr/\,|w|)w$. Then

$$x_t + w = \frac{|w|}{tr} z_t + \Big(1 - \frac{|w|}{tr}\Big) x_t$$

is on the line segment $[x_t, z_t]$ from the center to the boundary of the ball B_t.

The point $z_1 = x_1 + (r/\,|w|)w$ lies on the boundary of the ball B_1 and $z_t = x_0 + t(z_1 - x_0)$. By convexity and (12.5),

$$\begin{aligned} f(x_t + w) - f(x_t) &\le \frac{|w|}{tr}\big(f(z_t) - f(x_t)\big) \\ &= \frac{|w|}{tr}\big[f(x_0 + t(z_1 - x_0)) - f(x_0 + t(x_1 - x_0))\big] \le \frac{2C}{r}\,|w|\,. \end{aligned}$$

The opposite inequality

$$f(x_t + w) - f(x_t) \ge -(2C/r)\,|w|$$

comes the same way, by writing x_t as a convex combination of $x_t + w$ and $x_t - (tr/\,|w|)w$.

These inequalities combine to give

$$|f(x_t + w) - f(x_t)| \le (2C/r)\,|w| \quad \text{for } |w| \le tr.$$

For $0 < t \le 1$, $x_t \in (\operatorname{dom} f)^\circ$ and so f is differentiable at x_t. The bound above implies $|\nabla f(x_t)| \le 2C/r$ for $0 < t \le 1$. Since $x_t \to x_0$ as $t \searrow 0$, property (c) of the definition of essential smoothness implies that $x_0 \notin \operatorname{bd}(\operatorname{dom} f)$. Since $x_0 \in \operatorname{dom} f$, it must be that $x_0 \in (\operatorname{dom} f)^\circ$. □

We have the tools in place for the proof of the main theorem of the section.

Proof of Theorem 12.5. Let $y \in \operatorname{ri}(\operatorname{dom} f^*)$. By Theorem 12.10, $\exists x \in \partial f^*(y)$. By Theorem 4.21, $y \in \partial f(x)$. By Theorem 12.12, $x \in (\operatorname{dom} f)^\circ$ and $y = \nabla f(x)$. By Lemma 12.11, y is an exposed point of f^* with exposing hyperplane $x \in (\operatorname{dom} f)^\circ$. □

12.2. Gärtner-Ellis theorem

Let $\{\mu_n\}$ be a sequence of Borel probability measures on $\mathbb{R}^d$ and let $0 < r_n \nearrow \infty$ be a sequence of normalization constants. The basic assumption is that the limiting logarithmic moment generating function (or limiting cumulant generating function, or pressure) exists and is not entirely trivial.

Assumption 12.13. *The limit below exists in* $(-\infty, \infty]$ *for all* $\theta \in \mathbb{R}^d$*:*

$$p(\theta) = \lim_{n\to\infty} \frac{1}{r_n} \log \int_{\mathbb{R}^d} e^{r_n \langle \theta, x\rangle}\, \mu_n(dx). \tag{12.6}$$

Furthermore, the origin lies in the interior $(\operatorname{dom} p)^\circ$ *of the effective domain* $\operatorname{dom} p = \{\theta \in \mathbb{R}^d : p(\theta) < \infty\}$.

As in Theorem 4.24, p is convex as a consequence of Hölder's inequality.

Remark 12.14. The assumption that $0 \in (\operatorname{dom} p)^\circ$ is enough to imply that $p(\theta) > -\infty$ $\forall \theta \in \mathbb{R}^d$ even if this is not assumed to begin with. Let $r > 0$ be a radius such that $B(0,r) \subset \operatorname{dom} p$. Pick $\delta > 0$ small enough so that $-\delta\theta \in B(0,r)$. By convexity

$$p(0) \le \frac{1}{1+\delta} p(-\delta\theta) + \frac{\delta}{1+\delta} p(\theta).$$

Since $p(-\delta\theta) < \infty$, $p(\theta) = -\infty$ would force $p(0) = -\infty$. But $p(0) = 0$, so $p(\theta) > -\infty$ must hold.

The convex conjugate of p is, as before,

$$p^*(x) = \sup_{\theta \in \mathbb{R}^d} \{\langle \theta, x\rangle - p(\theta)\}.$$

Let $\mathcal{E}$ be the set of exposed points of p^* whose exposing hyperplane is an element of $(\operatorname{dom} p)^\circ$.

Gärtner-Ellis theorem. *Let Assumption 12.13 hold. Then p^* is a tight convex rate function, and we have the following large deviation bounds.*

(i) *For any closed set $F \subset \mathbb{R}^d$,*

$$\overline{\lim_{n\to\infty}} \frac{1}{r_n} \log \mu_n(F) \le - \inf_{x\in F} p^*(x).$$

(ii) *For any open set $G \subset \mathbb{R}^d$,*

$$\underline{\lim_{n\to\infty}} \frac{1}{r_n} \log \mu_n(G) \ge - \inf_{x\in G\cap\mathcal{E}} p^*(x).$$

(iii) *Assume additionally that p is lower semicontinuous and essentially smooth. Then* $\mathrm{LDP}(\mu_n, r_n, p^*)$ *holds.*

Proof. Abbreviate

$$p_n(\theta) = \log \int e^{\langle\theta,x\rangle}\, \mu_n(dx)$$

so that $r_n^{-1} p_n(r_n\theta) \to p(\theta)$.

Step 1. We show that p^* is a tight convex rate function.

As a convex conjugate, p^* is lower semicontinuous and convex, and $p(0) = 0$ implies $p^*(x) \ge 0$.

Let $e_1, \dots, e_d$ denote the standard basis vectors of $\mathbb{R}^d$. Pick $t > 0$ small enough so that $\pm te_i \in (\operatorname{dom} p)^\circ$ for $1 \le i \le d$. The set $K = \{\theta \in \mathbb{R}^d : |\theta|_1 \le t\}$ is the convex hull of $\{\pm te_i\}_{1\le i\le d}$. Hence an arbitrary $\theta \in K$ is of the form $\theta = \sum_{i=1}^d (\alpha_i te_i + \alpha_{d+i}(-te_i))$ with coefficients $\alpha_i \ge 0$ such that $\sum_1^{2d} \alpha_i = 1$. By convexity

$$p(\theta) \le \sum_{i=1}^d \bigl(\alpha_i p(te_i) + \alpha_{d+i} p(-te_i)\bigr).$$

Since the values $p(\pm te_i)$ are finite, there exists a constant c_0 such that $p(\theta) \le c_0$ $\forall \theta \in K$. (As a real-valued convex function, p is in fact continuous on the convex open set $(\operatorname{dom} p)^\circ$ [**68**, Theorem 10.1] and hence bounded on K. But we do not need this further fact.) For $x \in \mathbb{R}^d$

$$p^*(x) = \sup_\theta \{\langle\theta, x\rangle - p(\theta)\} \ge td^{-1}\,|x| - c_0$$

by choosing $\theta = td^{-1}|x|^{-1}x \in K$. Consequently

$$\{p^* \le a\} \subset \{|x| \le t^{-1}d(a + c_0)\}.$$

Thus $\{p^* \le a\}$ is a closed, bounded subset of $\mathbb{R}^d$, and such sets are compact.

Step 2. Proof of the upper bound in part (i).

The upper bound for compact sets comes from Theorem 4.24. To get the upper bound for closed sets we check exponential tightness.

Let t be as above. Then by virtue of the limits (12.6) there is a constant $c < \infty$ such that

$$r_n^{-1} p_n(\pm r_n t e_i) \le c \quad \text{for all } n \in \mathbb{N} \text{ and } i = 1, \dots, d.$$

Let $b > 0$. We bound the tail in a single coordinate direction e_i:

$$\begin{aligned} \mu_n\{x : x_i \ge b\} &\le e^{-r_n tb} \int_{x_i \ge b} e^{r_n \langle te_i, x\rangle}\, \mu_n(dx) \\ &\le e^{-r_n[tb - r_n^{-1} p_n(r_n t e_i)]} \le e^{-r_n(tb-c)}. \end{aligned}$$

The same bound works for $\mu_n\{x : x_i \le -b\}$. Next we have a union bound:

$$\begin{aligned} \mu_n\{x : |x| \ge b\sqrt{d}\,\} &\le \sum_{i=1}^{d} \bigl(\mu_n\{x : x_i \ge b\} + \mu_n\{x : x_i \le -b\}\bigr) \\ &\le 2de^{-r_n(tb-c)}. \end{aligned}$$

Since c, t, d are fixed, we can take b as large as needed to meet the requirement for exponential tightness.

Step 3. Proof of the lower bound in part (ii). This is a change of measure argument.

Fix $z \in G \cap \mathcal{E}$. (This automatically implies $p^*(z) < \infty$.) Let $\sigma \in (\operatorname{dom} p)^\circ$ be an exposing hyperplane for z. Since $p(\sigma)$ is finite, $p_n(r_n \sigma)$ is finite for large enough n. For these n define the new probability measure ν_n on $\mathbb{R}^d$ by

$$\nu_n(A) = e^{-p_n(r_n \sigma)} \int_A e^{r_n \langle \sigma, x\rangle}\, \mu_n(dx), \quad A \in \mathscr{B}_{\mathbb{R}^d}.$$

Choose $\delta > 0$ small enough so that $B = B(z, \delta) \subset G$. The key step of the change of measure argument is that $\nu_n(B) \to 1$. In our earlier applications of this idea (lower bound proofs for Cramér's theorem on page 32 and Sanov's theorem on page 75) we had i.i.d. variables and $\nu_n(B) \to 1$ came from the law of large numbers. This time we appeal to the upper bound just proved. To do that we verify that measures $\{\nu_n\}$ satisfy Assumption 12.13.

The logarithmic moment generating function of ν_n is

$$q_n(\theta) = \log \int e^{\langle \theta, x\rangle}\, \nu_n(dx) = p_n(r_n\sigma + \theta) - p_n(r_n \sigma).$$

For all $\theta \in \mathbb{R}^d$ the limits

$$q(\theta) = \lim_{n\to\infty} r_n^{-1} q_n(r_n \theta) = p(\sigma + \theta) - p(\sigma)$$

exist in $(-\infty, \infty]$, and $q(\theta)$ is finite for θ in an open neighborhood of 0. These conclusions follow because $\sigma \in (\operatorname{dom} p)^\circ$ implies that $\sigma + \theta \in (\operatorname{dom} p)^\circ$ for small enough θ. Assumption 12.13 has been verified for $\{\nu_n\}$.

The convex conjugate is

$$\begin{aligned} q^*(x) &= \sup_\theta\{\langle \theta, x\rangle - q(\theta)\} = \sup_\theta\{\langle \theta, x\rangle - p(\sigma+\theta)\} + p(\sigma) \\ &= p^*(x) - \langle \sigma, x\rangle + p(\sigma). \end{aligned}$$

It is a tight rate function by Step 1 of the proof. The upper bound gives

$$\varlimsup_{n\to\infty} r_n^{-1} \log \nu_n(B^c) \le -\inf_{x\in B^c} q^*(x).$$

To have the desired limit $\nu_n(B) \to 1$, we argue that $\inf_{B^c} q^* > 0$. If this infimum is finite, compact sublevel sets and lower semicontinuity imply the existence of $\bar{x} \in B^c$ such that $q^*(\bar{x}) = \inf_{B^c} q^*$. Since z is an exposed point of p^* with exposing hyperplane σ,

$$p^*(\bar{x}) > p^*(z) + \langle \sigma, \bar{x} - z\rangle \iff q^*(\bar{x}) > q^*(z).$$

This guarantees $q^*(\bar{x}) > 0$ and we have $\nu_n(B) \to 1$.

The remaining steps of the lower bound proof are familiar. We have

$$\begin{aligned} r_n^{-1} \log \mu_n(B) &= r_n^{-1} \log \int_B e^{p_n(r_n\sigma) - r_n\langle \sigma, y\rangle}\, \nu_n(dy) \\ &= r_n^{-1} p_n(r_n\sigma) - \langle \sigma, z\rangle + r_n^{-1} \log \int_B e^{r_n\langle \sigma, z-y\rangle}\, \nu_n(dy) \\ &\ge r_n^{-1} p_n(r_n\sigma) - \langle \sigma, z\rangle - |\sigma|\,\delta + r_n^{-1}\log \nu_n(B). \end{aligned}$$

From this,

$$\varliminf_{n\to\infty} r_n^{-1} \log \mu_n(G) \ge -\big[\langle \sigma, z\rangle - p(\sigma)\big] - |\sigma|\,\delta \ge -p^*(z) - |\sigma|\,\delta.$$

Let $\delta \searrow 0$. Take the supremum on the right over $z \in G \cap \mathcal{E}$. This completes the proof of part (ii).

Step 4. Full LDP of part (iii). The upper bound is already in part (i). To get the lower bound from part (ii) we need to show

$$\inf_{x\in G\cap\mathcal{E}} p^*(x) \le \inf_{x\in G} p^*(x).$$

We can assume that $G \cap (\operatorname{dom} p^*)$ is nonempty because otherwise the right-hand side above is infinite. Then $\operatorname{ri}(\operatorname{dom} p^*)$ is also nonempty, and by Theorem 12.5, $\operatorname{ri}(\operatorname{dom} p^*) \subset \mathcal{E}$. Now it suffices to show that

$$\inf_{x\in G\cap \operatorname{ri}(\operatorname{dom} p^*)} p^*(x) \le \inf_{x\in G\cap(\operatorname{dom} p^*)} p^*(x) \tag{12.7}$$

with the usual interpretation that the infimum of an empty set equals ∞. Let $x \in G \cap (\operatorname{dom} p^*)$ and $z \in \operatorname{ri}(\operatorname{dom} p^*)$. The convex combinations $(1-s)x + sz$

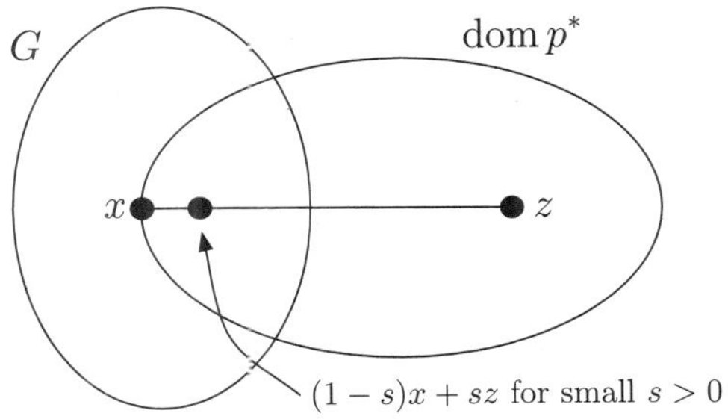

Figure 12.2

lie in ri(dom p^*) for all $s \in (0,1)$ by Lemma 12.7(b) and in G for small enough $s \in (0,1)$. See Figure 12.2. Consequently

$$\inf_{G\cap \mathrm{ri}(\mathrm{dom}\, p^*)} p^* \le \lim_{s\searrow 0} p^*((1-s)x+sz) \le \lim_{s\searrow 0}\big((1-s)p^*(x)+sp^*(z)\big)$$
$$\le p^*(x).$$

The infimum over $x \in G\cap(\mathrm{dom}\, p^*)$ proves (12.7). □

Exercise 12.15. A simple example of a lower semicontinuous convex function that satisfies essential smoothness except for the steepness assumption is

$$p(t) = \begin{cases} t^2/2, & t \le 1, \\ \infty, & t > 1. \end{cases}$$

(a) Find p^* and identify the set $\mathcal{E}$.

(b) Construct a sequence of probability measures $\{\mu_n\}$ with this limiting logarithmic moment generating function and such that the large deviation lower bound with rate p^* fails for some open sets.

Hint: Here is a possible course. First find a random variable Y such that $E(e^{tY}) < \infty$ if and only if $t \le 1$. (For example, consider a density function of the form $Ce^{-x}/(1+x)$, $x \ge 0$.) Let Z be a standard normal. Then

$$\xi_n = \begin{cases} n^{-1/2}Z & \text{with probability } 1-\delta_n, \\ n^{-1}Y & \text{with probability } \delta_n \end{cases}$$

for a suitable sequence $\delta_n \searrow 0$. Let μ_n be the distribution of ξ_n.

12.3. Large deviations for the current of particles

As an application of the Gärtner-Ellis theorem we derive an LDP for the current of particles executing independent random walks. Begin by placing one particle at each of the sites of the integer lattice $\mathbb{Z}$. Then let each particle independently execute a simple random walk on $\mathbb{Z}$, jumping with equal probability one step to the left and one step to the right.

To introduce notation, let X_m^k denote the position at time $m \in \mathbb{Z}_+$ of the particle that initially resided at position $k \in \mathbb{Z}$. Then $X_0^k = k$. We give a formula for the joint probability distribution of the walks. For $k \in \{-K, \dots, K\}$ let $(x_m^k)_{m=0}^N$ be an admissible random walk path of length N with initial position k; in other words $x_0^k = k$ and $x_m^k - x_{m-1}^k \in \{\pm 1\}$. Then, since the walks are independent,

$$\begin{aligned} &P\{X_m^k = x_m^k \ \forall k = -K, \dots, K \ \forall m = 0, \dots, N\} \\ &\quad = \prod_{k=-K}^{K} P\{X_m^k = x_m^k \ \forall k = 0, \dots, N\} = \prod_{k=-K}^{K} 2^{-N} = 2^{-N(2K+1)}. \end{aligned}$$

A quantity of interest is the net current J_n of particles that flow from right to left across the point $\frac{1}{2}$ on the real line, up to time n, defined by

$$J_n = \sum_{k=1}^{\infty} \mathbb{1}\{X_n^k \le 0\} - \sum_{k=-\infty}^{0} \mathbb{1}\{X_n^k > 0\}. \tag{12.8}$$

The sums are finite because by time n an individual particle can have traveled at most distance n.

Let $X_n = X_n^0$ denote the walk that starts from the origin. The marginal distribution of walk X_n^k is the same as that of $k + X_n$. Then

$$\begin{aligned} EJ_n &= \sum_{k=1}^{\infty} P(X_n^k \le 0) - \sum_{k=-\infty}^{0} P(X_n^k > 0) \\ &= \sum_{k=1}^{\infty} P(X_n \le -k) - \sum_{k=-\infty}^{0} P(X_n > -k) \\ &= \sum_{k=1}^{\infty} P(X_n \ge k) - \sum_{j=1}^{\infty} P(X_n \ge j) = 0 \end{aligned}$$

where on the last line we used the symmetry of a simple random walk and changed a summation index. Thus the average value of the current is zero. Our task is to find the correct scale on which large deviations take place, derive a rate function, and, if possible, understand how the system of particles behaves to produce an atypical current.

Definition (12.8) expresses J_n as a sum of independent Bernoulli random variables with $\pm$ signs. Thus it is not surprising that the Bernoulli rate function appears in the solution. We denote it and its convex dual by

$$H_p(x) = x \log \frac{x}{p} + (1-x) \log \frac{1-x}{1-p} \quad \text{and} \quad H_p^*(t) = \log(pe^t + 1 - p)$$

for $0 < p < 1$, $x \in [0,1]$, and $t \in \mathbb{R}$. Let $\Phi(x) = (2\pi)^{-1/2} \int_{-\infty}^{x} e^{-s^2/2}\, ds$ denote the standard normal cumulative distribution function.

The first result gives the limiting logarithmic moment generating function. Proofs follow after the theorems and discussion.

Theorem 12.16. *The limit*

$$p(\theta) = \lim_{n\to\infty} n^{-1/2} \log E(e^{\theta J_n}) \tag{12.9}$$

exists for all $\theta \in \mathbb{R}$ and is given by

$$p(\theta) = \int_0^\infty H^*_{\Phi(-x)}(\theta)\, dx + \int_{-\infty}^0 H^*_{\Phi(x)}(-\theta)\, dx. \tag{12.10}$$

That the integrals above converge for all θ can be checked by using the inequalities

$$\begin{aligned} &\log(1-x) \geq -x - x^2/(2\delta) \quad \text{for } 0 \leq x \leq 1-\delta < 1,\\ &\log(1+x) \leq x \quad \text{for } x \geq 0. \end{aligned} \tag{12.11}$$

Once we check that p is differentiable we know that it is essentially smooth. Then the Gärtner-Ellis theorem gives us the LDP for the distributions of $n^{-1/2} J_n$ with normalization $r_n = n^{1/2}$ and rate function p^*. Before formally stating the LDP we develop a more illuminating formula for the rate function.

Introduce the function

$$F_\theta(x) = \frac{e^{-\theta}\Phi(x)}{e^{-\theta}\Phi(x) + 1 - \Phi(x)}. \tag{12.12}$$

Differentiation (to be justified when we cover proofs below) gives

$$p'(\theta) = \int_0^\infty (1 - F_\theta(x))\, dx - \int_{-\infty}^0 F_\theta(x)\, dx. \tag{12.13}$$

The derivative p' is a strictly increasing bijection on $\mathbb{R}$, and hence for each $s \in \mathbb{R}$ there exists a unique $\alpha(s) \in \mathbb{R}$ such that

$$s = p'(\alpha(s)). \tag{12.14}$$

Define

$$I(s) = \int_{-\infty}^\infty H_{\Phi(x)}\big(F_{\alpha(s)}(x)\big)\, dx. \tag{12.15}$$

Once we show that $I = p^*$, the following theorem has been justified.

Theorem 12.17. *For closed sets $A \subset \mathbb{R}$ and open sets $G \subset \mathbb{R}$ we have the large deviation bounds*

$$\overline{\lim_{n\to\infty}}\, n^{-1/2} \log P\{n^{-1/2} J_n \in A\} \leq -\inf_{s\in A} I(s) \tag{12.16}$$

and

(12.17) $$\underline{\lim}_{n\to\infty} n^{-1/2} \log P\{n^{-1/2} J_n \in G\} \geq -\inf_{s\in G} I(s).$$

Why is the scale $n^{1/2}$ natural? By the central limit theorem, the typical fluctuations of a simple random walk are of order $n^{1/2}$. Therefore the number of particles that have probability at least $\delta > 0$ to cross the origin is of order $n^{1/2}$. The large deviations of $n^{1/2}$ Bernoulli variables occur on the scale $n^{1/2}$.

The rate function can be explained similarly. By the central limit theorem, the probability that a particle initially at $x\sqrt{n}$ crosses the origin by time n is approximately $\Phi(x)$ if $x < 0$ and $1 - \Phi(x)$ if $x > 0$. To produce an average current $s\sqrt{n}$, the system alters these crossing probabilities to $F_{\alpha(s)}(x)$ for $x < 0$ and $1 - F_{\alpha(s)}(x)$ for $x > 0$. Equations (12.13) and (12.14) express in macroscopic variables that the parameter $\alpha(s)$ is chosen so that the mean scaled current is s. The rate function $I(s)$ is a sum (which becomes an integral in macroscopic variables) of the entropies of the new crossing probabilities relative to the old crossing probabilities. The unique zero of I is at $I(0) = 0$ because $\alpha(0) = 0$ and $H_{\Phi(x)}(F_\theta(x)) = 0$ if and only if $\theta = 0$.

The reader may be surprised that the large deviations of the current are produced by diffusive fluctuations (described by the central limit theorem) of the individual walks, and not by large deviations of the walks. This point is made precise in Exercise 12.20.

In the remainder of this section we sketch the proof of Theorem 12.17. It is the kind of probability proof that inevitably requires some estimates. Such work is best done with pencil and paper and not read from a book. Hence we leave some steps as guided exercises.

In order to apply the Gärtner-Ellis theorem we must compute the limiting logarithmic moment generating function. Let $\theta \in \mathbb{R}$. Begin by using the independence of the particles:

$$\begin{aligned} E(e^{\theta J_n}) &= E\Big[\prod_{k=1}^{\infty} e^{\theta \mathbb{1}\{X_n^k \leq 0\}} \cdot \prod_{k=-\infty}^{0} e^{-\theta \mathbb{1}\{X_n^k > 0\}}\Big] \\ &= \prod_{k=1}^{\infty}\big(1 + (e^{\theta} - 1)P(X_n \leq -k)\big) \cdot \prod_{k=-\infty}^{0}\big(1 + (e^{-\theta} - 1)P(X_n > -k)\big). \end{aligned}$$

All but finitely many factors in the products are equal to 1.

The next task is to take the limit of

$$
\begin{aligned}
n^{-1/2}\log E(e^{\theta J_n}) &= n^{-1/2}\sum_{k=1}^{\infty}\log\Bigl(1+(e^{\theta}-1)P\Bigl\{\frac{X_n}{\sqrt{n}}\le\frac{-k}{\sqrt{n}}\Bigr\}\Bigr) \\
&\quad + n^{-1/2}\sum_{k=-\infty}^{0}\log\Bigl(1+(e^{-\theta}-1)P\Bigl\{\frac{X_n}{\sqrt{n}}>\frac{-k}{\sqrt{n}}\Bigr\}\Bigr).
\end{aligned}
\tag{12.18}
$$

The normalization by $\sqrt{n}$ was inserted in the probabilities to indicate that their convergence will come from the central limit theorem (Section B.5). The sums themselves converge to integrals. To help this argument it is useful to truncate the sums.

Exercise 12.18. Show that, given $\varepsilon > 0$, we can choose a constant $0 < a < \infty$ so that, for large enough n,

$$
\begin{aligned}
&n^{-1/2}\sum_{k\ge a\sqrt{n}}\Bigl|\log\Bigl(1+(e^{\theta}-1)P\Bigl\{\frac{X_n}{\sqrt{n}}\le\frac{-k}{\sqrt{n}}\Bigr\}\Bigr)\Bigr| \\
&\quad + n^{-1/2}\sum_{k\le -a\sqrt{n}}\Bigl|\log\Bigl(1+(e^{-\theta}-1)P\Bigl\{\frac{X_n}{\sqrt{n}}>\frac{-k}{\sqrt{n}}\Bigr\}\Bigr)\Bigr| \le \varepsilon.
\end{aligned}
$$

Hint: Through the inequalities in (12.11) this boils down to estimating sums $\sum_{k\ge a\sqrt{n}} P(X_n \ge k)$. Use exponential estimates: for $t > 0$,

$$
P(X_n \ge k) \le e^{-tk}E(e^{tX_n}) = \exp\bigl[-tk + n\log\bigl(\tfrac{e^t+e^{-t}}{2}\bigr)\bigr],
$$

approximate the logarithm and pick suitable t-values for different ranges of k.

After discarding part of the first sum in (12.18), we take the limit

$$
\begin{aligned}
&n^{-1/2}\sum_{1\le k<a\sqrt{n}}\log\Bigl(1+(e^{\theta}-1)P\Bigl\{\frac{X_n}{\sqrt{n}}\le\frac{-k}{\sqrt{n}}\Bigr\}\Bigr) \\
&= \int_0^a \log\Bigl(1+(e^{\theta}-1)P\Bigl\{\frac{X_n}{\sqrt{n}}\le -\frac{\lceil x\sqrt{n}\,\rceil}{\sqrt{n}}\Bigr\}\Bigr)\,dx \;+\; O(n^{-1/2}) \\
&\underset{n\to\infty}{\longrightarrow} \int_0^a \log\bigl(1+(e^{\theta}-1)\Phi(-x)\bigr)\,dx.
\end{aligned}
$$

The $O(n^{-1/2})$ error above comes from switching from the sum to the integral. The integrand converges by the central limit theorem and the integral by the dominated convergence theorem because the integrands are bounded uniformly over n and x. Applying the same argument to the second sum in

(12.18) proves the existence of the limit $p(\theta)$ and gives the formula

$$p(\theta) = \int_0^\infty \log\bigl(1 + (e^\theta - 1)\Phi(-x)\bigr)\,dx \tag{12.19}$$
$$+ \int_{-\infty}^0 \log\bigl(1 + (e^{-\theta} - 1)\Phi(x)\bigr)\,dx.$$

The formula above is the same as (12.10). This proves Theorem 12.16.

To check the differentiability of p, apply the mean value theorem to $h^{-1}(p(\theta + h) - p(\theta))$ and let $h \to 0$. Here is the derivation for the first integral in (12.19), where $\tilde{h}(x)$ is a number between 0 and h:

$$\int_0^\infty \frac{1}{h}\bigl[\log\bigl(1 + (e^{\theta+h} - 1)\Phi(-x)\bigr) - \log\bigl(1 + (e^\theta - 1)\Phi(-x)\bigr)\bigr]\,dx$$
$$= \int_0^\infty \frac{e^{\theta+\tilde{h}(x)}\Phi(-x)}{1 + (e^{\theta+\tilde{h}(x)} - 1)\Phi(-x)}\,dx$$
$$\underset{h\to 0}{\longrightarrow} \int_0^\infty \frac{e^\theta \Phi(-x)}{1 + (e^\theta - 1)\Phi(-x)}\,dx = \int_0^\infty (1 - F_\theta(x))\,dx.$$

The limit can be justified by dominated convergence. Combining with a similar argument for the second integral in (12.19) gives formula (12.13) for $p'(\theta)$.

At this stage we know that the LDP holds with rate function $I = p^*$ and it remains to derive formula (12.15). We shall do it indirectly by appeal to the next convex analysis fact.

Exercise 12.19. Let f and g be differentiable convex functions on $\mathbb{R}$ such that $f'(g'(y)) = y$ for all $y \in \mathbb{R}$. Then there exists a constant c such that $g = f^* + c$.

Hint: Show that f' is a bijection on $\mathbb{R}$. Use the properties in Theorem 4.21 and Exercise 4.23.

Next derive the formulas

$$p''(\theta) = -\int_{-\infty}^\infty \frac{d}{d\theta} F_\theta(x)\,dx \quad \text{and} \quad \alpha'(s) = \frac{1}{p''(\alpha(s))}.$$

Notice that $H'_{\Phi(x)}\bigl(F_\theta(x)\bigr) = -\theta$ and then after some simplification $I'(s) = \alpha(s)$. By (12.14) α is strictly increasing and hence I is convex. We have the identity $p'(I'(s)) = s$ and the exercise above implies $I(s) = p^*(s) + c$. From $I(0) = 0 = p^*(0)$ we conclude $c = 0$. $I = p^*$ has been verified and Theorem 12.17 thereby proved.

Exercise 12.20. Let $Y_n = \sum_{k \ge b(n)} \mathbb{1}\{X_n^k \le 0\}$ denote the contribution to the current from walks that begin at points $k \ge b(n)$. Show that if

$b(n)n^{-1/2} \to \infty$, then for every $\varepsilon > 0$

$$\lim_{n\to\infty} n^{-1/2} \log P\{n^{-1/2} Y_n \geq \varepsilon\} = -\infty.$$

In other words, particles that lie beyond a distance of order $n^{1/2}$ from the origin do not contribute to the deviations of the current on the scale $n^{1/2}$.

Hint: Following the proof in the text show that $n^{-1/2} \log E(e^{\theta Y_n}) \to 0$ and then apply an exponential Chebyshev inequality.

Literature notes. The convex analysis of Section 12.1 is developed more fully in Rockafellar [**68**]. For example, there is a precise characterization for the conjugate of an essentially smooth convex function. The argument in the proof of Theorem 12.12 is adapted from Lemma A.5 of [**15**], which credits the idea to A. Ioffe. The example in Section 12.3 comes from [**56**].

Chapter 13

Large deviations for Markov chains

This chapter extends Sanov's theorem to Markov chains. We restrict the treatment to a countable state space where fairly general elegant results are accessible, especially if we restrict ourselves even further to a finite state space. We begin with a general theorem about relative entropy under kernels. This result will be useful both for Markov chains in this chapter and for nonstationary independent random variables in Chapter 15.

13.1. Relative entropy for kernels

Let $\mathcal{X}$ and $\mathcal{Y}$ be Polish spaces, let $\kappa \in \mathcal{M}_1(\mathcal{Y})$, and let $y \mapsto \rho(y)$ be a measurable map (in other words, a stochastic kernel) from $\mathcal{Y}$ into $\mathcal{M}_1(\mathcal{X})$. We write $\rho(y, A)$ for the value that measure $\rho(y)$ assigns to the Borel set $A \subset \mathcal{X}$. Define two probability measures on $\mathcal{X} \times \mathcal{Y}$ by $r(y) = \rho(y) \otimes \delta_y$ and $Q(dx, dy) = \rho(y, dx)\, \kappa(dy)$. The expectation of a bounded measurable function f on $\mathcal{X} \times \mathcal{Y}$ under the measure $r(y)$ is denoted by

$$r(y, f) = E^{r(y)}[f] = \int_{\mathcal{X}} f(x, y)\, \rho(y, dx).$$

Define a functional Λ for $f \in \mathscr{C}_b(\mathcal{X} \times \mathcal{Y})$ by

$$\Lambda(f) = \int_{\mathcal{Y}} \log r(y, e^f)\, \kappa(dy) = \int_{\mathcal{Y}} \log\Big(\int_{\mathcal{X}} e^{f(x,y)}\, \rho(y, dx)\Big)\, \kappa(dy). \tag{13.1}$$

For $\nu \in \mathcal{M}_1(\mathcal{X} \times \mathcal{Y})$ and $\alpha \in \mathcal{M}_1(\mathcal{X})$ define

$$I(\nu) = \sup_{f \in \mathscr{C}_b(\mathcal{X} \times \mathcal{Y})} \{E^\nu[f] - \Lambda(f)\} \tag{13.2}$$

and

$$I_{\mathcal{X}}(\alpha) = \sup_{g \in \mathscr{C}_b(\mathcal{X})} \{E^{\alpha}[g] - \Lambda(g)\}. \tag{13.3}$$

The defining formula (13.1) of Λ works just as well for functions that depend only on the variable x, and so Λ can be regarded also as a function on $\mathscr{C}_b(\mathcal{X})$. This is the sense in which Λ appears in definition (13.3). Let $\nu_{\mathcal{X}}$ and $\nu_{\mathcal{Y}}$ denote the $\mathcal{X}$- and $\mathcal{Y}$-marginals of a measure $\nu \in \mathcal{M}_1(\mathcal{X} \times \mathcal{Y})$. That is, $\nu_{\mathcal{X}}(A) = \nu(A \times \mathcal{Y})$ for $A \in \mathscr{B}_{\mathcal{X}}$ and similarly for $\nu_{\mathcal{Y}}$.

Theorem 13.1. *We have the following identities for $\nu \in \mathcal{M}_1(\mathcal{X} \times \mathcal{Y})$ and $\alpha \in \mathcal{M}_1(\mathcal{X})$:*

$$I(\nu) = \begin{cases} H(\nu \,|\, Q) & \text{if } \nu_{\mathcal{Y}} = \kappa, \\ \infty & \text{if } \nu_{\mathcal{Y}} \neq \kappa, \end{cases} \tag{13.4}$$

and

$$I_{\mathcal{X}}(\alpha) = \inf\{I(\mu) : \ \mu \in \mathcal{M}_1(\mathcal{X} \times \mathcal{Y}) \text{ and } \mu_{\mathcal{X}} = \alpha\}. \tag{13.5}$$

Before embarking on the proof of the theorem let us apply the result to a Markov kernel. Let $\mathscr{C}_b^+(\mathcal{S})$ be the space of functions $f \in \mathscr{C}_b(\mathcal{S})$ that are strictly positive and bounded away from 0. (This means that there exists a constant $0 < c < \infty$ such that $c^{-1} \le f(x) \le c$.) Let p be a stochastic kernel on $\mathcal{S}$, in other words a Markov transition probability kernel. As before in Section 7.3, for a function f and a measure α, define a function pf and a measure αp by $pf(x) = \int f(y)p(x, dy)$ and $\alpha p(A) = \int p(x, A)\, \alpha(dx)$. Define also a joint measure $\alpha \times p$ on $\mathcal{S} \times \mathcal{S}$ as

$$\alpha \times p(A \times B) = \int_{\mathcal{S}} \mathbb{1}_A(x) p(x, B)\, \alpha(dx).$$

In the next section we generalize Sanov's theorem to Markov chains with the rate function given by the *Donsker-Varadhan entropy*

$$J(\alpha) = \sup_{f \in \mathscr{C}_b^+(\mathcal{S})} \int_{\mathcal{S}} \log \frac{f}{pf}\, d\alpha, \qquad \alpha \in \mathcal{M}_1(\mathcal{S}). \tag{13.6}$$

The next theorem gives this function an entropy representation. Convexity and lower semicontinuity of J are immediate from the definition.

Theorem 13.2. *Let $\mathcal{S}$ be a Polish space and let p be a stochastic kernel on $\mathcal{S}$. Then for $\alpha \in \mathcal{M}_1(\mathcal{S})$*

$$J(\alpha) = \inf_{q : \alpha q = \alpha} H(\alpha \times q \,|\, \alpha \times p). \tag{13.7}$$

The infimum is over stochastic kernels q on $\mathcal{S}$ that fix α. $J(\alpha) = 0$ if and only if $\alpha = \alpha p$.

Proof. Apply Theorem 13.1 with the following choices: $\mathcal{X} = \mathcal{Y} = \mathcal{S}$, $\rho = p$, and $\kappa = \alpha$. Then (13.5) says

$$J(\alpha) = \inf_{\nu} H(\nu \,|\, Q)$$

with infimum over $\nu \in \mathcal{M}_1(\mathcal{S} \times \mathcal{S})$ with both marginals equal to α, and $Q = \alpha \times p$. (Think of the space $\mathcal{S} \times \mathcal{S}$ as $\mathcal{Y} \times \mathcal{X}$ rather than $\mathcal{X} \times \mathcal{Y}$.) If $\nu_{\mathcal{Y}} = \nu_{\mathcal{X}} = \alpha$, the conditional probability distribution of x under ν, given y, defines a kernel q that fixes α. Conversely, any such kernel defines a ν.

If $\alpha = \alpha p$, we can take $\nu = \alpha \times p = Q$ and get $H(\nu \,|\, Q) = 0$. Conversely, suppose $J(\alpha) = 0$. Since the sublevel sets of relative entropy are compact (Proposition 5.7) and the space of probability measures with marginals α is closed, there exists a minimizer ν. Then $H(\nu \,|\, Q) = 0$ implies $\nu = Q$. Hence both marginals of Q are α, which is the same as saying $\alpha = \alpha p$. This completes the proof of Theorem 13.2. □

The remainder of this section is devoted to the proof of Theorem 13.1.

Proof of Theorem 13.1. We prove first (13.4). By Jensen's inequality and variational formula (5.2),

$$I(\nu) \ge \sup_{f \in \mathscr{C}_b(\mathcal{X}\times\mathcal{Y})} \{E^{\nu}[f] - \log E^{Q}[e^f]\} = H(\nu \,|\, Q). \tag{13.8}$$

Write ν^y for the conditional probability measure of ν, given $y \in \mathcal{Y}$. (See Theorem B.5.) If $\nu_{\mathcal{Y}} = \kappa$,

$$\begin{aligned} I(\nu) &= \sup_{f \in \mathscr{C}_b(\mathcal{X}\times\mathcal{Y})} \int_{\mathcal{Y}} \big(E^{\nu^y}[f] - \log r(y, e^f) \big)\, \kappa(dy) \\ &\le \int_{\mathcal{Y}} H(\nu^y \,|\, r(y))\, \kappa(dy) = H(\nu \,|\, Q). \end{aligned}$$

The last equality used the conditional entropy formula (Exercise 5.13). Taking $f \in \mathscr{C}_b(\mathcal{Y})$ in (13.2) shows that $I(\nu) = \infty$ if $\nu_{\mathcal{Y}} \neq \kappa$. We have proved (13.4).

We prove (13.5) first for the case of compact $\mathcal{X}$ and $\mathcal{Y}$. By dominated convergence, Λ is a strongly continuous function on the Banach space $\mathscr{C}(\mathcal{X} \times \mathcal{Y})$, in other words, continuous when $\mathscr{C}(\mathcal{X} \times \mathcal{Y})$ is equipped with the supremum norm. By Hölder's inequality it is a convex function. It follows that Λ is a weakly lower semicontinuous function on $\mathscr{C}(\mathcal{X} \times \mathcal{Y})$. Here is the argument for this step. By continuity and convexity, the set $U = \{f : \Lambda(f) \le s\}$ is a strongly closed, convex set. By the Hahn-Banach separation theorem a function $g \in U^c$ can be strictly separated from U with a linear functional. Thus each $g \in U^c$ lies in the weak interior of U^c, which makes U^c weakly open, and thereby U is weakly closed. That sets of type U are weakly closed is exactly the definition of weak lower semicontinuity.

As a convex, weakly lower semicontinuous function, Λ is equal to its convex biconjugate. Let $g \in \mathscr{C}(\mathcal{X})$. Below we can think of g also as a function on $\mathcal{X} \times \mathcal{Y}$, by composing it with the projection $(x, y) \mapsto x$:

$$\begin{aligned}\Lambda(g) &= \Lambda^{**}(g) = I^*(g) \\ &= \sup\{E^{\nu_{\mathcal{X}}}[g] - I(\nu) : \nu \in \mathcal{M}_1(\mathcal{X} \times \mathcal{Y})\} \\ &= \sup\big\{E^{\alpha}[g] - \inf\{I(\nu) : \nu_{\mathcal{X}} = \alpha\} : \alpha \in \mathcal{M}_1(\mathcal{X})\big\}.\end{aligned}$$

The third equality above hides a couple of steps. The dual of $\mathscr{C}(\mathcal{X} \times \mathcal{Y})$ is the space $\mathcal{M}(\mathcal{X} \times \mathcal{Y})$ of finite signed Borel measures on $\mathcal{X} \times \mathcal{Y}$. (This is one of the Riesz representation theorems; see Section 7.3 in [**34**].) Definition (13.2) defines I as a convex and lower semicontinuous function on the space $\mathcal{M}(\mathcal{X} \times \mathcal{Y})$. By the proof of Theorem 5.4, $I(\nu) = \infty$ unless ν is a probability measure. Thus the supremum in the definition of I^* can be restricted to probability measures ν as done above.

The function $\alpha \mapsto \inf\{I(\nu) : \nu_{\mathcal{X}} = \alpha\}$ is again convex and lower semicontinuous in the $\mathscr{C}(\mathcal{X})$-generated weak topology of $\mathcal{M}(\mathcal{X})$. Since $\mathcal{X}$ and $\mathcal{Y}$ are assumed compact I is automatically tight, and so lower semicontinuity follows from part (b) of the contraction principle (page 35). Thus taking convex conjugates once more gives

$$I_{\mathcal{X}}(\alpha) = \Lambda^*(\alpha) = \inf\{I(\nu) : \nu_{\mathcal{X}} = \alpha\}$$

and completes the proof of (13.5) for compact $\mathcal{X}$ and $\mathcal{Y}$.

To prove (13.5) for Polish $\mathcal{X}$ and $\mathcal{Y}$, we begin by observing that we can assume that $\mathcal{X}$ and $\mathcal{Y}$ are dense Borel subsets of compact metric spaces $\bar{\mathcal{X}}$ and $\bar{\mathcal{Y}}$. Here is the argument. By separability $\mathcal{X}$ has a totally bounded metric $\bar{d}$, and then the completion $(\bar{\mathcal{X}}, \bar{d})$ of $(\mathcal{X}, \bar{d})$ is compact (details can be found in Theorem 2.8.2 in [**26**]). As a Polish space, $\mathcal{X}$ also has a complete metric; in other words, $(\mathcal{X}, \bar{d})$ is a topologically complete metric space. By Theorem 2.5.4 in [**26**] a metric space is topologically complete if and only if it is a countable intersection of open sets (that is, a G_δ set) in its completion.

Definition (13.1) of Λ works just as well for functions f on $\bar{\mathcal{X}} \times \bar{\mathcal{Y}}$ because we can think of ρ^y and κ as probability measures on $\bar{\mathcal{X}}$ and $\bar{\mathcal{Y}}$ that happen to satisfy $\kappa(\mathcal{Y}) = 1$ and $\rho(y, \mathcal{X}) = 1$ κ-almost surely. Let $\bar{I}$ and $\bar{I}_{\mathcal{X}}$ denote the functions defined by (13.2) and (13.3) on $\mathcal{M}_1(\bar{\mathcal{X}} \times \bar{\mathcal{Y}})$ and $\mathcal{M}_1(\bar{\mathcal{X}})$, respectively, with the suprema now over $\mathscr{C}(\bar{\mathcal{X}} \times \bar{\mathcal{Y}})$ and $\mathscr{C}(\bar{\mathcal{X}})$. A probability measure α on $\mathcal{X}$ can be thought of as a measure on $\bar{\mathcal{X}}$, and hence the proof for compact spaces gives

$$\bar{I}_{\mathcal{X}}(\alpha) = \inf\{\bar{I}(\bar{\nu}) : \bar{\nu} \in \mathcal{M}_1(\bar{\mathcal{X}} \times \bar{\mathcal{Y}}), \bar{\nu}_{\bar{\mathcal{X}}} = \alpha\}. \tag{13.9}$$

By Lemma A.11 the supremum in definition (13.3) of $I_{\mathcal{X}}(\alpha)$ can just as well be taken over $g \in \mathscr{U}_{b,\bar{d}}(\mathcal{X})$, where $\mathscr{U}_{b,\bar{d}}(\mathcal{X})$ is the space of bounded uniformly continuous functions on $(\mathcal{X}, \bar{d})$. Functions in $\mathscr{U}_{b,\bar{d}}(\mathcal{X})$ and $\mathscr{C}(\bar{\mathcal{X}})$ correspond to each other bijectively via restriction and unique extension; hence $\bar{I}_{\mathcal{X}}(\alpha) = I_{\mathcal{X}}(\alpha)$. Since $\mathcal{X} \times \mathcal{Y}$ is dense in $\bar{\mathcal{X}} \times \bar{\mathcal{Y}}$ this same argument gives $\bar{I}(\nu) = I(\nu)$ for probability measures ν on $\mathcal{X} \times \mathcal{Y}$. Since $Q(\mathcal{X} \times \mathcal{Y}) = 1$ and $H(\bar{\nu} \,|\, Q)$ is finite only if $\bar{\nu} \ll Q$, (13.4) shows that $\bar{I}(\bar{\nu}) = \infty$ unless $\bar{\nu}(\mathcal{X} \times \mathcal{Y}) = 1$ too. These facts combine to imply that for $\alpha \in \mathcal{M}_1(\mathcal{X})$, (13.9) is the same as (13.5). □

Exercise 13.3. Use (13.7) to show that J is convex.

Hint: Check that if $\mu = t\mu_1 + (1-t)\mu_2$ and if $\mu_i q_i = \mu_i$ and $g_i = \frac{d\mu_i}{d\mu}$ for $i = 1, 2$, then $q(x, dy) = tg_1(x)q_1(x, dy) + (1-t)g_2(x)q_2(x, dy)$ fixes μ.

Exercise 13.4. Let $\mathcal{S}$ be compact and let p be a stochastic kernel on $\mathcal{S}$. For $f, h \in \mathscr{C}(\mathcal{S})$ define

$$K_h(f) = \sup_{x \in \mathcal{S}} \log \int e^{f(x) - h(x) + h(y)}\, p(x, dy) \tag{13.10}$$

and then

$$K(f) = \inf_{h \in \mathscr{C}(\mathcal{S})} K_h(f). \tag{13.11}$$

With J as defined in (13.6), show that $K = J^*$.

Hint: Use the minimax theorem.

The alternative version $J = K^*$ of the rate function has been useful in random medium models. See the literature notes at the end of Chapter 16.

13.2. Countable Markov chains

Let $\mathcal{S}$ be a countable state space and p a Markov transition kernel (stochastic kernel) on $\mathcal{S}$. On a countable space, p is given by a stochastic matrix $p = \{p(x, y)\}_{x, y \in \mathcal{S}}$. $\mathcal{S}$ endowed with the discrete topology is Polish and all functions $f : \mathcal{S} \to \mathbb{R}$ are continuous.

We use the canonical construction to define the $\mathcal{S}$-valued Markov chain, with path space $\Omega = \mathcal{S}^{\mathbb{Z}_+}$ and coordinate random variables $(X_n)_{n \geq 0}$, as was done for the i.i.d. process on page 74. The distribution of the Markov chain $(X_n)_{n \geq 0}$ with kernel p and initial state $x \in \mathcal{S}$ is the probability measure P_x on Ω determined by the identities

$$P_x\{X_0 = x_0, X_1 = x_1, \ldots, X_n = x_n\} = \mathbb{1}\{x_0 = x\} \prod_{i=0}^{n-1} p(x_i, x_{i+1})$$

for $n \in \mathbb{Z}_+$ and states $x_0, \ldots, x_n \in \mathcal{S}$. E_x denotes expectation under P_x. Assume p irreducible; that is, for any $x, y \in \mathcal{S}$ there exists $n \geq 0$ such that

$P_x\{X_n = y\} > 0$. Let L_n denote the empirical measure:

$$L_n = n^{-1} \sum_{k=0}^{n-1} \delta_{X_k}.$$

The average of a function is denoted by $L_n(f) = E^{L_n}[f] = n^{-1} \sum_{k=0}^{n-1} f(X_k)$. Recall the Donsker-Varadhan entropy introduced in (13.6):

$$J(\alpha) = \sup_{f \in \mathscr{C}_b^+(\mathcal{S})} \int_{\mathcal{S}} \log \frac{f}{pf}\, d\alpha, \qquad \alpha \in \mathcal{M}_1(\mathcal{S}). \tag{13.12}$$

Theorem 13.5. *Assume the transition matrix p is irreducible. Then the weak LDP holds for the distribution of L_n under P_x for all $x \in \mathcal{S}$, with normalization n and rate function J given by (13.12). More precisely, for compact $K \subset \mathcal{M}_1(\mathcal{S})$*

$$\overline{\lim_{n\to\infty}} \frac{1}{n} \log P_x\{L_n \in K\} \le - \inf_{\nu \in K} J(\nu) \tag{13.13}$$

and for open $G \subset \mathcal{M}_1(\mathcal{S})$

$$\underline{\lim_{n\to\infty}} \frac{1}{n} \log P_x\{L_n \in G\} \ge - \inf_{\nu \in G} J(\nu) \quad \forall x \in \mathcal{S}. \tag{13.14}$$

Proof. We begin with the quick upper bound proof, which uses no assumptions at all and gives bounds uniformly over the initial state x.

Upper bound. Given $\mu \in \mathcal{M}_1(\mathcal{S})$ and $c < J(\mu)$ find $f \in \mathscr{C}_b^+(\mathcal{S})$ such that $h = \log(f/pf)$ satisfies

$$\int h\, d\mu \ge c.$$

Fix $\varepsilon > 0$ and let

$$B_\mu = \{\nu \in \mathcal{M}_1(\mathcal{S}) : |E^\nu[h] - E^\mu[h]\,| < \varepsilon\}.$$

B_μ is a weak neighborhood of μ. By repeated application of the Markov property

$$\begin{aligned}
&E_x\Big[f(X_n)\, e^{\sum_{k=0}^{n-1} h(X_k)}\Big] = E_x\Big[pf(X_{n-1})\, e^{\sum_{k=0}^{n-1} h(X_k)}\Big] \\
&= E_x\Big[\frac{f(X_{n-1})}{pf(X_{n-1})} \cdot pf(X_{n-1})\, e^{\sum_{k=0}^{n-2} h(X_k)}\Big] = E_x\Big[f(X_{n-1})\, e^{\sum_{k=0}^{n-2} h(X_k)}\Big] \\
&= \cdots = f(x).
\end{aligned}$$

Next,

$$\begin{aligned}
\frac{f(x)}{\inf f} &\ge E_x\Big[e^{\sum_{k=0}^{n-1} h(X_k)}\Big] \ge E_x\big[e^{nL_n(h)} \mathbb{1}\{L_n \in B_\mu\}\big] \\
&\ge P_x\{L_n \in B_\mu\}\, e^{nc - n\varepsilon}
\end{aligned}$$

where the last inequality is true because

$$L_n \in B_\mu \text{ implies } L_n(h) \ge E^\mu[h] - \varepsilon \ge c - \varepsilon.$$

Since f is bounded and bounded away from 0, we get an intermediate upper bound for the neighborhood B_μ:

$$\overline{\lim_{n\to\infty}} \frac{1}{n} \log \sup_{x\in\mathcal{S}} P_x\{L_n \in B_\mu\} \le -c + \varepsilon.$$

Given a compact set K, let $c < \inf_{\nu\in K} J(\nu)$. For each $\mu \in K$ find h as above and define B_μ. By compactness there exist $\mu_1, \dots, \mu_m \in K$ so that $K \subset B_{\mu_1} \cup \cdots \cup B_{\mu_m}$. Then

$$\begin{aligned}\overline{\lim_{n\to\infty}} \frac{1}{n} \log \sup_{x\in\mathcal{S}} P_x\{L_n \in K\} &\le \max_{1\le i\le m} \overline{\lim_{n\to\infty}} \frac{1}{n} \log \sup_{x\in\mathcal{S}} P_x\{L_n \in B_{\mu_i}\} \\ &\le -c + \varepsilon.\end{aligned}$$

Let $\varepsilon \searrow 0$ and $c \nearrow \inf_{\nu\in K} J(\nu)$. The upper bound for compact sets is proved.

Lower bound: Change of measure argument. Let G be an open subset of $\mathcal{M}_1(\mathcal{S})$ and let $\mu \in G$ be such that $J(\mu) < \infty$. By (13.7) we can pick a stochastic matrix q that fixes μ (that is, $\mu q = \mu$) and satisfies $H(\mu \times q \,|\, \mu \times p) < \infty$. The finite entropy implies that if $\mu(x) > 0$ and $p(x, y) = 0$, then $q(x, y) = 0$.

For $x \in \mathcal{S}$ let Q_x be the distribution of the Markov chain with transition kernel q and initial state x. Let $P_x^{(n)}$ and $Q_x^{(n)}$ denote the distributions of $(X_1, \dots, X_n)$ under P_x and Q_x. By the invariance and the finite entropy $Q_x^{(n)} \ll P_x^{(n)}$ for μ-a.e. x (Exercise 13.8) with Radon-Nikodym derivative

$$f_n(\omega) = \frac{dQ_x^{(n)}}{dP_x^{(n)}}(\omega) = \prod_{k=0}^{n-1} \frac{q(X_k, X_{k+1})}{p(X_k, X_{k+1})}.$$

Next we perform the familiar change of measure argument but with an extra twist. We include a condition $X_n \in V$ for a set $V \subset \mathcal{S}$ that will help us in the sequel. Assume $Q_x\{L_n \in G, X_n \in V\} > 0$. Then

$$
\begin{aligned}
&n^{-1}\log P_x\{L_n \in G, X_n \in V\}\\
&\quad \ge n^{-1}\log P_x\{L_n \in G, X_n \in V, f_n > 0\}\\
&\quad = n^{-1}\log E^{Q_x}\big[\mathbb{1}\{L_n \in G, X_n \in V\} f_n^{-1}\big]\\
&\quad = n^{-1}\log \frac{E^{Q_x}\big[\mathbb{1}\{L_n \in G, X_n \in V\} f_n^{-1}\big]}{Q_x\{L_n \in G, X_n \in V\}}\\
&\qquad + n^{-1}\log Q_x\{L_n \in G, X_n \in V\}\\
&\quad \ge -\frac{1}{Q_x\{L_n \in G, X_n \in V\}}\cdot n^{-1}E^{Q_x}\big[\mathbb{1}\{L_n \in G, X_n \in V\}\log f_n\big]\\
&\qquad + n^{-1}\log Q_x\{L_n \in G, X_n \in V\}.
\end{aligned}
$$

Here is an intermediate bound:

$$
\begin{aligned}
&E^{Q_x}\big[\mathbb{1}\{L_n \in G, X_n \in V\}\log f_n\big]\\
&\quad = E^{Q_x}\big[\log f_n\big] - E^{P_x}\big[(1-\mathbb{1}\{L_n \in G, X_n \in V\}) f_n \log f_n\big]\\
&\quad \le E^{Q_x}\Big[\sum_{k=0}^{n-1}\log\frac{q(X_k, X_{k+1})}{p(X_k, X_{k+1})}\Big] + \frac{1}{e}.
\end{aligned}
$$

Substitute this back up:

$$
\begin{aligned}
&n^{-1}\log P_x\{L_n \in G, X_n \in V\}\\
&\quad \ge -\frac{1}{Q_x\{L_n \in G, X_n \in V\}}\cdot n^{-1}E^{Q_x}\Big[\sum_{k=0}^{n-1}\log\frac{q(X_k, X_{k+1})}{p(X_k, X_{k+1})}\Big]\\
&\qquad -\frac{1}{neQ_x\{L_n \in G, X_n \in V\}} + n^{-1}\log Q_x\{L_n \in G, X_n \in V\}.
\end{aligned}
\tag{13.15}
$$

As in our earlier uses of the change of measure argument, we need to establish that $Q_x\{L_n \in G, X_n \in V\}$ can be taken close to 1. For that purpose, let us recall facts about countable state space Markov chains.

Let $\{C_i : i \ge 1\}$ be the disjoint positive recurrent closed communicating classes of the q-Markov chain. (At least one exists because q has an invariant probability distribution μ.) On each C_i there is a unique invariant distribution π_i for q. Furthermore, $\pi_i(x) > 0$ for each $x \in C_i$ and on C_i the q-chain is ergodic. Lastly, μ is supported on $\bigcup C_i$ and hence there exist constants $\alpha_i \in [0,1]$ such that $\sum_i \alpha_i = 1$ and $\mu = \sum_i \alpha_i \pi_i$.

To achieve the lower bound we force L_n to approximate μ via this strategy: first $L_{\lfloor \alpha_1 n\rfloor} \approx \pi_1$, then $L_{\lfloor \alpha_2 n\rfloor}\circ\theta^{\lfloor \alpha_1 n\rfloor} \approx \pi_2$, and so on. We make this idea rigorous in four steps.

Lower bound: Step 1. Consider first the case where μ is supported on a single positive recurrent communicating class C of q, G is an open neighborhood of μ, and $x \in C$. In this case, the law of large numbers for Markov chains, together with the justification for (5.5), gives $L_n \to \mu$ Q_x-almost surely, and thereby

$$Q_x\{L_n \in G\} \to 1.$$

By the convergence theorem for irreducible Markov chains [**27**, Theorem 6.7.2], if d is the period of the Markov chain on C with transition kernel q, then

$$\lim_{m\to\infty} d^{-1} \sum_{k=0}^{d-1} Q_x\{X_{md+k} \in V\} = \mu(V).$$

This implies that

$$\varliminf_{n\to\infty} Q_x\{X_n \in V\} \ge 1 - d(1 - \mu(V)).$$

Consequently,

$$\begin{aligned}\varliminf_{n\to\infty} Q_x\{L_n \in G, X_n \in V\} &\ge \varliminf_{n\to\infty} \big(Q_x\{X_n \in V\} - Q_x\{L_n \notin G\}\big)\\ &\ge 1 - d(1 - \mu(V)).\end{aligned}$$

Also, by the ergodicity of the q-Markov chain

$$\begin{aligned}&\frac{1}{n} E^{Q_x}\Big[\sum_{k=0}^{n-1} \log \frac{dq(X_k, X_{k+1})}{dp(X_k, X_{k+1})}\Big]\\ &\qquad \longrightarrow \sum_{x,y} \mu(x) q(x,y) \log \frac{q(x,y)}{p(x,y)} = H(\mu \times q \,|\, \mu \times p).\end{aligned}$$

Provided $\mu(V) > 1 - d^{-1}$, the bounds above substituted back up in (13.15) combine to give for μ-a.e. x

$$\varliminf_{n\to\infty} \frac{1}{n} \log P_x\{L_n \in G, X_n \in V\} \ge -\frac{H(\mu \times q \,|\, \mu \times p)}{1 - d(1 - \mu(V))}. \tag{13.16}$$

Lower bound: Step 2. Consider next the case $\mu = \alpha\pi_1 + (1-\alpha)\pi_2$ for some $\alpha \in (0,1]$ where π_i is the unique invariant probability measure on positive recurrent class C_i of q, and C_1 and C_2 are two distinct classes. We can choose open neighborhoods G_i of π_i such that if $\mu_i' \in G_i$, $i = 1, 2$, then $\alpha\mu_1' + (1-\alpha)\mu_2' \in G$.

Let V_i be a finite subset of C_i. By the irreducibility of p, for any $u \in V_1$ and $v \in V_2$ there exists $j = j(u,v) \in \mathbb{N}$ such that $P_u\{X_j \in v\} > 0$. Let

$$m = \max\{j(u,v) : u \in V_1, v \in V_2\}$$

and

$$\delta = \min\{P_u(X_{j(u,v)} = v) : u \in V_1, v \in V_2\}.$$

Let $n_1 = \lfloor \alpha n \rfloor$ and $n_2 = n - n_1$. Given an integer $1 \le j \le m$ write

$$L_n = \frac{n_1 - j}{n} L_{n_1 - j} + \frac{n_2}{n} L_{n_2} \circ \theta^{n_1} + \frac{1}{n} \sum_{i=n_1-j}^{n_1-1} \delta_{X_i}.$$

Shrink the neighborhoods G_i of π_i a little further. Then, since m is fixed, for all large enough n, $L_{n_1-j} \in G_1$ and $L_{n_2} \circ \theta^{n_1} \in G_2$ together imply $L_n \in G$. To see this last point precisely, suppose that for a bounded function g on $\mathcal{S}$,

$$|E^{\pi_1}[g] - L_{n_1-j}(g)| < \gamma \quad \text{and} \quad |E^{\pi_2}[g] - L_{n_2}(g \circ \theta^{n_1})| < \gamma.$$

Then

$$\begin{aligned} |E^{\mu}[g] - L_n(g)| \le\ & \left| \alpha E^{\pi_1}[g] - \frac{n_1}{n} L_{n_1-j}(g) \right| \\ & + \left| (1-\alpha) E^{\pi_2}[g] - \frac{n_2}{n} L_{n_2}(g \circ \theta^{n_1}) \right| + \frac{2j}{n} \|g\|_\infty \\ & \le \gamma + \frac{2m+2}{n} \|g\|_\infty . \end{aligned}$$

With this inequality we can use the metric (B.3) for $\mathcal{M}_1(\mathcal{S})$ to construct neighborhoods G_i that meet the requirement.

We derive a lower bound. First take $x \in C_1$. Then

$$\begin{aligned} & \frac{1}{n} \log P_x\{L_n \in G\} \\ & \quad \ge \frac{1}{n} \log \sum_{u \in V_1, v \in V_2} P_x\{L_{n_1-j} \in G_1, X_{n_1-j} = u, X_{n_1} = v, L_{n_2} \circ \theta^{n_1} \in G_2\}, \end{aligned}$$

where inside the braces $j = j(u,v)$. By the Markov property

$$\begin{aligned} & \frac{1}{n} \log P_x\{L_n \in G\} \\ & \quad \ge \frac{1}{n} \log \sum_{u \in V_1, v \in V_2} P_x\{L_{n_1-j} \in G_1, X_{n_1-j} = u\} P_u\{X_j = v\} P_v\{L_{n_2} \in G_2\} \\ & \quad \ge \frac{n_1 - m}{n} \cdot \min_{1 \le j \le m} \frac{1}{n_1 - j} \log P_x\{L_{n_1-j} \in G_1, X_{n_1-j} \in V_1\} \\ & \qquad + \frac{n_2}{n} \cdot \min_{v \in V_2} \frac{1}{n_2} \log P_v\{L_{n_2} \in G_2\} + \frac{\log \delta}{n}. \end{aligned}$$

Apply the lower bound (13.16) from Step 1 to the two measures π_i with neighborhoods G_i:

$$\begin{aligned} & \varliminf_{n \to \infty} \frac{1}{n} \log P_x\{L_n \in G\} \\ & \quad \ge - \frac{\alpha H(\pi_1 \times q \,|\, \pi_1 \times p)}{1 - d_1(1 - \pi_1(V_1))} - (1-\alpha) H(\pi_2 \times q \,|\, \pi_2 \times p), \end{aligned}$$

where d_1 is the period of class C_1 under q. Increase V_1 up to C_1 to get

$$\underline{\lim}_{n\to\infty} \frac{1}{n} \log P_x\{L_n \in G\} \geq -\alpha H(\pi_1 \times q \,|\, \pi_1 \times p) - (1-\alpha)H(\pi_2 \times q \,|\, \pi_2 \times p).$$

Direct inspection shows that

$$\alpha H(\pi_1 \times q \,|\, \pi_1 \times p) + (1-\alpha)H(\pi_2 \times q \,|\, \pi_2 \times p) = H(\mu \times q \,|\, \mu \times p).$$

The lower bound above then becomes

$$\underline{\lim}_{n\to\infty} \frac{1}{n} \log P_x\{L_n \in G\} \geq -H(\mu \times q \,|\, \mu \times p).$$

We proved this bound for $x \in C_1$. Interchanging the roles of C_1 and C_2 shows that the bound holds for $x \in C_1 \cup C_2$, in other words for μ-almost every x.

Lower bound: Step 3. The argument of Step 2 generalizes naturally to the case $\mu = \sum_{i=1}^k \alpha_i \pi_i$ where μ is supported on k distinct positive recurrent classes $\{C_i : 1 \leq i \leq k\}$. We leave the details to the reader.

Lower bound: Step 4. Lastly, we treat the general case where q has positive recurrent classes $\{C_i : i \geq 1\}$, invariant distributions π_i on C_i for each i, and $\mu = \sum_{i\geq 1} \alpha_i \pi_i$ for $\alpha_i \geq 0$ and $\sum_{i\geq 1} \alpha_i = 1$. In this case, define

$$\mu_k = (\alpha_1 + \cdots + \alpha_k)^{-1} \sum_{i=1}^k \alpha_i \pi_i.$$

Then $\mu_k q = \mu_k$ for each k and

$$\lim_{k\to\infty} H(\mu_k \times q \,|\, \mu_k \times p) = H(\mu \times q \,|\, \mu \times p).$$

Since $\mu_k \to \mu$ weakly, $\mu_k \in G$ for large enough k. Thus, for k large enough, the result from Step 3 gives

$$\underline{\lim}_{n\to\infty} \frac{1}{n} \log P_x\{L_n \in G\} \geq -H(\mu_k \times q \,|\, \mu_k \times p)$$

for any large enough $k \geq 1$ and any $x \in C_1 \cup \cdots \cup C_k$. Take $k \to \infty$ to get

$$\underline{\lim}_{n\to\infty} \frac{1}{n} \log P_x\{L_n \in G\} \geq -H(\mu \times q \,|\, \mu \times p) \quad \text{for } \mu\text{-a.e. } x.$$

Irreducibility allows us to extend the lower bound to all initial states $x \in \mathcal{S}$. Given $z \in \mathcal{S}$, pick y in the support of μ and $m \in \mathbb{Z}_+$ such that $P_x\{X_m = y\} > 0$. Pick a smaller neighborhood G' of μ such that, if n is (deterministically) large enough, $L_{n-m} \circ \theta^m \in G'$ implies $L_n \in G$. Now use the Markov property and the lower bound above for P_y to write

$$\begin{aligned}\varliminf_{n\to\infty} \frac{1}{n} \log P_x\{L_n \in G\} &\geq \varliminf_{n\to\infty} \frac{1}{n} \log P_x\{L_{n-m} \circ \theta^m \in G', X_m = y\} \\ &= \varliminf_{n\to\infty} \frac{1}{n} \log P_x\{X_m = y\} P_y\{L_{n-m} \in G'\} \\ &\geq -H(\mu \times q \,|\, \mu \times p).\end{aligned}$$

Finally, take the supremum over kernels q that fix μ and then the supremum over $\mu \in G$ to get the desired lower bound in Theorem 13.5. □

The theorem above extends readily to reducible Markov chains. The rate function can then depend on the initial state. Write

$$p^k(x,y) = \sum_{z_1,\dots,z_{k-1}\in\mathcal{S}} p(x,z_1)p(z_1,z_2)\cdots p(z_{k-1},y)$$

for the k-step transition probability. Let us say a state y is accessible from x if $p^m(x,y) > 0$ for some $m > 0$ and abbreviate this by $x \stackrel{p}{\longrightarrow} y$. For a given state $x \in \mathcal{S}$, define the set of measures

$$\begin{aligned}\mathcal{A}_x = \{\mu \in \mathcal{M}_1(\mathcal{S}) :\ &\text{if } \mu(y) > 0, \text{ then } y = x \text{ or } x \stackrel{p}{\longrightarrow} y, \text{ and} \\ &\text{if } \mu(y) \wedge \mu(z) > 0 \text{ for } y \neq z, \text{ then } y \stackrel{p}{\longrightarrow} z \text{ or } z \stackrel{p}{\longrightarrow} y\}.\end{aligned} \tag{13.17}$$

Corollary 13.6. *Let p be an arbitrary transition matrix on $\mathcal{S}$ and fix the initial state $x \in \mathcal{S}$. Then the weak large deviation bounds* (13.13) *and* (13.14) *are valid when J is replaced with rate function*

$$I_x(\mu) = \begin{cases} J(\mu), & \mu \in \mathcal{A}_x, \\ \infty, & \mu \notin \mathcal{A}_x. \end{cases} \tag{13.18}$$

The proof of the corollary parallels closely that of Theorem 13.5, so we leave the details in Exercise 13.9 with thorough hints.

By Theorem 2.19, to upgrade a weak LDP to a full LDP with a tight rate function it suffices to show that the distributions $P_x\{L_n \in \cdot\}$ are exponentially tight. An easy way (and one that appears in much of the literature) is to make an assumption that allows us to rely on properties of i.i.d. processes. We present such a result next. Exercise 13.10 addresses the utility of admitting a k-step transition in the hypothesis on p used below.

Theorem 13.7. *Assume that the transition matrix p satisfies the following uniformity condition: there exist $k \in \mathbb{N}$, $1 < r < \infty$, and a probability measure λ supported on all of $\mathcal{S}$ such that the Radon-Nikodym derivatives*

$f_x(y) = p^k(x,y)/\lambda(y)$ satisfy

$$\sup_{x\in\mathcal{S}} \sum_{y\in\mathcal{S}} f_x(y)^r \lambda(y) < \infty. \tag{13.19}$$

Then for all $x \in \mathcal{S}$ the distributions $P_x\{L_n \in \cdot\}$ are exponentially tight, with normalization n. Consequently under this uniformity assumption the weak LDPs of Theorem 13.5 *and Corollary* 13.6 *can be upgraded to full LDPs with tight rate functions.*

Proof. The exponential tightness of $P_x\{L_n \in \cdot\}$ will come from the i.i.d. process with marginal distribution λ. The idea is simple but we need a few technical steps to execute it.

Step 1. We observe that it is enough to prove exponential tightness for the sequence of distributions $P_x\{L_{n+a(n)} \circ \theta_{\ell(n)} \in \cdot\}$ for any bounded sequences $a(n) \in \mathbb{Z}$ and $\ell(n) \in \mathbb{Z}_+$.

This can be seen conveniently from Lemma 2.20, applied to the Polish space $\mathcal{M}_1(\mathcal{S})$. Use the metric (B.3) on $\mathcal{M}_1(\mathcal{S})$. For a bounded function g on $\mathcal{S}$ and for large enough n,

$$\begin{aligned}
&\bigl|L_n(g) - L_{n+a(n)}(g \circ \theta_{\ell(n)})\bigr| \\
&\quad = \left| \frac{1}{n}\sum_{i=0}^{n-1} g(X_i) - \frac{1}{n+a(n)} \sum_{i=0}^{n+a(n)-1} g(X_{i+\ell(n)}) \right| \\
&\quad \le \frac{2\,|a(n)| + 2\ell(n)}{n - |a(n)|}\, \|g\|_\infty .
\end{aligned}$$

Hence if $B_1, \dots, B_m$ are δ-balls in $\mathcal{M}_1(\mathcal{S})$, for large enough n it is the case that $L_{n+a(n)} \circ \theta_{\ell(n)} \in \bigcup_{i=1}^m B_i$ implies $L_n \in \bigcup_{i=1}^m \widetilde{B}_i$ where $\widetilde{B}_i$ is the 2δ-ball centered at the center of B_i. Thus

$$P\left\{L_n \in \left(\bigcup_{i=1}^m \widetilde{B}_i\right)^c\right\} \le P\left\{L_{n+a(n)} \circ \theta_{\ell(n)} \in \left(\bigcup_{i=1}^m B_i\right)^c\right\}$$

and the condition of Lemma 2.20 passes from the distributions of $L_{n+a(n)} \circ \theta_{\ell(n)}$ to those of L_n.

Step 2. The next observation is that we can assume that the compact sets K_b in the definition (2.5) of exponential tightness are convex.

This is a consequence of the fact that in the present setting the closed convex hull $\overline{\mathrm{co}}\, K_b$ of K_b is compact, and if (2.5) holds for K_b, it certainly holds for the larger set $\overline{\mathrm{co}}\, K_b$. The compactness of $\overline{\mathrm{co}}\, K_b$ follows from general principles in locally convex spaces [**69**, Theorem 3.20(b)]. On $\mathcal{M}_1(\mathcal{S})$ we can also see it directly. Since $\mathcal{S}$ is Polish, given $\varepsilon > 0$ we can find a compact $A \subset \mathcal{S}$ such that $\nu(A^c) \le \varepsilon$ for all $\nu \in K_b$. This property is preserved under convex combinations and weak limits of probability measures ν, and thereby we have shown that $\overline{\text{co}}\, K_b$ is tight.

Step 3. We finish the proof by showing exponential tightness of the distributions of $L_{mk} \circ \theta_k$ as $m \to \infty$ and k is the parameter given in the statement of the theorem. Step 1 guarantees that this is enough (pick $a(n)$ so that $n + a(n)$ is a nearby multiple of k).

Decompose $L_{mk} \circ \theta_k$ into a convex combination of empirical measures of k-separated variables:

$$L_{mk} \circ \theta_k = \frac{1}{mk} \sum_{i=k}^{(m+1)k-1} \delta_{X_i} = \frac{1}{k} \sum_{j=k}^{2k-1} \frac{1}{m} \sum_{i=0}^{m-1} \delta_{X_{j+ik}} = \frac{1}{k} \sum_{j=k}^{2k-1} L_m^{(j)}$$

where we defined empirical measures $L_m^{(j)} = m^{-1} \sum_{i=0}^{m-1} \delta_{X_{j+ik}}$.

For a compact convex set $K \subset \mathcal{M}_1(\mathcal{S})$,

$$P_x\{L_{mk} \circ \theta_k \in K^c\} \le \sum_{j=k}^{2k-1} P_x\{L_m^{(j)} \in K^c\}. \tag{13.20}$$

Let $Q = \lambda^{\otimes \mathbb{N}}$ denote the i.i.d. product measure with marginal λ. Let $1 < s < \infty$ be the conjugate exponent of r: $r^{-1} + s^{-1} = 1$. Work on an individual probability from the right-hand side above and apply Hölder's inequality:

$$\begin{aligned} P_x\{L_m^{(j)} \in K^c\} &= \sum_{y_0} p^{j-k}(x, y_0) P_{y_0}\{L_m^{(k)} \in K^c\} \\ &= \sum_{y_0} p^{j-k}(x, y_0) \int \mathbb{1}\Big\{\frac{1}{m} \sum_{i=1}^{m} \delta_{y_i} \in K^c\Big\} \prod_{i=1}^{m} f_{y_{i-1}}(y_i) Q(dy_1, \dots, dy_m) \\ &\le \sum_{y_0} p^{j-k}(x, y_0) \Big(Q\{L_m \in K^c\}\Big)^{\frac{1}{s}} \Big(\int \prod_{i=1}^{m} f_{y_{i-1}}(y_i)^r Q(dy_1, \dots, dy_m)\Big)^{\frac{1}{r}}. \end{aligned}$$

Assumption (13.19) bounds inductively the last integral by a power of a constant, independently of y_0:

$$\begin{aligned} &\int \prod_{i=1}^{m} f_{y_{i-1}}(y_i)^r Q(dy_1, \dots, dy_m) \\ &\qquad = \int \prod_{i=1}^{m-1} f_{y_{i-1}}(y_i)^r \Big\{\sum_{z \in \mathcal{S}} f_{y_{m-1}}(z)^r \lambda(z)\Big\} Q(dy_1, \dots, dy_{m-1}) \\ &\qquad \le C \int \prod_{i=1}^{m-1} f_{y_{i-1}}(y_i)^r \; Q(dy_1, \dots, dy_{m-1}) \le \dots \le C^m. \end{aligned}$$

Go back up to (13.20) and collect the bounds:

$$P_x\{L_{mk}\circ\theta_k \in K^c\} \le C^{\frac{m}{r}} k\big(Q\{L_m \in K^c\}\big)^{\frac{1}{s}}.$$

Finally, with $n = mk$ and $n, m \to \infty$, given $0 < t < \infty$,

$$\begin{aligned}\overline{\lim_{m\to\infty}} \frac{1}{n}\log P_x\{L_{mk}\circ\theta_k \in K^c\} &\le \frac{\log C}{kr} + \overline{\lim_{m\to\infty}} \frac{1}{mks}\log Q\{L_m \in K^c\}\\ &\le -t\end{aligned}$$

where we use the exponential tightness of $Q\{L_m \in \cdot\}$ from Step 2 of the proof of Sanov's theorem (page 74) to choose the compact convex $K \subset \mathcal{M}_1(\mathcal{S})$ so that property (2.5) holds for $b = tks + (s/r)\log C$. Exponential tightness has been verified and the proof is completed. □

***Exercise 13.8.** In the countable state space setting of Theorem 13.5, show that $\mu q = \mu$ and $H(\mu\times q\,|\,\mu\times p) < \infty$ imply that $Q_x^{(n)} \ll P_x^{(n)}$ for μ-a.e. x.

Hint: Show by induction that $\mu(x_0)\prod_{i=0}^{n-1} q(x_i, x_{i+1}) > 0$ implies $\mu(x_n) > 0$ and $\prod_{i=0}^{n-1} p(x_i, x_{i+1}) > 0$.

***Exercise 13.9.** Prove Corollary 13.6. Below is an outline.

(a) $\mathcal{A}_x$ is closed and $P_x\{L_n \in \mathcal{A}_x\} = 1$. Upper bound (13.13) follows.

For the lower bound pick $\mu \in \mathcal{A}_x \cap G$ and a kernel q that fixes μ and satisfies $H(\mu\times q\,|\,\mu\times p) < \infty$.

(b) If C_1 and C_2 are distinct closed irreducible classes of q such that $\mu(C_1)\wedge\mu(C_2) > 0$, then either $w \xrightarrow{p} y$ $\forall(w,y)\in C_1\times C_2$ or $y \xrightarrow{p} w$ $\forall(w,y)\in C_1\times C_2$. Abbreviate the first case as $C_1 \xrightarrow{p} C_2$. The following observations lead to the proof. Since μ charges every point in $C_1\cup C_2$, finite entropy implies that $q \ll p$ on $C_1\cup C_2$. Hence p is irreducible on C_1 and on C_2. By the definition of $\mathcal{A}_x$, either $w \xrightarrow{p} y$ or $y \xrightarrow{p} w$ for every pair $(w,y)\in C_1\times C_2$.

(c) Suppose μ charges distinct closed irreducible classes $C_1, C_2, \dots, C_m$ of q. Then they can be ordered so that $C_{i_1} \xrightarrow{p} C_{i_2} \xrightarrow{p} \cdots \xrightarrow{p} C_{i_m}$. Use (b) and induction.

(d) After these preliminaries the lower bound argument of Theorem 13.5 can be repeated.

Exercise 13.10. Create an irreducible transition matrix p such that the uniformity condition of Theorem 13.7 holds with $k = 2$ for some $\lambda \in \mathcal{M}_1(\mathcal{S})$ but with $k = 1$ it fails for all $\lambda \in \mathcal{M}_1(\mathcal{S})$.

Hint: Let ν be a probability measure on $\mathcal{S} = \mathbb{N}$. Take $p(2x, 2x-1) = 1$ and $p(2x-1, 2y) = \nu(y)$ for $x, y \in \mathbb{N}$.

Exercise 13.11. Suppose z is a state such that $p^m(z,z) = 0$ for all $m > 0$. Show that if $\mu(z) > 0$, then $J(\mu) = \infty$. (Such states can satisfy $\mu(z) > 0$ for $\mu \in \mathcal{A}_x$ in (13.17).)

Exercise 13.12. Suppose $\mathcal{S}$ is finite. Show that $J(\mu) < \infty$ for all $\mu \in \mathcal{M}_1(\mathcal{S})$ if and only if $p_{x,x} > 0$ for all $x \in \mathcal{S}$.

Hint: Consider $\mu = \delta_x$. Which kernel trivially fixes every μ?

Exercise 13.13. Assume the same conditions as in Theorem 13.5. Assume the full large deviation principle holds with some lower semicontinuous rate function I. Prove that then $I = J$. In other words, J is the only candidate for the rate function of a full large deviations principle.

Hint: Use lower semicontinuity of I, lower bound (13.14), and upper bound (13.13) with rate I to deduce that $I \le J$. For the other direction modify the proof of the upper bound in Theorem 13.5 to get

$$\overline{\lim_{n\to\infty}} \frac{1}{n} \log \sup_{x\in\mathcal{S}} P_x\{L_n \in K\} \le - \sup_{f\in\mathscr{C}_b^+(\mathcal{S})} \inf_{\nu\in K} \int \log \frac{f}{pf}\, d\nu$$

for all measurable sets K.

The next two exercises demonstrate how the full large deviation principle may fail to hold even in very basic cases.

Exercise 13.14. Fix $p \in (0,1)$ and consider the Markov chain on $\mathcal{S} = \mathbb{Z}$ with kernel $p(x, x+1) = 1 - p(x, x-1) = p$ for all $x \in \mathbb{Z}$. Prove that the full large deviation principle cannot hold.

Hint: Consider the closed set $K = \{(2m+1)^{-1}\sum_{j=0}^{2m} \delta_j : m \ge 1\}$. Show that $P_0\{L_{2m+1} \in K\} \ge p^{2m}$. Next, using the test function $f(2j+1) = 1$ and $f(2j) = k$, $k > 1$, in (13.6), show that $J(\mu) = \infty$ for all $\mu \in K$. Use Theorem 13.5 and the previous exercise to conclude.

Exercise 13.15. Fix $p \in (0,1)$. Let $\mathcal{S} = \mathbb{Z}_+$ and consider the positive recurrent Markov chain with kernel $p(x, x+1) = p$, $p(x, 0) = 1 - p$ for all $x \in \mathbb{Z}_+$. Prove that the full large deviation principle cannot hold.

Hint: Consider the set $K = \{m^{-1}\sum_{j=\ell+1}^{\ell+m} \delta_j, \ell \ge 0, m \ge 1\}$ and proceed as in the exercise above.

Exercise 13.16. Prove a process level LDP for a finite state space Markov chain with a strictly positive transition matrix, as a corollary of the LDP of Gibbs measures in Theorem 8.6.

Hint: Recall Example 7.21.

Literature notes. Corollary 13.6 appears as Theorem 2.1 in [**52**]. Exercises 13.13–13.15 were inspired by [**7**]. LDPs hold for Markov chains on general

Polish spaces under sufficient assumptions of irreducibility and Feller continuity ([**25**], [**51**], [**14**]). A uniformity condition can also be formulated for Polish spaces ([**33**], [**25**], Section 6.3 of [**15**]).

Process level large deviations for Markov chains can be derived more generally than through the Gibbs connection utilized in Exercise 13.16. In our setting we could apply Theorems 13.5, Corollary 13.6, and Theorem 13.7 to the multivariate Markov chain $\bar{X}_n = (X_n, \dots, X_{n+\ell})$, for any $\ell \geq 1$. A *projective limit* argument extends these multivariate results to a process level LDP, similarly to Theorem 6.13. The rate function is a specific relative entropy similar to (6.4). See Section 4.6 of [**15**].

13.3. Finite Markov chains

A finite space is compact, and hence Theorem 13.5 gives a full LDP for every irreducible finite Markov chain. In the finite case we can profitably investigate the structure of the rate function J and its dual with calculus and linear algebra.

Let us denote the transition matrix also by $p_{x,y} = p(x, y)$. For a function f write $p(x, f) = pf(x) = \sum_y p_{x,y} f(y)$, and also write $p(x, A) = \sum_{y \in A} p_{x,y}$ for a set $A \subset \mathcal{S}$. We restate definition (13.6) of the rate function J and identity (13.7) in the present setting: for a probability measure μ on $\mathcal{S}$,

$$J(\mu) = \sup_{h \in \mathbb{R}^{\mathcal{S}}} \sum_{x \in \mathcal{S}} \mu(x) \Big[h(x) - \log \sum_{y \in \mathcal{S}} p_{x,y} e^{h(y)} \Big] \tag{13.21}$$

$$= \inf_{q : \mu q = \mu} \sum_{x,y \in \mathcal{S}} \mu(x) q_{x,y} \log \frac{q_{x,y}}{p_{x,y}}. \tag{13.22}$$

In the last formula the correct convention is $0 \log \frac{0}{0} = 0$ because the sum represents the relative entropy $\sum_x \mu(x) H(q_{x,\cdot} \,|\, p_{x,\cdot})$. In the case $J(\mu) < \infty$ a minimizing stochastic matrix q always exists in (13.22), by continuity and compactness. By the strict convexity of relative entropy (Exercise 5.5) the minimizing matrix is unique on the support $\mathcal{S}_\mu = \{x \in \mathcal{S} : \mu(x) > 0\}$ of μ.

We record two preliminary observations. Take $h(x) = -M \mathbb{1}_{\mathcal{S}_\mu^c}(x)$ in (13.21) and let $M \nearrow \infty$ to deduce that

$$\text{if } p(x, \mathcal{S}_\mu) = 0 \text{ for some } x \in \mathcal{S}_\mu, \text{ then } J(\mu) = \infty. \tag{13.23}$$

Second, if $\mathcal{S}_\mu \neq \mathcal{S}$, then for any $v \in \mathbb{R}^{\mathcal{S}}$ and $M > 0$, $h = v - M \mathbb{1}_{\mathcal{S}_\mu^c}$ gives a strictly better value on the right-hand side of (13.21) than $h = v$. Thus in (13.21) the optimal course is to let $h \searrow -\infty$ on $\mathcal{S} \smallsetminus \mathcal{S}_\mu$ and the maximum cannot be attained. From these considerations it is natural that we maximize the function

$$F_\mu(v) = \sum_{x\in\mathcal{S}_\mu} \mu(x)\Big[v(x) - \log \sum_{y\in\mathcal{S}_\mu} p_{x,y}e^{v(y)}\Big] \quad \text{for} \quad v \in \mathbb{R}^{\mathcal{S}_\mu}. \tag{13.24}$$

The next lemma shows how the μ-fixing kernel q arises naturally from calculus.

Lemma 13.17. *Let $\mu \in \mathcal{M}_1(\mathcal{S})$. Assume that $p(x,\mathcal{S}_\mu) = \sum_{y\in\mathcal{S}_\mu} p_{x,y} > 0$ for each $x \in \mathcal{S}_\mu$. Then for $v \in \mathbb{R}^{\mathcal{S}_\mu}$ the following statements are equivalent.*

(i) *F_μ is maximized by v.*

(ii) *$\nabla F_\mu(v) = 0$.*

(iii) *On the space $\mathcal{S}_\mu$, μ is invariant under the stochastic kernel*

$$q_{x,y} = \frac{p_{x,y}e^{v(y)}}{p(x, e^v \mathbb{1}_{\mathcal{S}_\mu})}, \quad x, y \in \mathcal{S}_\mu. \tag{13.25}$$

If such a maximizer v exists, then $J(\mu) = F_\mu(v)$ and q extends to a kernel on $\mathcal{S}$ that minimizes in (13.22).

Proof. Differentiation gives for $z \in \mathcal{S}_\mu$

$$\frac{\partial}{\partial v(z)} F_\mu(v) = \mu(z) - \sum_{x\in\mathcal{S}_\mu} \mu(x) \frac{p_{x,z}e^{v(z)}}{p(x, e^v \mathbb{1}_{\mathcal{S}_\mu})}.$$

This shows that for $v \in \mathbb{R}^{\mathcal{S}_\mu}$, $\nabla F_\mu(v) = 0$ if and only if μ restricted to $\mathcal{S}_\mu$ is invariant under the stochastic kernel (13.25). The equivalence of (ii) and (iii) follows.

(i)$\Longleftrightarrow$(ii) comes from calculus. (i)$\Longrightarrow$(ii) is immediate. For the other direction differentiate again:

$$\frac{\partial^2 F_\mu(v)}{\partial v(z)\partial v(w)} = \sum_{x\in\mathcal{S}_\mu} \mu(x) \frac{p_{x,z}e^{v(z)}p_{x,w}e^{v(w)}}{p(x, e^v \mathbb{1}_{\mathcal{S}_\mu})^2} - \mathbb{1}\{z = w\} \sum_{x\in\mathcal{S}_\mu} \mu(x) \frac{p_{x,z}e^{v(z)}}{p(x, e^v \mathbb{1}_{\mathcal{S}_\mu})}.$$

For a vector $u = (u_z)_{z\in\mathcal{S}_\mu}$ we compute the quadratic form:

$$\begin{aligned} \langle D^2 F_\mu(v)u, u\rangle &= \sum_{z,w\in\mathcal{S}_\mu} u_z u_w \frac{\partial^2 F_\mu(v)}{\partial v(z)\partial v(w)} \\ &= \sum_{x\in\mathcal{S}_\mu} \mu(x) \bigg\{ \bigg(\frac{\sum_{z\in\mathcal{S}_\mu} p_{x,z}u_z e^{v(z)}}{p(x, e^v \mathbb{1}_{\mathcal{S}_\mu})} \bigg)^2 - \frac{\sum_{z\in\mathcal{S}_\mu} p_{x,z}u_z^2 e^{v(z)}}{p(x, e^v \mathbb{1}_{\mathcal{S}_\mu})} \bigg\} \le 0. \end{aligned} \tag{13.26}$$

The expression in braces is ≤ 0 as the negative of a variance. This sign of the quadratic form gives us the maximization property. For suppose $\nabla F_\mu(v) = 0$. Let $u \in \mathbb{R}^{\mathcal{S}_\mu}$ and set $\psi(t) = F_\mu(v + tu)$ for $t \in \mathbb{R}$. Then $\psi'(0) = \langle \nabla F_\mu(v), u\rangle = 0$ and $\psi''(t) = \langle D^2 F_\mu(v + tu)u, u\rangle \le 0$, which implies that $\psi(0) \ge \psi(t)$. Thus v is a maximizer of F_μ.

Now suppose v is a maximizer. Then q of (13.25) can be extended to a kernel on $\mathcal{S}$ that fixes μ, namely by setting $q_{x,y} = 0$ for $x \in \mathcal{S}_\mu$, $y \in \mathcal{S} \smallsetminus \mathcal{S}_\mu$, and by taking $q_{x,\cdot}$ to be an arbitrary probability measure for $x \in \mathcal{S} \smallsetminus \mathcal{S}_\mu$. Then q is a legitimate kernel in (13.22). Using the invariance $\mu q = \mu$, we bound as follows:

$$\begin{aligned} J(\mu) &\le \sum_{x,y \in \mathcal{S}_\mu} \mu(x) q_{x,y} \log \frac{q_{x,y}}{p_{x,y}} \\ &= \sum_{x \in \mathcal{S}_\mu} \mu(x) \Big[v(x) - \log \sum_{y \in \mathcal{S}_\mu} p_{x,y} e^{v(y)} \Big] \\ &= \lim_{\substack{h(x) = v(x) \text{ on } \mathcal{S}_\mu \\ h(y) \to -\infty \text{ for } y \in \mathcal{S}_\mu^c}} \sum_{x \in \mathcal{S}} \mu(x) \Big[h(x) - \log \sum_{y \in \mathcal{S}} p_{x,y} e^{h(y)} \Big] \le J(\mu). \qquad \square \end{aligned}$$

The best situation is the one where p is positive.

Theorem 13.18. *Assume $p_{x,y} > 0$ for all $x, y \in \mathcal{S}$. Then for each $\mu \in \mathcal{M}_1(\mathcal{S})$ there exists a maximizer $v \in \mathbb{R}^{\mathcal{S}_\mu}$ of F_μ. In particular, $J(\mu) < \infty$ and all the statements of Lemma 13.17 are in force, the maximizer of F_μ is unique up to an additive constant, and J is a strictly convex function on $\mathcal{M}_1(\mathcal{S})$.*

Proof. Step 1. Adding a constant to v does not alter $F_\mu(v)$, so we can restrict to $\min_x v(x) = 0$. Taking $v = 0$ gives $\sup F_\mu(v) \ge 0$, so any v for which $F_\mu(v) < 0$ is irrelevant for $\sup F_\mu(v)$. So consider v such that $F_\mu(v) \ge 0$. Pick any $x_0, x_1 \in \mathcal{S}_\mu$. Let $y(x_0) = x_1$ and $y(x) = x$ for $x \in \mathcal{S}_\mu \smallsetminus \{x_0\}$. Then

$$\begin{aligned} 0 \le F_\mu(v) &= \sum_{x \in \mathcal{S}_\mu} \mu(x) \Big[v(x) - \log \sum_{y \in \mathcal{S}_\mu} p_{x,y} e^{v(y)} \Big] \\ &\le \sum_{x \in \mathcal{S}_\mu} \mu(x) [v(x) - \log p_{x,y(x)} e^{v(y(x))}] \\ &= \sum_{x \in \mathcal{S}_\mu} \mu(x) v(x) - \sum_{x \in \mathcal{S}_\mu} \mu(x) v(y(x)) - \sum_{x \in \mathcal{S}_\mu} \mu(x) \log p_{x,y(x)} \\ &\le \mu(x_0)(v(x_0) - v(x_1)) - \log \min_{x,y} p_{x,y}. \end{aligned}$$

If $v(x_0) = 0$, we conclude that $v(x_1) \le C = -\log \min_{x,y} p_{x,y} / \min_{x \in \mathcal{S}_\mu} \mu(x)$. Consequently it is enough to consider $v \in [0, C]^{\mathcal{S}_\mu}$ for $\sup F_\mu(v)$. By continuity and compactness, there is a maximizer.

Step 2. Suppose v and $v + u$ are maximizers. Then ψ defined in the paragraph below (13.26) satisfies $\psi''(t) = \langle D^2 F_\mu(v + tu) u, u \rangle = 0$. Since p is now strictly positive, the expression in braces on the second line in (13.26) cannot vanish unless u is constant on $\mathcal{S}_\mu$.

We have shown that F_μ possesses a maximizer that is unique up to an additive constant. Hence the statements of Lemma 13.17 are in force.

Step 3. It remains to argue that J is strictly convex. Convexity was observed above Theorem 13.2. Let $\nu = t\mu + (1-t)\tilde{\mu}$ for some $0 < t < 1$ but $J(\nu) = tJ(\mu) + (1-t)J(\tilde{\mu})$. Let q and $\tilde{q}$ be minimizing kernels for $J(\mu)$ and $J(\tilde{\mu})$ in (13.22). Define the Radon-Nikodym derivatives $b(x) = \frac{\mu(x)}{\nu(x)}$ and $\tilde{b}(x) = \frac{\tilde{\mu}(x)}{\nu(x)}$ for $x \in \mathcal{S}_\nu = \mathcal{S}_\mu \cup \mathcal{S}_{\tilde{\mu}}$. Then

$$\big(tb(x) + (1-t)\tilde{b}(x)\big)\nu(x) = t\mu(x) + (1-t)\tilde{\mu}(x) = \nu(x)$$

implies that $tb + (1-t)\tilde{b} = 1$ on $\mathcal{S}_\nu$. The stochastic matrix

$$\bar{q}_{x,y} = tb(x)q_{x,y} + (1-t)\tilde{b}(x)\tilde{q}_{x,y}$$

fixes ν. By the convexity of relative entropy

$$\begin{aligned} J(\nu) &\le \sum_{x\in\mathcal{S}_\nu} \nu(x) H(\bar{q}_{x,\cdot} \,|\, p_{x,\cdot}) \\ &\le \sum_{x\in\mathcal{S}_\nu} \nu(x)\big[tb(x)H(q_{x,\cdot} \,|\, p_{x,\cdot}) + (1-t)\tilde{b}(x)H(\tilde{q}_{x,\cdot} \,|\, p_{x,\cdot})\big] \\ &= t\sum_{x\in\mathcal{S}_\mu} \mu(x)H(q_{x,\cdot} \,|\, p_{x,\cdot}) + (1-t)\sum_{x\in\mathcal{S}_{\tilde{\mu}}} \tilde{\mu}(x)H(\tilde{q}_{x,\cdot} \,|\, p_{x,\cdot}) \\ &= tJ(\mu) + (1-t)J(\tilde{\mu}) = J(\nu). \end{aligned}$$

The inequalities above are forced to be equalities. Hence $\bar{q}$ is the minimizing kernel for ν. Consequently all three kernels q, $\tilde{q}$, and $\bar{q}$ are of the type (13.25), and so in particular strictly positive on their domains.

Suppose $x \in \mathcal{S}_\mu$. If $x \notin \mathcal{S}_{\tilde{\mu}}$, then $\tilde{b}(x) = 0$ and $\bar{q}_{x,\cdot} = q_{x,\cdot}$, while if $x \in \mathcal{S}_{\tilde{\mu}}$, then $b(x) \wedge \tilde{b}(x) > 0$ and strict convexity of relative entropy (Exercise 5.5) forces $\bar{q}_{x,\cdot} = q_{x,\cdot} = \tilde{q}_{x,\cdot}$. We have that $\bar{q}_{x,\cdot} = q_{x,\cdot}$ for all $x \in \mathcal{S}_\mu$. This forces $\mathcal{S}_\nu = \mathcal{S}_\mu$ because for $y \in \mathcal{S}_\nu \smallsetminus \mathcal{S}_\mu$ we would have $\bar{q}_{x,y} = q_{x,y} = 0$, contradicting $\bar{q} > 0$ on $\mathcal{S}_\nu$. We conclude that $\bar{q}$ and q are the same kernel. By irreducibility this kernel has a unique invariant distribution. This implies $\nu = \mu$, and thereby also $\tilde{\mu} = \mu$. Strict convexity of J has been verified. □

The next example shows how the ideal properties can fail once p is not positive. This example, a simple random walk with reflecting boundaries, shows that $J(\mu) = \infty$ can happen, and even if $J(\mu) < \infty$, functional F_μ may not have a maximizer or, even if it does have one, the maximizers are not necessarily unique up to additive constants. It also shows that the minimizing kernel can have $q_{x,y} = 0$ for some $x, y \in \mathcal{S}_\mu$ even though $p_{x,y} > 0$. Furthermore, it demonstrates that J may fail strict convexity.

Example 13.19. Let $\mathcal{S} = \{1, 2, 3, 4\}$ and let the transition matrix be

$$p = \begin{bmatrix} 0 & 1 & 0 & 0 \\ \frac{1}{2} & 0 & \frac{1}{2} & 0 \\ 0 & \frac{1}{2} & 0 & \frac{1}{2} \\ 0 & 0 & 1 & 0 \end{bmatrix}.$$

Let

$$\alpha = [\tfrac{1}{2}, \tfrac{1}{2}, 0, 0], \quad \beta = [0, 0, \tfrac{1}{2}, \tfrac{1}{2}],$$
$$\gamma = t\alpha + (1-t)\beta = [\tfrac{t}{2}, \tfrac{t}{2}, \tfrac{1-t}{2}, \tfrac{1-t}{2}], \quad \text{and} \quad \delta = [\tfrac{1}{4}, \tfrac{3}{4}, 0, 0].$$

F_α and F_β have multiple maximizers. Indeed, after cancellation

$$F_\alpha(v) = F_\beta(v) = \tfrac{1}{2}\log 2 \quad \text{for all } v.$$

$F_\gamma(v)$ has no maximizer but the supremum is $\frac{1}{2}\log 2$. Thus we get

$$J(\alpha) = J(\beta) = J(\gamma) = \tfrac{1}{2}\log 2.$$

This shows that strict convexity of J fails. Also

$$F_\delta(v) = \tfrac{3}{4}\log 2 + \tfrac{1}{2}(v_2 - v_1),$$

which leads to $J(\delta) = \infty$.

Let

$$q = \begin{bmatrix} 0 & 1 & 0 & 0 \\ 1 & 0 & 0 & 0 \\ 0 & 0 & 0 & 1 \\ 0 & 0 & 1 & 0 \end{bmatrix}.$$

For $\mu \in \{\alpha, \beta, \gamma\}$ the restriction of q to $\mathcal{S}_\mu$ is the unique kernel absolutely continuous with respect to p that fixes μ. Consequently q is the minimizing kernel for all three measures. This example also shows that reducible minimizing kernels q appear naturally and consequently had to be considered in the lower bound proof of Theorem 13.5.

We turn now to study the dual of J. Continue to assume that the transition matrix p is irreducible. For $g \in \mathbb{R}^{\mathcal{S}}$, the LDP and Varadhan's theorem give the existence of the limiting free energy or pressure

$$\Lambda(g) = \lim_{n\to\infty} n^{-1} \log E_x[e^{\sum_{k=0}^{n-1} g(X_k)}],$$

together with the representation

$$\Lambda(g) = \sup_{\mu \in \mathcal{M}_1(\mathcal{S})} \{E^\mu[g] - J(\mu)\}. \tag{13.27}$$

We describe a recipe for the maximizer in (13.27) in terms of the Perron-Frobenius theory of nonnegative matrices. (See Appendix D for the necessary linear algebra facts.)

Let $g \in \mathbb{R}^{\mathcal{S}}$ be given. Define the nonnegative irreducible matrix

$$R = (r_{x,y})_{x,y\in\mathcal{S}} = (e^{g(x)}p_{x,y})_{x,y\in\mathcal{S}}. \tag{13.28}$$

Let $\rho(R)$ be the Perron-Frobenius eigenvalue of R and let s and σ be the strictly positive right and left Perron-Frobenius eigenvectors normalized so that $\sigma s = \sum_x s(x)\sigma(x) = 1$. Define the stochastic kernel π and the probability measure μ on $\mathcal{S}$ by

$$\pi_{x,y} = \frac{p_{x,y}s(y)}{\rho(R)e^{-g(x)}s(x)} = \frac{p_{x,y}e^{\log s(y)}}{p(x, e^{\log s})} \quad \text{and} \quad \mu(x) = s(x)\sigma(x). \tag{13.29}$$

Theorem 13.20. *Given $g \in \mathbb{R}^{\mathcal{S}}$ define R by (13.28). Then $\Lambda(g) = \log \rho(R)$. Define π and μ by (13.29). Then π is a stochastic kernel that fixes μ: $\mu\pi = \mu$. Furthermore, $q = \pi$ is the unique minimizing kernel in (13.22), $\log s$ is a maximizer of F_μ, and μ is the unique maximizer in (13.27).*

Proof. The right eigenvector equation

$$\sum_y e^{g(x)}p_{x,y}s(y) = \rho(R)s(x)$$

gives the stochastic kernel

$$\pi_{x,y} = \frac{p_{x,y}s(y)}{\rho(R)e^{-g(x)}s(x)} = \frac{p_{x,y}e^{\log s(y)}}{p(x, e^{\log s})} \tag{13.30}$$

and the equality of the two denominators. The left eigenvector equation is

$$\begin{aligned}\sum_x \sigma(x)e^{g(x)}p_{x,y} = \rho(R)\sigma(y) &\iff \sum_x s(x)\sigma(x)\frac{p_{x,y}s(y)}{\rho(R)e^{-g(x)}s(x)} = s(y)\sigma(y)\\ &\iff \sum_x \mu(x)\pi_{x,y} = \mu(y).\end{aligned}$$

In other words, probability measure $\mu(x) = s(x)\sigma(x)$ is invariant under π.

To derive $\Lambda(g)$ write

$$\begin{aligned}n^{-1}\log E_x\big[e^{\sum_{k=0}^{n-1} g(X_k)}\big] &= n^{-1}\log \sum_y r^n_{x,y} = n^{-1}\log \sum_y r^n_{x,y}\frac{s(y)}{s(y)}\\ &\le n^{-1}\log(R^n s)_x - n^{-1}\log \min_x s(x)\\ &= \log \rho(R) + n^{-1}\log s(x) - n^{-1}\log \min_y s(y)\end{aligned}$$

with a matching lower bound with $\max_y s(y)$. From this, the free energy is (independently of initial state x)

$$\Lambda(g) = \lim_{n\to\infty} n^{-1}\log E_x\big[e^{\sum_{k=0}^{n-1} g(X_k)}\big] = \log \rho(R).$$

By Lemma 13.17 the invariance of μ under π guarantees that $\log s$ maximizes F_μ, $q = \pi$ is the minimizing kernel for $J(\mu)$ in (13.22), and $J(\mu) = F_\mu(\log s)$. We have

$$J(\mu) = \sum_x \mu(x)\big[\log s(x) - \log \sum_y p_{x,y} s(y)\big] = \sum_x \mu(x) \log \frac{s(x)}{p(x,s)}$$

$$= \sum_x \mu(x) \log \frac{e^{g(x)}}{\rho(R)} = E^\mu[g] - \Lambda(g). \tag{13.31}$$

This tells us that μ is a maximizer.

Suppose $\tilde\mu$ is another maximizer in (13.27). To show that $\tilde\mu = \mu$ we basically rerun the convexity argument from Step 3 on page 206. Put $\nu = t\mu + (1-t)\tilde\mu$ for some $0 < t < 1$. By the convexity of J, ν is also a maximizer in (13.27), and then it must be that $J(\nu) = tJ(\mu) + (1-t)J(\tilde\mu)$. Let $\tilde q$ be the minimizing kernel for $J(\tilde\mu)$ in (13.22). Define the Radon-Nikodym derivatives $b(x) = \frac{\mu(x)}{\nu(x)}$ and $\tilde b(x) = \frac{\tilde\mu(x)}{\nu(x)}$. Since $\nu(x) > 0$ for all x,

$$\big(tb(x) + (1-t)\tilde b(x)\big)\nu(x) = t\mu(x) + (1-t)\tilde\mu(x) = \nu(x)$$

implies that $tb + (1-t)\tilde b = 1$. The stochastic matrix

$$\bar q_{x,y} = tb(x) q_{x,y} + (1-t)\tilde b(x) \tilde q_{x,y}$$

fixes ν. By the convexity of relative entropy

$$\begin{aligned}
J(\nu) &\le \sum_x \nu(x) H(\bar q_{x,\cdot} \,|\, p_{x,\cdot}) \\
&\le \sum_x \nu(x)\big[tb(x) H(q_{x,\cdot} \,|\, p_{x,\cdot}) + (1-t)\tilde b(x) H(\tilde q_{x,\cdot} \,|\, p_{x,\cdot})\big] \\
&= t\sum_x \mu(x) H(q_{x,\cdot} \,|\, p_{x,\cdot}) + (1-t)\sum_x \tilde\mu(x) H(\tilde q_{x,\cdot} \,|\, p_{x,\cdot}) \\
&= tJ(\mu) + (1-t)J(\tilde\mu) = J(\nu).
\end{aligned}$$

The inequalities above are forced to be equalities. Hence $\bar q$ is the minimizing kernel for ν. By the strict convexity of relative entropy (Exercise 5.5), whenever $\tilde b(x) > 0$, we must have $\tilde q_{x,\cdot} = q_{x,\cdot}$ (because $b(x) > 0$ for all x). Hence $\bar q = q$. From (13.30) we see that $\bar q = q = \pi$ is irreducible and therefore has a unique invariant distribution. This implies $\nu = \mu$, and thereby also $\tilde\mu = \mu$. We have proved that $\mu(x) = s(x)\sigma(x)$ is the unique maximizer in (13.27). ∎

Exercise 13.4 gives an alternate variational formula

$$\Lambda(g) = K(g) = \inf_{h \in \mathbb{R}^S} \max_x \log \sum_y p_{x,y} e^{g(x)+h(y)-h(x)}. \tag{13.32}$$

For finite $\mathcal{S}$ this formula is the same as the characterization of the Perron-Frobenius eigenvalue given in the second formula on the right of (D.1). The minimizer of this formula is unique up to an additive constant and is given by $h(x) = \log s(x)$.

Exercise 13.21. Assume transition matrix p is irreducible and aperiodic. Let $g \in \mathbb{R}^{\mathcal{S}}$. Prove that the minimizer of (13.32) (unique up to an additive constant) is given by

$$h(x) = \lim_{n\to\infty} \bigl\{ \log E_x[e^{\sum_{k=0}^{n-1} g(X_k)}] - n\Lambda(g)\bigr\}.$$

Hint: Apply (D.2) to the matrix in (13.28).

The next exercise shows how kernel π in (13.29) arises from a *tilt* of the Markov chain by the *potential* g, much in the spirit of Gibbs measures.

Exercise 13.22. Assume transition matrix p is irreducible and aperiodic. Let $g \in \mathbb{R}^{\mathcal{S}}$ and let π and μ be defined by (13.29). Fix $m \geq 1$ and $x \in \mathcal{S}$. Let $Q_{x,n}^{(m)}$ be the probability distribution of $(X_0, \dots, X_m)$ given by $Q_{x,n}^{(m)}\{X_0 = x\} = 1$ and for $x_0 = x$

$$Q_{x,n}^{(m)}(x_0, \dots, x_m) = \frac{E_x[e^{\sum_{k=0}^{n-1} g(X_k)} \mathbb{1}\{X_1 = x_1, \dots, X_m = x_m\}]}{E_x[e^{\sum_{k=0}^{n-1} g(X_k)}]}. \tag{13.33}$$

Let Q_x be the Markov chain, starting at x, with transition kernel π. Prove that as $n \to \infty$ measures $Q_{x,n}^{(m)}$ converge weakly to the marginal of Q_x on $(X_0, \dots, X_m)$. Show also that for all x, y

$$\mu(y) = \lim_{n\to\infty} \frac{E_x[e^{\sum_{k=0}^{n-1} g(X_k)} \mathbb{1}\{X_{\lfloor n/2\rfloor} = y\}]}{E_x[e^{\sum_{k=0}^{n-1} g(X_k)}]}.$$

Hint: Divide the numerator and denominator of (13.33) by $e^{n\Lambda(g)}$ and use the previous exercise. For the second claim express powers of π in terms of (13.29).

Since J is convex and lower semicontinuous, (13.27) and convex duality give for $\mu \in \mathcal{M}_1(\mathcal{S})$

$$J(\mu) = \sup_{g\in\mathbb{R}^{\mathcal{S}}} \{E^{\mu}[g] - \Lambda(g)\}. \tag{13.34}$$

The next three exercises address the maximizing g.

Exercise 13.23. Assume p is irreducible. Let $\mu \in \mathcal{M}_1(\mathcal{S})$ be such that $\mu(x) > 0$ for all x and $J(\mu) < \infty$.

(a) Prove that if there exists a v that maximizes F_μ, as described in Lemma 13.17, then

$$g(x) = v(x) - \log p(x, e^v)$$

is a maximizer in (13.34).

(b) Prove that if $p > 0$, then the maximizing g exists and is unique up to an additive constant.

Hint: For the uniqueness apply Lemma D.1 to prove strict convexity of Λ, that is, that $\Lambda(g) < \theta\Lambda(g_1) + (1-\theta)\Lambda(g_2)$ for $g = \theta g_1 + (1-\theta)g_2$, $0 < \theta < 1$, except when $g_1 - g_2$ is constant.

In certain situations, even when p has zeros, one can still find maximizers for F_μ, and hence a maximizing g. Uniqueness up to additive constants may fail in this case (recall Example 13.19).

Exercise 13.24. Consider a Markov chain with finite state space $\mathcal{S}$ and irreducible transition matrix p. Let $\mu \in \mathcal{M}_1(\mathcal{S})$ be such that $p(x, \mathcal{S}_\mu) > 0$ for each $x \in \mathcal{S}_\mu$. Let q be the unique minimizer of (13.22). Assume q is equivalent to p, that is, $q_{x,y} > 0$ if and only if $p_{x,y} > 0$. Prove that there exists $v_0 \in \mathbb{R}^{\mathcal{S}_\mu}$ that maximizes F_μ. Furthermore,

$$q_{x,y} = \frac{p_{x,y} e^{v_0(y)}}{p(x, e^{v_0} \mathbb{1}_{\mathcal{S}_\mu})}, \qquad x, y \in \mathcal{S}_\mu, \tag{13.35}$$

for any such v_0. Conclude the existence of maximizers in (13.34).

Hint: Use Lagrange multipliers to prove that the minimizer of (13.22) has the form (13.35) for some $v \in \mathbb{R}^{\mathcal{S}_\mu}$ and by a direct computation show that the corresponding v maximizes F_μ. Conversely, use Lemma 13.17 to show that if v_0 is a maximizer and kernel $\tilde{q}$ is defined by the right-hand side of (13.35), then $\mu\tilde{q} = \mu$ and $\tilde{q}$ minimizes (13.22). Conclude that $\tilde{q} = q$ and thus (13.35) holds.

When p has zeros and when the kernel q that maximizes (13.22) is not equivalent to p, (13.34) may fail to have a maximizing g.

Exercise 13.25. Show that for $\mu = \alpha$ from Example 13.19 there is no maximizing g for (13.34).

Hint: Assume otherwise and use that to compute $\Lambda(g)$ from the duality. Prove that this cannot be the logarithm of the Perron-Frobenius eigenvalue of matrix R defined by (13.28).

Chapter 14

Convexity criterion for large deviations

In this chapter we prove an abstract large deviation theorem whose hypotheses involve convexity. The result comes from an unpublished 1991 manuscript of Baxter and Jain [**8**]. In the next two chapters we apply this result to a nonstationary independent process and to a walk in a random medium.

Let $\mathcal{X}$ and $\mathcal{L}$ be two real vector spaces in duality, endowed with their weak topologies as described in the beginning of Chapter 4. Assumption 4.1 is in force so that spaces $\mathcal{X}$ and $\mathcal{L}$ are Hausdorff spaces. We prove a large deviation theorem for a sequence of Borel probability measures $\{\mu_n\}_{n\in\mathbb{N}}$ on a convex, compact subset $\mathcal{X}_0$ of $\mathcal{X}$. The concrete example to keep in mind is the one where $\mathcal{X} = \mathcal{M}(\mathcal{S})$ is the space of real-valued Borel measures on a compact metric space $\mathcal{S}$, $\mathcal{L} = \mathscr{C}(\mathcal{S}) = \mathscr{C}_b(\mathcal{S})$, and $\mathcal{X}_0 = \mathcal{M}_1(\mathcal{S})$ is the space of Borel probability measures on $\mathcal{S}$. When we apply this result we will be able to remove the compactness assumption. In Chapter 15 we compactify the underlying space and use exponential tightness. In Chapter 16 we generalize the theorem to a bounded set $\mathcal{X}_0$ with additional assumptions of exponential tightness and compact sublevel sets.

Let $0 < r_n \nearrow \infty$ be a normalizing sequence. For $\varphi \in \mathcal{L}$ define the (upper) pressure by

$$\bar{p}(\varphi) = \varlimsup_{n\to\infty} \frac{1}{r_n} \log \int_{\mathcal{X}_0} e^{r_n\langle x,\varphi\rangle}\, \mu_n(dx). \tag{14.1}$$

This function is finite because each φ is bounded on the compact space $\mathcal{X}_0$. By Hölder's inequality $\bar{p}$ is convex. Let $J : \mathcal{X} \to [0, \infty]$ denote the convex conjugate of $\bar{p}$:

$$J(x) = \sup_{\varphi \in \mathcal{L}} \{\langle x, \varphi \rangle - \bar{p}(\varphi)\}, \quad x \in \mathcal{X}. \tag{14.2}$$

If the limit in (14.1) exists, we drop the bar and write

$$p(\varphi) = \lim_{n \to \infty} \frac{1}{r_n} \log \int_{\mathcal{X}_0} e^{r_n \langle x, \varphi \rangle} \, \mu_n(dx). \tag{14.3}$$

For $x \in \mathcal{X}$ define upper and lower local rate functions by

$$\overline{\kappa}(x) = - \inf_{G \subset \mathcal{X} \,:\, x \in G,\, G \text{ open}} \ \overline{\lim_{n \to \infty}} \frac{1}{r_n} \log \mu_n(G) \tag{14.4}$$

and

$$\underline{\kappa}(x) = - \inf_{G \subset \mathcal{X} \,:\, x \in G,\, G \text{ open}} \ \underline{\lim}_{n \to \infty} \frac{1}{r_n} \log \mu_n(G). \tag{14.5}$$

Since the measures μ_n are supported on $\mathcal{X}_0$, for $x \in \mathcal{X} \smallsetminus \mathcal{X}_0$ we have $\overline{\kappa}(x) = \underline{\kappa}(x) = \infty$. The functions $\overline{\kappa}$ and $\underline{\kappa}$ are $[0, \infty]$-valued and lower semicontinuous. By Exercise 2.24 and the compactness of the space $\mathcal{X}_0$, if $\overline{\kappa} = \underline{\kappa} = \kappa$, then the LDP holds with rate function κ. The main theorem of this section asserts that this is the case provided the pressure (14.3) exists and $\underline{\kappa}$ is convex.

Baxter-Jain theorem. *Let $\{\mu_n\}_{n \in \mathbb{N}}$ be a sequence of Borel probability measures on the compact convex subset $\mathcal{X}_0$ of $\mathcal{X}$. Assume that the limit $p(\varphi)$ in (14.3) exists for all $\varphi \in \mathcal{L}$ and that $\underline{\kappa}$ is convex. Then the LDP holds for $\{\mu_n\}$ with rate function $J = \overline{\kappa} = \underline{\kappa}$ and normalization $\{r_n\}$.*

We proceed in the proof without the two assumptions of the existence of (14.3) and the convexity of $\underline{\kappa}$, as far as possible. When these two assumptions are used, they are stated explicitly in the hypotheses of the lemma.

Lemma 14.1. *$J = \infty$ on $\mathcal{X} \smallsetminus \mathcal{X}_0$. For all $x \in \mathcal{X}_0$*

$$J(x) = \overline{\kappa}^{**}(x) \le \overline{\kappa}(x) \le \underline{\kappa}(x). \tag{14.6}$$

The set $\{\overline{\kappa} = 0\}$ is not empty. Consequently J is a lower semicontinuous, proper convex function.

Proof. Hahn-Banach separation implies that $J = \infty$ on $\mathcal{X} \smallsetminus \mathcal{X}_0$. This is left as Exercise 14.6 at the end of the chapter.

In (14.6) the last inequality is evident from definitions (14.4) and (14.5). The middle inequality is part (b) of Proposition 4.10.

We claim that

$$\bar{p}(\varphi) \geq \langle x, \varphi \rangle - \overline{\kappa}(x) \qquad \forall x \in \mathcal{X},\ \forall \varphi \in \mathcal{L}. \tag{14.7}$$

Given x and φ and $c < \langle x, \varphi \rangle$, let $G = \{y \in \mathcal{X} : \langle y, \varphi \rangle > c\}$. Then

$$\int_{\mathcal{X}_0} e^{r_n \langle y, \varphi \rangle}\, \mu_n(dy) \geq \int_G e^{r_n \langle y, \varphi \rangle}\, \mu_n(dy) \geq e^{r_n c} \mu_n(G)$$

from which

$$\bar{p}(\varphi) \geq c + \varlimsup_{n\to\infty} \frac{1}{r_n} \log \mu_n(G) \geq c - \overline{\kappa}(x).$$

Letting $c \nearrow \langle x, \varphi \rangle$ verifies (14.7).

Taking the supremum over x on the right in (14.7) gives $\bar{p} \geq \overline{\kappa}^*$, and another round of convex conjugation gives

$$J(x) \leq \overline{\kappa}^{**}(x). \tag{14.8}$$

Next we claim that

$$\bar{p}(\varphi) \leq \overline{\kappa}^*(\varphi) = \sup_{x \in \mathcal{X}_0} \{\langle x, \varphi \rangle - \overline{\kappa}(x)\}. \tag{14.9}$$

The equality above is simply the definition. The inequality completes the proof of (14.6) because it implies $J(x) \geq \overline{\kappa}^{**}(x)$. The argument for (14.9) will be needed shortly again so we separate it into the next lemma.

The claim that $\{\overline{\kappa} = 0\} \neq \varnothing$ is left as Exercise 14.8. Thus J is not identically infinite. J is convex and lower semicontinuous by its definition (14.2). ∎

Lemma 14.2. *Let $g : \mathcal{X}_0 \to [-\infty, \infty)$ be upper semicontinuous and extend the use of the notation $\bar{p}(g)$ to such functions by defining*

$$\bar{p}(g) = \varlimsup_{n\to\infty} \frac{1}{r_n} \log \int_{\mathcal{X}_0} e^{r_n g(x)}\, \mu_n(dx). \tag{14.10}$$

Then

$$\bar{p}(g) \leq \sup_{x \in \mathcal{X}_0} \{g(x) - \overline{\kappa}(x)\}. \tag{14.11}$$

Proof. Denote the right-hand side of (14.11) by $A(g)$. We can assume $A(g) < \infty$. Let $a > A(g)$ and $\varepsilon > 0$ such that $a - \varepsilon > A(g)$. (We have not ruled out $A(g) = -\infty$ but a is real.) We claim that each $x \in \mathcal{X}_0$ has an open neighborhood G_x such that for large enough n

$$\int_{G_x} e^{r_n g}\, d\mu_n \leq e^{r_n a}. \tag{14.12}$$

This suffices for the proof because by compactness we can cover $\mathcal{X}_0$ with a finite collection $G_{x_1}, \dots, G_{x_m}$ of these neighborhoods. Then for large enough n,

$$\int_{\mathcal{X}_0} e^{r_n g}\, d\mu_n \le \sum_{i=1}^m \int_{G_{x_i}} e^{r_n g}\, d\mu_n \le m e^{r_n a}.$$

This implies $\bar{p}(g) \le a$, and letting $a \searrow A(g)$ verifies (14.11).

Now we show (14.12). If $g(x) = -\infty$, take $G_x = \{g < a\}$, which is open by upper semicontinuity. Note that $g(x) = \infty$ is not allowed.

If $-\infty < g(x) < \infty$, pick G_x so that $g(y) < g(x) + \varepsilon/4$ for $y \in G_x$. Since

$$-\overline{\kappa}(x) \le A(g) - g(x) < a - \varepsilon - g(x),$$

looking at the definition of $\overline{\kappa}(x)$ we can shrink G_x to ensure that

$$\overline{\lim_{n\to\infty}} \frac{1}{r_n} \log \mu_n(G_x) \le a - \varepsilon/2 - g(x).$$

Consequently, for large enough n

$$\mu_n(G_x) \le \exp\{r_n(a - \varepsilon/4 - g(x))\}$$

and (14.12) follows. □

Combined with the observation that $\overline{\kappa} = \underline{\kappa}$ implies the LDP, Lemma 14.1 tells us that $J = \underline{\kappa}$ implies the LDP with rate function J. Our goal will now be to show that if the pressure exists and $\underline{\kappa}$ is convex, then $J = \underline{\kappa}$. The next definition is the key.

Definition 14.3. Let $f : \mathcal{X}_0 \to (-\infty, \infty]$ be a lower semicontinuous convex function. A point $z \in \mathcal{X}_0$ is an *almost exposed point* of f if, for any $\delta > 0$ and any open set $G \ni z$, there exist $\psi \in \mathcal{L}$ and an open set U such that $z \in U \subset G$ and the following inequalities hold:

$$\inf_{x \in \mathcal{X}_0 \smallsetminus U} \{f(x) - \langle x, \psi \rangle\} > f(z) - \langle z, \psi \rangle \tag{14.13}$$

and

$$|\langle x, \psi \rangle - \langle z, \psi \rangle| < \delta \qquad \text{for } x \in U. \tag{14.14}$$

Exercises 14.9 and 14.10 justify the term *almost exposed.* The strict inequality in (14.13) is important. We illustrate the usefulness of this notion by showing that $J(z) = \underline{\kappa}(z)$ at almost exposed points z of J. We assume that the limit (14.3) defining the pressure actually exists.

Lemma 14.4. *Assume that the limit in* (14.3) *exists for all* $\varphi \in \mathcal{L}$. *Then* $J(z) = \underline{\kappa}(z)$ *at each almost exposed point* z *of* J.

Proof. Fix an almost exposed point z, pick $\delta > 0$ and a neighborhood G of z, and let $\psi \in \mathcal{L}$ and neighborhood U be given by Definition 14.3. Define the upper semicontinuous $g : \mathcal{X} \to [-\infty, \infty)$ by

$$g(x) = \langle x, \psi \rangle \cdot \mathbb{1}_{U^c}(x) - \infty \cdot \mathbb{1}_U(x).$$

By Exercise 14.7,

$$p(\psi) \le \Big(\varliminf_{n\to\infty} \frac{1}{r_n} \log \int_U e^{r_n \langle x, \psi \rangle}\, \mu_n(dx) \Big) \vee \bar{p}(g). \tag{14.15}$$

From the definition (14.2),

$$p(\psi) \ge \langle z, \psi \rangle - J(z). \tag{14.16}$$

On the other hand, combining Lemma 14.2, the definition of g, inequality (14.6), and property (14.13) of almost exposed points gives

$$\begin{aligned} \bar{p}(g) &\le \sup_{x \in U^c} \{ \langle x, \psi \rangle - \overline{\kappa}(x) \} \le \sup_{x \in U^c} \{ \langle x, \psi \rangle - J(x) \} \\ &< \langle z, \psi \rangle - J(z). \end{aligned}$$

Consequently the second member of the maximum in (14.15) is irrelevant. Starting with (14.16) and using the second property (14.14) of almost exposed points,

$$\begin{aligned} \langle z, \psi \rangle - J(z) \le p(\psi) &\le \varliminf_{n\to\infty} \frac{1}{r_n} \log \int_U e^{r_n \langle x, \psi \rangle}\, \mu_n(dx) \\ &\le \langle z, \psi \rangle + \delta + \varliminf_{n\to\infty} \frac{1}{r_n} \log \mu_n(U), \end{aligned}$$

from which, since $U \subset G$,

$$\varliminf_{n\to\infty} \frac{1}{r_n} \log \mu_n(G) \ge -J(z) - \delta.$$

Since G is an arbitrary neighborhood of z and $\delta > 0$ is arbitrary, $-\underline{\kappa}(z) \ge -J(z)$ follows. □

We can now state the main technical result of this section, which shows that almost exposed points are plentiful. Recall that a point z is an *extreme point* of a convex set K if $z \in K$ and it has the following property: if $z = sx + (1-s)y$ for some $x, y \in K$ and $0 < s < 1$, then $x = y = z$. In other words, z cannot lie on a line segment inside K unless it is an endpoint. The set of extreme points of K is denoted by $\mathrm{ex}(K)$.

Theorem 14.5. *Let $f : \mathcal{X}_0 \to (-\infty, \infty]$ be a lower semicontinuous convex function. Let $\varphi \in \mathcal{L}$, $c \in \mathbb{R}$, and define the affine function $g(x) = \langle x, \varphi \rangle + c$. Assume that $g \le f$ on $\mathcal{X}_0$ and that the convex set $A = \{x \in \mathcal{X}_0 : f(x) = g(x)\}$ is nonempty. Then every extreme point of A is an almost exposed point of f.*

A is a closed subset of the compact space $\mathcal{X}_0$, and hence it is a compact convex set. By the Krein-Milman theorem [**69**, Theorem 3.23] a compact convex set is the closed convex hull of its extreme points. Thus A has extreme points, and consequently f has almost exposed points. Before proving Theorem 14.5 let us see how it completes the LDP.

Proof of the Baxter-Jain theorem. By its definition $\underline{\kappa}$ is lower semicontinuous, and the assumption is that it is also convex. Hence $\underline{\kappa} = \underline{\kappa}^{**}$ and by Exercise 4.11 for $z \in \mathcal{X}$,

$$\underline{\kappa}(z) = \sup\{\langle z, \varphi\rangle + c : \varphi \in \mathcal{L},\, c \in \mathbb{R},\, \langle x, \varphi\rangle + c \le \underline{\kappa}(x)\,\forall x \in \mathcal{X}\}.$$

Suppose $\langle x, \varphi\rangle + c \le \underline{\kappa}(x)$ for all $x \in \mathcal{X}$. We claim that then also

$$\langle x, \varphi\rangle + c \le J(x) \qquad \text{for all } x \in \mathcal{X}. \tag{14.17}$$

This implies that $\underline{\kappa} \le J$, and thereby $\underline{\kappa} = J$ since the opposite inequality was in Lemma 14.1. Then we have $J = \overline{\kappa} = \underline{\kappa}$, and as already pointed out, this is enough for the LDP with rate J.

Inequality (14.17) needs to be verified only on $\mathcal{X}_0$ because $J \equiv \infty$ on $\mathcal{X} \setminus \mathcal{X}_0$. Let c_1 be the (finite) infimum of the lower semicontinuous function $J(x) - \langle x, \varphi\rangle$ on the compact set $\mathcal{X}_0$. Since the sets $\{x : J(x) - \langle x, \varphi\rangle \le c_1 + m^{-1}\}$, $m \in \mathbb{N}$, are a nested family of nonempty compact sets, their intersection $A = \{x : J(x) - \langle x, \varphi\rangle = c_1\}$ is not empty. Let $g(x) = \langle x, \varphi\rangle + c_1$. We are exactly in the situation of Theorem 14.5 and so $\mathrm{ex}(A)$ consists of almost exposed points of J. By Lemma 14.4, $J(z) = \underline{\kappa}(z)$ for all $z \in \mathrm{ex}(A)$.

To summarize, we have now for $z \in \mathrm{ex}(A)$ that

$$\langle z, \varphi\rangle + c \le \underline{\kappa}(z) = J(z) = \langle z, \varphi\rangle + c_1.$$

Consequently $c \le c_1$ and, by the definition $c_1 = \inf_x(J(x) - \langle x, \varphi\rangle)$, (14.17) follows. ☐

As the last item of this section we prove Theorem 14.5.

Proof of Theorem 14.5. Let $\varphi \in \mathcal{L}$, $g = \varphi + c$, and $A = \{f = g\}$ as given in the statement of the theorem. Let $z \in \mathrm{ex}(A)$, and let $\delta > 0$ and an open neighborhood G of z be given. We need to produce $\psi \in \mathcal{L}$ and a neighborhood U of z such that (14.13) and (14.14) are satisfied.

Pick an open neighborhood $U \subset G$ of z such that

$$|g(x) - g(z)| < \delta/2 \qquad \text{for } x \in U.$$

Let K be the closed convex hull of the compact set $A \setminus U$. Since A is compact and convex, $K \subset A$, and thereby K is also compact and convex. By Milman's theorem [**69**, Theorem 3.25], $\mathrm{ex}(K) \subset A \setminus U$. It follows that $z \notin K$. (If $z \in K$, then z must be extreme in K since it is extreme in the larger set A. But $z \in U$ and $\operatorname{ex}(K) \subset A \smallsetminus U$.) By the Hahn-Banach separation theorem there exists $\lambda \in \mathcal{L}$ such that

$$\sup_{x \in K} \langle x, \lambda \rangle < \langle z, \lambda \rangle.$$

By continuity, we can find an open set $V \supset K$ and $\eta > 0$ such that

$$\sup_{x \in V} \langle x, \lambda \rangle < \langle z, \lambda \rangle - \eta. \tag{14.18}$$

Since $U \cup V$ is an open set and contains A, by compactness there exists $\varepsilon > 0$ such that $f - g \geq \varepsilon$ on $\mathcal{X}_0 \smallsetminus (U \cup V)$. Again by compactness we can fix $a > 0$ small enough so that

$$a\,|\langle x - z, \lambda \rangle| \leq \tfrac{1}{2}(\varepsilon \wedge \delta) \quad \text{for all } x \in \mathcal{X}_0.$$

Now set

$$\psi = \varphi + a\lambda.$$

We check that (14.13) and (14.14) are satisfied. For (14.14), recalling that $g = \varphi + c$,

$$|\langle x, \psi \rangle - \langle z, \psi \rangle| \leq |g(x) - g(z)| + a\,|\langle x - z, \lambda \rangle| < \delta \quad \text{for } x \in U.$$

Condition (14.13) is checked in two parts. For $x \in \mathcal{X}_0 \smallsetminus (U \cup V)$, recalling that $f(z) = g(z)$,

$$f(x) - g(x) \geq \varepsilon \geq a\langle x - z, \lambda \rangle + \varepsilon/2 + f(z) - g(z).$$

This rearranges to

$$f(x) - \langle x, \psi \rangle \geq f(z) - \langle z, \psi \rangle + \varepsilon/2.$$

For $x \in V$, by (14.18) and $f \geq g$,

$$f(x) - g(x) - a\langle x, \lambda \rangle \geq f(z) - g(z) - a\langle z, \lambda \rangle + a\eta$$

and thus

$$f(x) - \langle x, \psi \rangle \geq f(z) - \langle z, \psi \rangle + a\eta.$$

This completes the proof of Theorem 14.5. □

***Exercise 14.6.** Use Hahn-Banach separation to show that $J = \infty$ on $\mathcal{X} \smallsetminus \mathcal{X}_0$.

***Exercise 14.7.** For $a_n, b_n \geq 0$,

$$\varlimsup_{n \to \infty} \frac{1}{r_n} \log(a_n + b_n) \leq \Big(\varlimsup_{n \to \infty} \frac{1}{r_n} \log a_n \Big) \vee \Big(\varlimsup_{n \to \infty} \frac{1}{r_n} \log b_n \Big).$$

***Exercise 14.8.** Show that $\overline{\kappa}$ is a rate function that satisfies the upper large deviation bound and that the set $\{\overline{\kappa} = 0\}$ is not empty.

Exercise 14.9. Let f be a finite convex function on (a,b). Show that $z \in (a,b)$ is almost exposed if and only if z is not an interior point of an interval (c,d) on which f is of the form $f(x) = \alpha x + \beta$.

Exercise 14.10. Let f be as in Definition 14.3. Recall that z is an exposed point of f if there exists $\varphi \in \mathcal{L}$ such that

$$f(x) > f(z) + \langle x - z, \varphi \rangle \quad \forall x \neq z.$$

(a) Show that an exposed point is almost exposed.

(b) Construct an example of a point z that is almost exposed but not exposed and such that $\partial f(z) \neq \varnothing$.

Hint: Use the characterization of Exercise 14.9 for inspiration.

Exercise 14.11. This exercise shows that an almost exposed point can have empty subdifferential. Let $a < b$ and $f : \mathbb{R} \to (-\infty, \infty]$ convex. Assume f is finite and continuous on $[a,b]$ and $f = \infty$ on (b, ∞). Show the following.

(a) If $f'(b-) < \infty$, then b is an exposed point of f.

(b) If $f'(b-) = \infty$, then b is an almost exposed point of f and $\partial f(b) = \varnothing$. In particular, b is not an exposed point of f.

Exercise 14.12. Let $\mathcal{X}_0$ be the unit interval $[0,1]$, $0 < \alpha < \infty$, and fix a subinterval $A = [a,b]$ of $(0,1)$. Let μ_n be the probability measure on $\mathcal{X}_0$ with Radon-Nikodym derivative

$$f_n(x) = \frac{1 - e^{-n\alpha}}{1 - m(A)} \mathbb{1}_{\mathcal{X}_0 \smallsetminus A}(x) + \frac{e^{-n\alpha}}{m(A)} \mathbb{1}_A(x)$$

relative to Lebesgue measure m.

(a) Prove the LDP by applying Exercise 2.24.

(b) Show that Bryc's theorem gives the same LDP.

(c) Check which hypotheses of the Gärtner-Ellis theorem and the Baxter-Jain theorem fail. These theorems must fail in this example because they give an LDP with a convex rate function.

Exercise 14.13. Re-prove Sanov's theorem with the Baxter-Jain theorem.

Chapter 15

Nonstationary independent variables

15.1. Generalization of relative entropy and Sanov's theorem

In this chapter we generalize Sanov's theorem to a sequence of independent but not identically distributed random variables. As in Section 5.2 the state space of the random variables $\{X_k\}$ is a Polish space $\mathcal{S}$. We assume the random variables are defined as coordinate variables on the sequence space $\Omega = \mathcal{S}^{\mathbb{N}}$. In this section we need to discuss even more frequently than before probability measures on the space $\mathcal{M}_1(\mathcal{S})$ of probability measures on $\mathcal{S}$, so let us abbreviate $\mathcal{M}_1 = \mathcal{M}_1(\mathcal{S})$ and $\mathcal{M}_1(\mathcal{M}_1) = \mathcal{M}_1(\mathcal{M}_1(\mathcal{S}))$. As throughout, $\mathcal{M}_1$ is given the weak topology generated by $\mathscr{C}_b(\mathcal{S})$. Generic elements of $\mathcal{M}_1$, that is, probability measures on $\mathcal{S}$, will be denoted by α, β, and γ, while κ will be a probability measure on $\mathcal{M}_1$ and μ and ν are probability measures on the product space $\mathcal{S} \times \mathcal{M}_1$.

Let $\boldsymbol{\lambda} = (\lambda_k)_{k\geq 1}$ be a sequence of probability measures on $\mathcal{S}$, that is, an element of $\mathcal{M}_1^{\mathbb{N}}$. Let $P^{\boldsymbol{\lambda}} = \bigotimes \lambda_k$ denote the product probability measure on Ω with marginals (λ_k), uniquely defined by the requirement that

$$P^{\boldsymbol{\lambda}}\{X_1 \in B_1, X_2 \in B_2, \ldots, X_n \in B_n\} = \prod_{k=1}^{n} \lambda_k(B_k)$$

for Borel subsets $B_k \subset \mathcal{S}$. Our interest lies in the large deviations under $P^{\boldsymbol{\lambda}}$ of the empirical measure

$$L_n = \frac{1}{n}\sum_{k=1}^{n} \delta_{X_k}.$$

We begin by developing the relevant generalization of relative entropy. Assume we are given a probability measure Ψ on $\mathcal{M}_1$. Define a joint distribution μ on $\mathcal{S} \times \mathcal{M}_1$ by

$$\mu(dx, d\alpha) = \alpha(dx)\, \Psi(d\alpha). \tag{15.1}$$

The $\mathcal{S}$-marginal $\mu_{\mathcal{S}}$ of μ can be regarded as the mean of Ψ:

$$\mu_{\mathcal{S}}(B) = \int_{\mathcal{M}_1} \alpha(B)\, \Psi(d\alpha) \quad \text{for Borel subsets } B \subset \mathcal{S}. \tag{15.2}$$

Recall Definition 5.2 of relative entropy H.

Definition 15.1. Given $\Psi \in \mathcal{M}_1(\mathcal{M}_1)$ and μ defined by (15.1), define the following entropy for $\gamma \in \mathcal{M}_1$:

$$K(\gamma \,|\, \Psi) = \inf_{\nu} H(\nu \,|\, \mu) \tag{15.3}$$

where the infimum is over probability measures ν on $\mathcal{S} \times \mathcal{M}_1$ with marginals γ and Ψ.

We now give the duality and other basic properties of K.

Theorem 15.2. (a) *For* $\gamma \in \mathcal{M}_1$

$$K(\gamma \,|\, \Psi) = \sup_{f \in \mathscr{C}_b(\mathcal{S})} \Big\{ E^{\gamma}[f] - \int_{\mathcal{M}_1} \log E^{\alpha}[e^f]\, \Psi(d\alpha) \Big\}. \tag{15.4}$$

(b) $K(\gamma \,|\, \Psi) \ge H(\gamma \,|\, \mu_{\mathcal{S}})$ *for all* $\gamma \in \mathcal{M}_1$. $K(\gamma \,|\, \Psi) = H(\gamma \,|\, \mu_{\mathcal{S}})$ *for all* $\gamma \in \mathcal{M}_1$ *if and only if* Ψ *is a point mass.*

(c) K *is a convex, tight rate function on* $\mathcal{M}_1$ *and* $K(\gamma \,|\, \Psi) = 0$ *if and only if* $\gamma = \mu_{\mathcal{S}}$.

Proof. (a) Apply Theorem 13.1 with the choices $\mathcal{X} = \mathcal{S}$, $\mathcal{Y} = \mathcal{M}_1$, and $\rho^{\alpha} = \alpha$ for $\alpha \in \mathcal{M}_1$.

(b) For the inequality apply Jensen's inequality inside the supremum in (15.4), as in (13.8). If Ψ is a point mass, then $\Psi = \delta_{\mu_{\mathcal{S}}}$ and $K(\gamma \,|\, \Psi) = H(\gamma \,|\, \mu_{\mathcal{S}})$ follows from (15.4).

Suppose now that $K(\gamma \,|\, \Psi) = H(\gamma \,|\, \mu_{\mathcal{S}})$ for all $\gamma \in \mathcal{M}_1$. Let $f \in \mathscr{C}_b(\mathcal{S})$ such that $\int e^f \, d\mu_{\mathcal{S}} = 1$. By Jensen's inequality

$$\int_{\mathcal{M}_1} \log E^{\alpha}[e^f]\, \Psi(d\alpha) \le \log \int_{\mathcal{M}_1} E^{\alpha}[e^f]\, \Psi(d\alpha) = 0. \tag{15.5}$$

Define the probability measure γ by $d\gamma = e^f \, d\mu_{\mathcal{S}}$. Then

$$E^{\gamma}[f] = H(\gamma \,|\, \mu_{\mathcal{S}}) = K(\gamma \,|\, \Psi) \ge E^{\gamma}[f] - \int_{\mathcal{M}_1} \log E^{\alpha}[e^f]\, \Psi(d\alpha)$$

from which

$$\int_{\mathcal{M}_1} \log E^{\alpha}[e^f]\, \Psi(d\alpha) \geq 0.$$

Thus we have equality in the Jensen inequality in (15.5). Then by the strict concavity of log, the random variable $\alpha \mapsto E^{\alpha}[e^f]$ must be degenerate, in other words $E^{\alpha}[e^f] = 1 = E^{\mu_{\mathcal{S}}}[e^f]$ for Ψ-a.e. α. Since $\mathcal{S}$ is Polish there exists a countable family $\mathcal{F}$ of bounded functions such that $\{e^f : f \in \mathcal{F}\}$ separates measures, and so this is enough for concluding that Ψ-a.e. α equals $\mu_{\mathcal{S}}$.

(c) Convexity is evident from either (15.3) or (15.4). $K(\mu_{\mathcal{S}} \,|\, \Psi) = 0$ follows from observing that for $\gamma = \mu_{\mathcal{S}}$, the right-hand side of (15.4) is ≤ 0 for all f, by Jensen's inequality. The rest follows from $K(\gamma \,|\, \Psi) \geq H(\gamma \,|\, \mu_{\mathcal{S}})$. □

Definition 15.3. A sequence $\boldsymbol{\lambda} = (\lambda_k)_{k\geq 1} \in \mathcal{M}_1^{\mathbb{N}}$ is called *regular* if the limit

$$\Psi(\boldsymbol{\lambda}) = \lim_{n\to\infty} \frac{1}{n} \sum_{k=1}^{n} \delta_{\lambda_k} \tag{15.6}$$

exists in the weak topology of $\mathcal{M}_1(\mathcal{M}_1)$.

The main theorem of this chapter asserts that regularity of the sequence $\boldsymbol{\lambda}$ is necessary and sufficient for Sanov's theorem to hold under the measure $P^{\boldsymbol{\lambda}}$.

Theorem 15.4. *The distributions $P^{\boldsymbol{\lambda}}\{L_n \in \cdot\}$ satisfy an LDP with normalization $\{n\}$ if and only if $\boldsymbol{\lambda}$ is regular. In this case the rate function is $K(\cdot \,|\, \Psi(\boldsymbol{\lambda}))$.*

15.2. Proof of the large deviation principle

We begin with a small generalization of the LDP proved in Chapter 14. As in that chapter, let $\mathcal{X}$ and $\mathcal{L}$ be two real vector spaces in duality, endowed with their weak topologies, assumed Hausdorff, and let $\mathcal{X}_0$ be a convex, compact subset of $\mathcal{X}$. Let $\mathcal{I}$ be an arbitrary index set, and for each $i \in \mathcal{I}$ let $\{\mu_n^{(i)} : n \in \mathbb{N}\}$ be a sequence of Borel probability measures on $\mathcal{X}_0$. Let $0 < r_n \nearrow \infty$ be a normalizing sequence.

For $x \in \mathcal{X}$ define the upper and lower local rate functions by

$$\overline{\kappa}^{(i)}(x) = -\inf_{G \subset \mathcal{X}\,:\, x \in G,\, G \text{ open}} \; \overline{\lim_{n\to\infty}} \frac{1}{r_n} \log \mu_n^{(i)}(G) \tag{15.7}$$

and

$$\underline{\kappa}^{(i)}(x) = -\inf_{G \subset \mathcal{X}\,:\, x \in G,\, G \text{ open}} \; \underline{\lim_{n\to\infty}} \frac{1}{r_n} \log \mu_n^{(i)}(G). \tag{15.8}$$

Assume that the sequences $\{\mu_n^{(i)} : n \in \mathbb{N}\}$ all determine the *same* pressure: for all $\varphi \in \mathcal{L}$ and $i \in \mathcal{I}$,

$$p(\varphi) = \lim_{n\to\infty} \frac{1}{r_n} \log \int_{\mathcal{X}_0} e^{r_n\langle x,\varphi\rangle}\, \mu_n^{(i)}(dx). \tag{15.9}$$

Let $J : \mathcal{X} \to [0,\infty]$ denote the convex conjugate of p:

$$J(x) = \sup_{\varphi\in\mathcal{L}}\{\langle x,\varphi\rangle - p(\varphi)\}, \quad x \in \mathcal{X}. \tag{15.10}$$

Define

$$\kappa(x) = \sup_{i\in\mathcal{I}} \underline{\kappa}^{(i)}(x). \tag{15.11}$$

Theorem 15.5. *Assume that the limit $p(\varphi)$ in (15.9) exists for all $\varphi \in \mathcal{L}$ and $i \in \mathcal{I}$ and that κ is convex. Then for each $i \in \mathcal{I}$ the sequence $\{\mu_n^{(i)}\}_{n\in\mathbb{N}}$ satisfies the LDP with rate function $J = \overline{\kappa}^{(i)} = \underline{\kappa}^{(i)} = \kappa$ and normalization $\{r_n\}$.*

***Exercise 15.6.** Go through Chapter 14 to check what more needs to be said to prove Theorem 15.5. Not much, you should discover.

To start the proof of Theorem 15.4, assume first that limit (15.6) holds and also temporarily that $\mathcal{S}$ is compact. We apply Theorem 15.5 to prove the LDP. Take $\mathcal{X}$ to be the vector space of real-valued Borel measures on $\mathcal{S}$, $\mathcal{X}_0 = \mathcal{M}_1$, and $\mathcal{L} = \mathscr{C}(\mathcal{S})$. The normalization sequence is naturally $r_n = n$. The index set is $\mathcal{I} = \mathbb{Z}_+ = \{0, 1, 2, \dots\}$, and the measures $\mu_n^{(i)}$ are distributions of shifted empirical measures:

$$\mu_n^{(i)}(B) = P^{\boldsymbol{\lambda}}\Big\{\frac{1}{n}\sum_{k=in+1}^{in+n} \delta_{X_k} \in B\Big\}, \qquad B \subset \mathcal{M}_1 \text{ measurable}. \tag{15.12}$$

We begin by verifying assumption (15.9). A linear functional $\varphi \in \mathcal{L}$ is now identified with a function $f \in \mathscr{C}(\mathcal{S})$ and the duality is given by integration: $\langle \alpha, \varphi\rangle = \int f\,d\alpha$ for $\alpha \in \mathcal{X}$. As a result, as $n \to \infty$, by (15.6),

$$\begin{aligned}
\frac{1}{n}\log\int_{\mathcal{X}_0} e^{n\langle\alpha,\varphi\rangle}\,\mu_n^{(i)}(d\alpha) &= \frac{1}{n}\log\int_\Omega \exp\Big\{\sum_{k=in+1}^{in+n} f(X_k)\Big\}\,dP^{\boldsymbol{\lambda}}\\
&= \frac{1}{n}\sum_{k=in+1}^{in+n} \log E^{\lambda_k}[e^f]\\
&= \frac{i+1}{(i+1)n}\sum_{k=1}^{in+n} \log E^{\lambda_k}[e^f] \;-\; \frac{i}{in}\sum_{k=1}^{in} \log E^{\lambda_k}[e^f]\\
&\longrightarrow \int_{\mathcal{M}_1} \log E^{\alpha}[e^f]\,\Psi(\boldsymbol{\lambda}, d\alpha) = p(f).
\end{aligned} \tag{15.13}$$

Convexity of κ will follow from verifying that, for all $i \in \mathbb{Z}_+$ and probability measures α, β, and $\gamma \in \mathcal{M}_1$ such that $\gamma = (\alpha+\beta)/2$,

$$\underline{\kappa}^{(i)}(\gamma) \le \tfrac{1}{2}\underline{\kappa}^{(2i)}(\alpha) + \tfrac{1}{2}\underline{\kappa}^{(2i+1)}(\beta). \tag{15.14}$$

To complete the argument for convexity of κ from (15.14), note that the right-hand side is bounded above by $\big(\kappa(\alpha)+\kappa(\beta)\big)/2$, and then take the supremum over i on the left and apply Exercise 4.7.

To prove (15.14), let $c > -\underline{\kappa}^{(i)}(\gamma)$, and pick an open neighborhood G of γ such that

$$c > \varlimsup_{n\to\infty} \frac{1}{n} \log P^{\boldsymbol{\lambda}}\Big\{\frac{1}{n}\sum_{k=in+1}^{in+n} \delta_{X_k} \in G\Big\}. \tag{15.15}$$

We can assume that G is of the form

$$G = \{\rho \in \mathcal{M}_1 : |E^\rho[f_\ell] - E^\gamma[f_\ell]| < 4\varepsilon,\ \ell = 1,\dots,L\}$$

for some functions $f_1,\dots,f_L \in \mathscr{C}(\mathcal{S})$ and $\varepsilon > 0$. Define neighborhoods of α and β by

$$G_1 = \{\rho \in \mathcal{M}_1 : |E^\rho[f_\ell] - E^\alpha[f_\ell]| < \varepsilon,\ \ell = 1,\dots,L\}$$

and

$$G_2 = \{\rho \in \mathcal{M}_1 : |E^\rho[f_\ell] - E^\beta[f_\ell]| < \varepsilon,\ \ell = 1,\dots,L\}.$$

Let $m = \lfloor n/2 \rfloor$. Do a quick calculation to check that, if n is large enough,

$$\begin{aligned} \bigg| \frac{1}{n}\sum_{k=in+1}^{in+n} f_\ell(X_k) &- \frac{1}{2}\cdot\frac{1}{m}\sum_{k=2im+1}^{(2i+1)m} f_\ell(X_k) \\ &- \frac{1}{2}\cdot\frac{1}{m}\sum_{k=(2i+1)m+1}^{(2i+2)m} f_\ell(X_k)\bigg| \le \varepsilon \end{aligned} \tag{15.16}$$

uniformly for all values $\{X_k\}$. (Since f_ℓ is bounded, the issue is only which terms appear in the sums.) This and independence of the $\{X_k\}$ under $P^{\boldsymbol{\lambda}}$ imply that

$$\begin{aligned} &P^{\boldsymbol{\lambda}}\Big\{\frac{1}{n}\sum_{k=in+1}^{in+n} \delta_{X_k} \in G\Big\} \\ &\quad\ge P^{\boldsymbol{\lambda}}\Big\{\frac{1}{m}\sum_{k=2im+1}^{(2i+1)m} \delta_{X_k} \in G_1,\ \frac{1}{m}\sum_{k=(2i+1)m+1}^{(2i+2)m} \delta_{X_k} \in G_2\Big\} \\ &\quad= P^{\boldsymbol{\lambda}}\Big\{\frac{1}{m}\sum_{k=2im+1}^{(2i+1)m} \delta_{X_k} \in G_1\Big\} P^{\boldsymbol{\lambda}}\Big\{\frac{1}{m}\sum_{k=(2i+1)m+1}^{(2i+2)m} \delta_{X_k} \in G_2\Big\}. \end{aligned}$$

Apply $n^{-1}\log$ and let $n\to\infty$ to continue from (15.15):

$$
\begin{aligned}
c &> \varliminf_{n\to\infty}\frac{1}{n}\log P^{\boldsymbol{\lambda}}\Big\{\frac{1}{n}\sum_{k=in+1}^{in+n}\delta_{X_k}\in G\Big\}\\
&\ge \frac{1}{2}\cdot\varliminf_{m\to\infty}\frac{1}{m}\log P^{\boldsymbol{\lambda}}\Big\{\frac{1}{m}\sum_{k=2im+1}^{(2i+1)m}\delta_{X_k}\in G_1\Big\}\\
&\quad+\frac{1}{2}\cdot\varliminf_{m\to\infty}\frac{1}{m}\log P^{\boldsymbol{\lambda}}\Big\{\frac{1}{m}\sum_{k=(2i+1)m+1}^{(2i+2)m}\delta_{X_k}\in G_2\Big\}\\
&\ge -\frac{1}{2}\underline{\kappa}^{(2i)}(\alpha)-\frac{1}{2}\underline{\kappa}^{(2i+1)}(\beta).
\end{aligned}
$$

Letting $c\searrow -\underline{\kappa}^{(i)}(\gamma)$ completes the proof of (15.14).

We have verified the hypotheses of Theorem 15.5. Consequently we have the LDP for the sequence $\{\mu_n^{(0)}\}$ with the rate function $J=p^*$ defined by (15.10). Calculation (15.13) identifies the pressure $p(f)$, and then a glance at (15.4) confirms that $J=K(\,\cdot\mid\Psi(\boldsymbol{\lambda}))$ as claimed in Theorem 15.4. We have completed the proof of the "if" part of Theorem 15.4 under the assumption that $\mathcal{S}$ is compact.

It remains to lift the compactness assumption. Now let $\mathcal{S}$ be Polish. We check exponential tightness.

Lemma 15.7. *Assume* (15.6). *Then for each* $b>0$ *there exists a compact subset* $K_b\subset\mathcal{M}_1$ *such that for all* $n\in\mathbb{N}$

$$
P^{\boldsymbol{\lambda}}\{L_n\in K_b^c\}\le e^{-bn}. \tag{15.17}
$$

***Exercise 15.8.** Show that $\Gamma\mapsto\rho_\Gamma(\cdot)=\int_{\mathcal{M}_1}\alpha(\cdot)\,\Gamma(d\alpha)$ defines a continuous mapping from $\mathcal{M}_1(\mathcal{M}_1)$ to $\mathcal{M}_1$. This follows fairly directly from the definition of weak topology on probability measures.

Proof of Lemma 15.7. Pick a positive sequence $\varepsilon_j\searrow 0$ and define $c_j=\varepsilon_j^{-1}(j+1+\log 2)$. By the exercise above and assumption (15.6) the probability measures

$$
\bar{\lambda}_n=\frac{1}{n}\sum_{k=1}^{n}\lambda_k
$$

converge weakly. A convergent sequence is tight (Exercise B.17), so we may pick compact sets $A_j\subset\mathcal{S}$ such that $\bar{\lambda}_n(A_j^c)\le e^{-c_j}$ for all n. By an exponential Chebyshev inequality and the arithmetic-geometric inequality (a special case of Jensen's inequality) and the definition of c_j,

$$P^{\boldsymbol{\lambda}}\{L_n(A_j^c) \ge \varepsilon_j\} = P^{\boldsymbol{\lambda}}\Big\{\sum_{k=1}^{n} \mathbb{1}_{A_j^c}(X_k) \ge n\varepsilon_j\Big\}$$

$$\le e^{-n\varepsilon_j c_j} \prod_{k=1}^{n} E^{\lambda_k}\big[e^{c_j \mathbb{1}_{A_j^c}}\big] \le e^{-n\varepsilon_j c_j}\Big(\frac{1}{n}\sum_{k=1}^{n} E^{\lambda_k}\big[e^{c_j \mathbb{1}_{A_j^c}}\big]\Big)^n$$

$$= e^{-n\varepsilon_j c_j}\Big(E^{\bar{\lambda}_n}\big[e^{c_j \mathbb{1}_{A_j^c}}\big]\Big)^n \le e^{-n\varepsilon_j c_j} 2^n = e^{-n(j+1)}.$$

Let $K_b = \{\alpha \in \mathcal{M}_1 : \alpha(A_j^c) \le \varepsilon_j \text{ for } j \ge b\}$. K_b is closed by the portmanteau theorem (page 279) and compact because the defining condition forces tightness. And finally,

$$P^{\boldsymbol{\lambda}}\{L_n \in K_b^c\} \le \sum_{j \ge b} P^{\boldsymbol{\lambda}}\{L_n(A_j^c) > \varepsilon_j\} \le e^{-bn}. \qquad \square$$

For the final step of the proof of the LDP we compactify, obtain the LDP on the compactification, and then argue that the LDP can be restricted to the original space.

Let $\bar{\mathcal{S}}$ be the compactification of $\mathcal{S}$ obtained by completing $\mathcal{S}$ under a totally bounded metric $\bar{d}$, as was done in the proof of Theorem 13.1. $\mathcal{S}$ is a dense Borel subset of $\bar{\mathcal{S}}$. $\mathcal{M}_1$ can be considered as a subset of $\mathcal{M}_1(\bar{\mathcal{S}})$ because every measure on $\mathcal{S}$ is also a measure on $\bar{\mathcal{S}}$. As a subspace of $\mathcal{M}_1(\bar{\mathcal{S}})$ the space $\mathcal{M}_1$ has its original topology. This is because the weak topology of $\mathcal{M}_1(\bar{\mathcal{S}})$ is generated by $\mathscr{C}(\bar{\mathcal{S}})$, which is in bijective correspondence with $\mathscr{U}_{b,\bar{d}}(\mathcal{S})$ via restriction and unique extension of functions. And in fact we can write $\overline{\mathcal{M}_1}$ for $\mathcal{M}_1(\bar{\mathcal{S}})$ because it is the closure of $\mathcal{M}_1$ (Exercise 15.13).

From these considerations it follows that the limit in assumption (15.6) is also valid in the space $\mathcal{M}_1(\overline{\mathcal{M}_1})$. The first part of the proof then gives us an LDP on the space $\overline{\mathcal{M}_1}$ with rate function $K(\cdot\,|\Psi(\boldsymbol{\lambda}))$. Let G be an open subset of $\mathcal{M}_1$ and let A be a closed subset of $\mathcal{M}_1$. By the definition of relative topology, there exist an open set $G_1 \subset \overline{\mathcal{M}_1}$ and a closed set $A_1 \subset \overline{\mathcal{M}_1}$ such that $G = G_1 \cap \mathcal{M}_1$ and $A = A_1 \cap \mathcal{M}_1$. Since $L_n \in \mathcal{M}_1$ $P^{\boldsymbol{\lambda}}$-a.s., the LDP on the larger space $\bar{\mathcal{M}}_1$ gives us the bounds

$$\underline{\lim}_{n\to\infty} \frac{1}{n} \log P^{\boldsymbol{\lambda}}\{L_n \in G\} \ge -\inf_{\alpha \in G_1} K(\alpha\,|\Psi(\boldsymbol{\lambda})) \tag{15.18}$$

and

$$\overline{\lim_{n\to\infty}} \frac{1}{n} \log P^{\boldsymbol{\lambda}}\{L_n \in A\} \le -\inf_{\alpha \in A_1} K(\alpha\,|\Psi(\boldsymbol{\lambda})). \tag{15.19}$$

The lower bound on the space $\mathcal{M}_1$ follows now because the right-hand side of (15.18) cannot increase if G_1 is replaced by G. To replace A_1 with A on the right-hand side of (15.19) we need to show that

$$K(\alpha\,|\,\Psi(\boldsymbol{\lambda})) = \infty \quad \text{for } \alpha \in \overline{\mathcal{M}_1} \setminus \mathcal{M}_1. \tag{15.20}$$

This is a consequence of exponential tightness: (15.17) and the lower bound give

$$\inf_{\alpha\in K_b^c} K(\alpha\,|\,\Psi(\boldsymbol{\lambda})) \geq -\varlimsup_{n\to\infty} \frac{1}{n} \log P^{\boldsymbol{\lambda}}\{L_n \in K_b^c\} \geq b.$$

Since $K_b \subset \mathcal{M}_1$ we can let $b \nearrow \infty$ to get (15.20).

To summarize, (15.18) and (15.19) turn into the LDP on the space $\mathcal{M}_1$ claimed in Theorem 15.4, and the proof of the "if" part is complete.

Next we prove the "only if" part of Theorem 15.4, namely that existence of an LDP implies the existence of the limit in (15.6). First we give a technical lemma.

Lemma 15.9. *Let $\mathscr{X}$ be a compact metric space, and let Φ and Ψ be Borel probability measures on $\mathcal{M}_1(\mathscr{X})$. Assume that*

$$\int_{\mathcal{M}_1(\mathscr{X})} \log E^\alpha[e^g]\,\Phi(d\alpha) = \int_{\mathcal{M}_1(\mathscr{X})} \log E^\alpha[e^g]\,\Psi(d\alpha) \tag{15.21}$$

for all $g \in \mathscr{C}(\mathscr{X})$. Then $\Phi = \Psi$.

Remark 15.10. By a compactification argument this lemma can be proved for Polish spaces but we need it only for compact spaces.

Proof. Let $h \in \mathscr{C}(\mathscr{X})$. Let $\delta > 0$ be small enough so that $\delta\,\|h\|_\infty < 1$. Then for $u \in (-\delta, \delta)$ let $g(x) = \log(1 + uh(x))$, and expand:

$$\begin{aligned}\int_{\mathcal{M}_1(\mathscr{X})} \log E^\alpha[e^g]\,\Phi(d\alpha) &= \int_{\mathcal{M}_1(\mathscr{X})} \log(1 + uE^\alpha[h])\,\Phi(d\alpha)\\ &= -\sum_{k=1}^{\infty} \frac{(-u)^k}{k} \int_{\mathcal{M}_1(\mathscr{X})} (E^\alpha[h])^k\,\Phi(d\alpha).\end{aligned}$$

The last expression is an analytic function of $u \in (-\delta, \delta)$, and consequently the coefficients of its power series expansion are uniquely determined. We can conclude that

$$\int_{\mathcal{M}_1(\mathscr{X})} (E^\alpha[h])^k\,\Phi(d\alpha) = \int_{\mathcal{M}_1(\mathscr{X})} (E^\alpha[h])^k\,\Psi(d\alpha) \quad \forall h \in \mathscr{C}(\mathscr{X}).$$

From this and the power series expansion for the exponential, we deduce the equality of the following characteristic functions:

$$\int_{\mathcal{M}_1(\mathscr{X})} e^{iE^\alpha[h]}\,\Phi(d\alpha) = \int_{\mathcal{M}_1(\mathscr{X})} e^{iE^\alpha[h]}\,\Psi(d\alpha) \quad \forall h \in \mathscr{C}(\mathscr{X})$$

where $i = \sqrt{-1}$ is the imaginary unit. Take linear combinations $h = \sum_{k=1}^m t_k g_k$ for $g_1, \dots, g_m \in \mathscr{C}(\mathscr{X})$ and vary the vectors $(t_1, \dots, t_m) \in \mathbb{R}^m$

to conclude that any vector of the form $(E^\alpha[g_1], \ldots, E^\alpha[g_m])$ has the same distribution under $\Phi(d\alpha)$ and $\Psi(d\alpha)$.

Integrating term by term, it follows that

$$\int_{\mathcal{M}_1(\mathscr{X})} p(E^\alpha[g_1], \ldots, E^\alpha[g_m])\,\Phi(d\alpha) = \int_{\mathcal{M}_1(\mathscr{X})} p(E^\alpha[g_1], \ldots, E^\alpha[g_m])\,\Psi(d\alpha)$$

for any polynomial p in m variables and $g_1, \ldots, g_m \in \mathscr{C}(\mathscr{X})$. Finally we use compactness. By the Stone-Weierstrass theorem [**34**, Section 4.7] this class of functions is dense among continuous functions on $\mathcal{M}_1(\mathscr{X})$. Consequently we have

$$\int_{\mathcal{M}_1(\mathscr{X})} F(\alpha)\,\Phi(d\alpha) = \int_{\mathcal{M}_1(\mathscr{X})} F(\alpha)\,\Psi(d\alpha)$$

for all continuous F on $\mathcal{M}_1(\mathscr{X})$, which implies the result. □

Now assume that distributions $P^{\boldsymbol{\lambda}}\{L_n \in \cdot\}$ satisfy an LDP with normalization $\{n\}$ and some rate function I on $\mathcal{M}_1$. With $\boldsymbol{\lambda}$ fixed, our goal is to prove that the probability measures

$$\Psi_n = \frac{1}{n}\sum_{k=1}^{n} \delta_{\lambda_k}$$

converge weakly on the space $\mathcal{M}_1(\mathcal{M}_1)$.

To get limit points for $\{\Psi_n\}$ for free we transfer the discussion to the compact space $\mathcal{M}_1(\bar{\mathcal{S}})$. By the contraction principle, the LDP holds for the distributions $P^{\boldsymbol{\lambda}}\{L_n \in \cdot\}$ on the space $\mathcal{M}_1(\bar{\mathcal{S}})$ with rate function

$$J(\alpha) = \begin{cases} I(\alpha), & \alpha \in \mathcal{M}_1, \\ \underline{\lim}_{\mathcal{M}_1 \ni \beta \to \alpha} I(\beta), & \alpha \in \mathcal{M}_1(\bar{\mathcal{S}}) \setminus \mathcal{M}_1. \end{cases} \tag{15.22}$$

(This situation was addressed in Exercise 3.3. See also Exercise 15.13 below.)

Since $\mathcal{M}_1(\bar{\mathcal{S}})$ is compact, so is $\mathcal{M}_1(\mathcal{M}_1(\bar{\mathcal{S}}))$. By standard metric space arguments convergence of the sequence $\{\Psi_n\}$ follows if we can show that there is a unique limit point. So let $\Psi = \lim_{j\to\infty} \Psi_{n_j}$ be the limit of some subsequence $\{\Psi_{n_j}\}$. Let $g \in \mathscr{C}(\bar{\mathcal{S}})$. Integrating the bounded, continuous function $F(\alpha) = \log E^\alpha[e^g]$ against the measure Ψ_{n_j} and taking the weak limit gives

$$\begin{aligned} \int_{\mathcal{M}_1(\bar{\mathcal{S}})} \log E^\alpha[e^g]\,\Psi(d\alpha) &= \lim_{j\to\infty} \frac{1}{n_j} \sum_{k=1}^{n_j} \log E^{\lambda_k}[e^g] \\ &= \lim_{j\to\infty} \frac{1}{n_j} \log E^{\boldsymbol{\lambda}}\Big[\exp\Big\{\sum_{k=1}^{n_j} g(X_k)\Big\}\Big]. \end{aligned} \tag{15.23}$$

On the other hand, Varadhan's theorem (page 38) together with the assumed LDP gives

$$\lim_{n\to\infty}\frac{1}{n}\log E^{\boldsymbol{\lambda}}\Big[\exp\Big\{\sum_{k=1}^{n} g(X_k)\Big\}\Big] = \sup_{\gamma\in\mathcal{M}_1(\bar{\mathcal{S}})}\{E^{\gamma}[g]-J(\gamma)\} = J^*(g).$$

Thus the quantities in (15.23) are uniquely defined and equal $J^*(g)$, for all limit points Ψ, and then by Lemma 15.9 there can be only one limit point.

We have proved that there is a limit $\Psi_n \to \Psi$ in the space $\mathcal{M}_1(\mathcal{M}_1(\bar{\mathcal{S}}))$. It remains to argue that this limit is actually an element of the space $\mathcal{M}_1(\mathcal{M}_1)$, or in other words that $\Psi\{\alpha : \alpha(\mathcal{S}) = 1\} = 1$. Let $\bar{\alpha}$ denote the mean of Ψ:

$$\bar{\alpha}(B) = \int_{\mathcal{M}_1(\bar{\mathcal{S}})} \alpha(B)\,\Psi(d\alpha) \quad \text{for } B\subset\mathcal{S} \text{ measurable.}$$

It is enough to show that $\bar{\alpha}(\mathcal{S}) = 1$ because this implies

$$0 = \bar{\alpha}(\bar{\mathcal{S}}\setminus\mathcal{S}) = \int_{\mathcal{M}_1(\bar{\mathcal{S}})} \alpha(\bar{\mathcal{S}}\setminus\mathcal{S})\,\Psi(d\alpha).$$

To show this last point, start with

$$J(\gamma) \ge J^{**}(\gamma) = \sup_{f\in\mathscr{C}(\bar{\mathcal{S}})}\Big\{E^{\gamma}(f) - \int_{\mathcal{M}_1(\bar{\mathcal{S}})} \log E^{\alpha}[e^f]\,\Psi(d\alpha)\Big\} \ge H(\gamma\,|\,\bar{\alpha})$$

where the last inequality came from part (b) of Theorem 15.2. Now rate function I cannot be identically infinite on $\mathcal{M}_1$ because the upper bound of the LDP implies $\inf I = 0$. Then $\bar{\alpha}(\mathcal{S}) > 0$ because otherwise every γ supported on $\mathcal{S}$ would fail $\gamma \ll \bar{\alpha}$ and the inequality above would force $I(\gamma) = \infty$. Combining Exercise 5.14 with the inequality above gives for $\gamma\in\mathcal{M}_1$

$$I(\gamma) = J(\gamma) \ge H(\gamma\,|\,\bar{\alpha}) \ge -\log\bar{\alpha}(\mathcal{S}) \ge 0.$$

Choose a sequence $\gamma_j \in \mathcal{M}_1$ such that $I(\gamma_j) \searrow 0$. This forces $\bar{\alpha}(S) = 1$ and completes the proof of Theorem 15.4.

Exercise 15.11 ([**73**]). Find the necessary and sufficient condition on $\boldsymbol{\lambda}$ under which $P^{\boldsymbol{\lambda}}\{L_n \in \cdot\}$ satisfies the LDP with rate given by relative entropy $H(\cdot\,|\,\rho)$ with respect to some probability measure ρ.

Exercise 15.12 ([**73**]). With notation as in (15.1) and (15.2), let p be the stochastic kernel from $\mathcal{S}$ into $\mathcal{M}_1$ defined by

$$\int_{\mathcal{S}} \mathbb{1}_A(x)p(x,B)\,\mu_{\mathcal{S}}(dx) = \mu(A\times B)$$

for Borel sets $A\subset\mathcal{S}$ and $B\subset\mathcal{M}_1$. Show that for $\gamma\in\mathcal{M}_1$,

$$K(\gamma\,|\,\Psi) = H(\gamma\,|\,\mu_{\mathcal{S}}) + K(\Psi\,|\,\gamma\circ p^{-1}).$$

Note that $\gamma \circ p^{-1}(D) = \gamma\{x : p(x) \in D\}$ is a probability measure on Borel sets $D \subset \mathcal{M}_1(\mathcal{M}_1)$ so the second K-entropy makes sense. In particular, $K(\gamma \,|\, \Psi) = H(\gamma \,|\, \mu_{\mathcal{S}})$ if and only if $\gamma p = \Psi$.

Hint: Apply (15.3) to $K(\Psi \,|\, \gamma \circ p^{-1})$ and use the $\mathcal{M}_1(\mathcal{M}_1)$-marginal $\gamma \circ p^{-1}$ to transport the joint entropy to the space $\mathcal{S} \times \mathcal{M}_1$. Use the conditional entropy formula (Exercise 5.13).

***Exercise 15.13.** Show that probability measures on $\mathcal{S}$ are dense among probability measures on $\bar{\mathcal{S}}$.

Hint: Convex combinations of point masses are dense among probability measures. This provides justification for formula (15.22).

Literature notes. The LDP of Theorem 15.4 appeared in [**6**] and was generalized and further elucidated in [**72**], [**73**]. In particular, process level versions exist. Applications to statistical mechanics, in the spirit of our Sections 3.4 and 5.3 and Chapter 8, appear in [**74**], [**75**].

Chapter 16

Random walk in a dynamical random environment

This chapter provides an introduction to a topic of contemporary research, namely motion in a random medium. It presents us with another opportunity for a nontrivial application of the Baxter-Jain theorem, which we generalize to a noncompact space. We discuss mainly a version of *random walk in a random environment* (RWRE) and briefly another model called *directed polymer in a random environment*, or *directed path in a random medium* (DPRM).

In the field of random media, results on probability distributions come in two flavors, *quenched* and *averaged.* (Depending on the context, the latter type may also be called *annealed.*) Quenched means that the random environment (the "medium") is fixed, and one seeks results valid for almost every realization of the environment. Averaged means that we average also over the environment, that is, the environment is not fixed but randomly fluctuating.

A simple example illustrates this difference. We have two coins, one fair and another one that gives heads with probability $3/4$. We pick one coin at random and start flipping it, without ever changing coins. Let $(X_k)_{0\le k<\infty}$ denote the sequence of coin flips. Once the coin is fixed, (X_k) is an i.i.d. sequence with success probability either $1/2$ or $3/4$, depending on which coin (which environment) we picked. This is the quenched distribution of (X_k), and it is a function of the environment. By contrast, the averaged distribution of (X_k) is an exchangeable distribution with success probability $\frac{1}{2}\times\frac{1}{2}+\frac{1}{2}\times\frac{3}{4}=\frac{5}{8}$.

The results we prove here are quenched large deviation principles, that is, LDPs valid under almost every fixed environment.

16.1. Quenched large deviation principles

We describe a particular RWRE model in two dimensions, where one dimension corresponds to time. Hence the model is 1+1-dimensional in the jargon of the field. The restriction to one space dimension is entirely to reduce technicalities that come from manipulating paths in higher dimensions. We adopt the notation of the random media field, which differs in some respects from the notation of the rest of the book.

RWRE on $\mathbb{Z}^2$ is a Markov chain whose transition probability matrix $\{\pi_{x,y}(\omega)\}_{x,y\in\mathbb{Z}^2}$ depends on a parameter ω called the *environment*. The environment itself is a sample point from a probability space $(\Omega,\mathfrak{S},\mathbb{P})$. Let $e_1=(1,0)$ and $e_2=(0,1)$ denote the basis vectors in $\mathbb{Z}^2$. We make two simplifying assumptions: (i) only nearest-neighbor jumps in coordinate directions are allowed and (ii) under the probability measure $\mathbb{P}$, the jump probability vectors $\{(\pi_{x,x+e_1}(\omega),\pi_{x,x+e_2}(\omega)) : x\in\mathbb{Z}^2\}$ are i.i.d. random vectors with values in the set $\{(p,1-p):0\le p\le 1\}$.

To realize the description above, we take the environment space as $\Omega=\mathcal{S}^{\mathbb{Z}^2}$ for some Polish space $\mathcal{S}$ and denote a generic environment configuration by $\omega=(\omega_x)_{x\in\mathbb{Z}^2}$. $\mathfrak{S}$ is the Borel σ-algebra, which is the same as the product σ-algebra on Ω. $\mathbb{P}$ is a product probability measure on Ω under which the coordinates $\{\omega_x\}$ are i.i.d. $\mathcal{S}$-valued random variables. Denote the shift on Ω by $(T_x\omega)_y=\omega_{x+y}$ for $x,y\in\mathbb{Z}^2$. $\mathcal{R}=\{e_1,e_2\}$ is the set of admissible steps of the walk.

To specify the transition probability, assume given a continuous function $p:\mathcal{S}\to[0,1]$ and then define $(\pi_{x,x+e_1}(\omega),\pi_{x,x+e_2}(\omega))=(p(\omega_x),1-p(\omega_x))$. For $z\notin\mathcal{R}$ set $\pi_{x,x+z}(\omega)=0$. This just means that steps z other than e_1 and e_2 have probability zero. The transition probability behaves naturally under the shifts: $\pi_{x,x+z}(\omega)=\pi_{0,z}(T_x\omega)$ for $x,z\in\mathbb{Z}^2$. Now at each vertex $x\in\mathbb{Z}^2$ we have a random jump probability $\{\pi_{x,y}(\omega):y\in\mathbb{Z}^d\}$ that is independent of the random jump probabilities at other vertices.

The *quenched path distribution* for walks that start at $x\in\mathbb{Z}^2$ is defined by

$$P_x^\omega(X_0=x_0,X_1=x_1,\dots,X_n=x_n)=\mathbb{1}\{x_0=x\}\cdot\prod_{i=1}^n\pi_{x_{i-1},x_i}(\omega) \tag{16.1}$$

for any sequence $(x_0,\dots,x_n)\in(\mathbb{Z}^2)^{n+1}$, $n\in\mathbb{N}$. Thus, in a fixed environment ω, the random walk in a random environment is the Markov chain with transition matrix $\{\pi_{x,y}(\omega)\}_{x,y\in\mathbb{Z}^2}$.

Throughout the chapter we make the basic assumption of *uniform ellipticity*:

$$\exists\eta>0 \text{ such that } \pi_{0,e_i}(\omega)\ge\eta \text{ for } \omega\in\Omega \text{ and } i=1,2. \tag{16.2}$$

We could also work with a moment assumption on $\log\pi_{0,e_i}(\omega)$, but uniform ellipticity is convenient for making some technical proofs cleaner.

Example 16.1. The simple "canonical" choice for the construction would be $\Omega=[\eta,1-\eta]^{\mathbb{Z}^2}$ with each $\omega_x\in[\eta,1-\eta]$ and transition probability $\pi_{x,x+e_1}(\omega)=\omega_x$ and $\pi_{x,x+e_2}(\omega)=1-\omega_x$.

We denote an infinite path by $x_{0,\infty}=(x_k)_{0\le k<\infty}$ and its steps by $z_k=x_k-x_{k-1}$. If the initial point x_0 is given, the path is equivalently represented by $x_{0,\infty}$ and $z_{1,\infty}=(z_k)_{1\le k<\infty}$. A path is admissible if each $z_k\in\mathcal{R}$. The steps of the random walk are $Z_k=X_k-X_{k-1}\in\mathcal{R}$. We write Z_1 for the z-variable on the space $\Omega\times\mathcal{R}$; that is, $Z_1(\omega,z)=z$.

Let $\mathbf{P}(d\omega,dz_{1,\infty})=P_0^\omega(dz_{1,\infty})\mathbb{P}(d\omega)$ denote the joint distribution of the random walk path started at the origin and the environment. $\mathbf{P}$ is a probability measure on the space $\Omega_+=\Omega\times\mathcal{R}^{\mathbb{N}}$. It is convenient to think of the quenched measure P^ω also as a probability measure on Ω_+ with Ω-marginal equal to δ_ω. Then P^ω is the conditional probability measure of $\mathbf{P}$, given ω. The variables X_k in (16.1) are defined on Ω_+ by $X_k=z_1+\cdots+z_k$.

The random environment is called *dynamical* because the walk sees a new environment at every time point. We can think of the environment as evolving in time with the walk, with a fresh new collection of environments $\{\omega_{i,j}:i+j=n\}$ sampled at time n. However, we should warn that it does not follow that the quenched walk behaves as if it evolved in an averaged environment. Under the averaged measure $\mathbf{P}$ the walk X_n is a standard random walk with jump probability $p(z)=\mathbb{E}[\pi_{0,z}(\omega)]$.

An empirical measure that records the evolution of the environment seen from the position of the moving particle and the steps of the walk is defined by

$$L_n=\frac{1}{n}\sum_{k=0}^{n-1}\delta_{(T_{X_k}\omega,\,Z_{k+1})}. \tag{16.3}$$

L_n is a function from Ω_+ to the space $\mathcal{M}_1=\mathcal{M}_1(\Omega\times\mathcal{R})$ of Borel probability measures on $\Omega\times\mathcal{R}$. We will study large deviations of L_n under the quenched measures P^ω. Average relative to L_n will be denoted by $L_n(f)=E^{L_n}[f]=n^{-1}\sum_{k=0}^{n-1}f(T_{X_k}\omega,Z_{k+1})$.

The larger vector space of all real-valued Borel measures on $\Omega \times \mathcal{R}$ is denoted by $\mathcal{M}$. We need a separable space of functions that generates the standard weak topology of probability measures on $\mathcal{M}_1$. To arrange that, pick a totally bounded metric for $\Omega\times\mathcal{R}$ (one exists by virtue of separability) and let $\mathscr{U}_b$ be the space of bounded, local, uniformly continuous functions on $\Omega\times\mathcal{R}$ under this metric. A local function depends on only finitely many coordinates ω_x of $\omega = (\omega_x)_{x\in\mathbb{Z}^2}$. (See Exercises 6.1 and 6.2 and Section B.4 for the background assumed here.)

We shall show that the limiting quenched pressure, or free energy, exists for bounded measurable functions $g : \Omega\times\mathcal{R}\to\mathbb{R}$ as the $\mathbb{P}$-a.s. limit

$$p(g) = \lim_{n\to\infty} \frac{1}{n} \log E_0^\omega\big[e^{\sum_{k=0}^{n-1} g(T_{X_k}\omega\,,\,Z_{k+1})}\big] \tag{16.4}$$

where the limit value $p(g)$ is nonrandom. Its convex conjugate is

$$J(\nu) = \sup_{g\in\mathscr{U}_b}\{E^\nu[g] - p(g)\}, \quad \nu\in\mathcal{M}. \tag{16.5}$$

The definition implies that $J(\nu)=\infty$ for $\nu\in\mathcal{M}\smallsetminus\mathcal{M}_1$.

We also consider the *point-to-point* model where both endpoints of the path are pinned down. For this purpose we introduce paths with a particular asymptotic velocity. The set of possible limiting velocities of admissible walks is

$$\mathcal{U} = \{\xi = (\xi_1, 1-\xi_1) : 0\le\xi_1\le 1\}.$$

(That is, any limit point of x_n/n for an admissible path $x_{0,\infty}$ lies in $\mathcal{U}$.) We use the following notation throughout: ξ_1 is the first coordinate of $\xi\in\mathcal{U}$. By setting $\hat{x}_n(\xi) = (\lfloor n\xi_1\rfloor\,,\, n-\lfloor n\xi_1\rfloor)$ we define a lattice path $\{\hat{x}_n(\xi)\}_{n\in\mathbb{Z}_+}$ such that $\hat{x}_n(\xi)$ is n steps away from the origin and approximates $n\xi$. When ξ is understood we may drop it from the notation and write $\hat{x}_n = \hat{x}_n(\xi)$.

The limiting quenched point-to-point pressure is

$$p(g,\xi) = \lim_{n\to\infty}\frac{1}{n}\log E_0^\omega\big[e^{\sum_{k=0}^{n-1} g(T_{X_k}\omega\,,\,Z_{k+1})}, X_n = \hat{x}_n(\xi)\big] \tag{16.6}$$

and its convex conjugate is

$$J(\nu,\xi) = \sup_{g\in\mathscr{U}_b}\{E^\nu[g] - p(g,\xi)\}, \quad \nu\in\mathcal{M},\ \xi\in\mathcal{U}. \tag{16.7}$$

As before, $J(\nu,\xi)=\infty$ for $\nu\in\mathcal{M}\smallsetminus\mathcal{M}_1$. To highlight the contrast with point-to-point, the pressure in (16.4) can be called *point-to-line* because the admissible paths go from the origin to the line $\{(i,j) : i+j=n\}$. In (16.6) we used the simple notation $E[X,A] = E[X\mathbb{1}_A]$.

The first theorem collects basic properties of these functions.

Theorem 16.2. (a) *Fix a bounded measurable local function $g:\Omega\times\mathcal{R}\to\mathbb{R}$. Then limits* (16.4) *and* (16.6) *exist both $\mathbb{P}$-a.s. and in $L^1(\mathbb{P})$. In particular,* *there is an event of full $\mathbb{P}$-probability on which limit* (16.6) *exists simultaneously for all $\xi \in \mathcal{U}$. The limit values $p(g)$ and $p(g,\xi)$ are nonrandom, and they satisfy*

$$p(g) = \sup_{\xi \in \mathcal{U}} p(g,\xi) = \sup_{\xi \in \mathcal{U}\cap\mathbb{Q}} p(g,\xi). \tag{16.8}$$

Functionals $p(g)$ and $p(g,\xi)$ are convex in g, while $p(g,\xi)$ is concave and continuous in $\xi \in \mathcal{U}$.

(b) *$J(\nu,\xi)$ is a lower semicontinuous convex function of (ν,ξ) from $\mathcal{M}_1 \times \mathcal{U}$ into $[0,\infty]$. If $E^\nu[Z_1] \neq \xi$, then $J(\nu,\xi) = \infty$. Similarly, $J(\nu)$ is a lower semicontinuous convex function from $\mathcal{M}_1$ into $[0,\infty]$. These functions satisfy*

$$J(\nu) = J(\nu, E^\nu[Z_1]) = \inf_{\xi} J(\nu,\xi). \tag{16.9}$$

The sublevel sets $\{\nu : J(\nu) \le c\}$ and $\{\nu : J(\nu,\xi) \le c\}$ are compact for all $\xi \in \mathcal{U}$ and $c \in \mathbb{R}$.

(c) *In the duality of $\mathscr{U}_b$ and $\mathcal{M}$, $p(g)$ and $J(\nu)$ are a pair of convex conjugates, and for each fixed ξ the functionals $p(g,\xi)$ and $J(\nu,\xi)$ are also convex conjugates of each other.*

Next we record LDPs, first for point-to-point distributions of L_n and then point-to-line.

Theorem 16.3. *Fix $\xi \in \mathcal{U}$. Then there exists an event Ω_0 such that $\mathbb{P}(\Omega_0) = 1$ and for all $\omega \in \Omega_0$ we have the following bounds. For closed $A \subset \mathcal{M}_1$ and open $G \subset \mathcal{M}_1$*

$$\overline{\lim_{n\to\infty}} \frac{1}{n} \log P_0^\omega\{L_n \in A, X_n = \hat{x}_n(\xi)\} \le -\inf_{\nu \in A} J(\nu,\xi) \tag{16.10}$$

and

$$\underline{\lim_{n\to\infty}} \frac{1}{n} \log P_0^\omega\{L_n \in G, X_n = \hat{x}_n(\xi)\} \ge -\inf_{\nu \in G} J(\nu,\xi). \tag{16.11}$$

Theorem 16.4. *There exists an event Ω_0 such that $\mathbb{P}(\Omega_0) = 1$ and for all $\omega \in \Omega_0$ the quenched distributions $P_0^\omega\{L_n \in \cdot\}$ on $\mathcal{M}_1$ satisfy an LDP with normalization $\{n\}$ and tight rate function J.*

Remark 16.5. A level 1 quenched LDP for the walk (an analogue of Cramér's theorem) follows from (16.6) with $g = 0$, and then we can use the duality of p and J:

$$\begin{aligned} \lim_{n\to\infty} \frac{1}{n} \log P_0^\omega\{X_n = \hat{x}_n(\xi)\} &= p(0,\xi) = -\inf_{\nu} J(\nu,\xi) \\ &= -\inf_{\nu : E^\nu[Z_1] = \xi} J(\nu). \end{aligned} \tag{16.12}$$

The last formula shows that this LDP is a contraction from Theorem 16.4, as it should. Once we have the limit (16.12), we can see that the LDP given by the bounds (16.10) and (16.11) is equivalent to an LDP for the conditional distributions $P_0^\omega\{L_n \in \cdot \mid X_n = \hat{x}_n(\xi)\}$.

We have merely scratched the surface of the topic in this section, but we must leave further study to the reader. For example, the rate functions have entropy representations, and the path component in the empirical measure can be extended to the entire process $Z_{1,\infty}$. The special feature of the RWRE treated here is that the walk never visits the same vertex twice. This is not necessary for the results. In fact, in the history of the subject the large deviations of such strictly directed walks were treated after other types of RWREs. See the literature notes at the end of the chapter.

Directed paths in a random medium.

We describe another model of motion in a random medium, namely the *directed path in a random medium* (DPRM), also called the *directed polymer in a random environment.* The environment space Ω is the same as in Section 16.1 with an i.i.d. probability measure $\mathbb{P}$. Admissible directed paths $(x_k)_{k=0}^n$ lie in $\mathbb{Z}^2$ and satisfy $x_k - x_{k-1} \in \mathcal{R} = \{e_1, e_2\}$. Let P_x be the standard random walk measure with initial point x:

$$P_x(X_0 = x_0, X_1 = x_1, \dots, X_n = x_n) = \mathbb{1}\{x_0 = x\} \cdot 2^{-n} \tag{16.13}$$

for an admissible path $(x_k)_{k=0}^n$. The expectation under P_x is denoted by E_x.

In a fixed environment ω, a sequence of quenched path measures Q_n^ω is constructed with a Gibbs-type recipe, in terms of Radon-Nikodym derivatives with respect to P_0. For this purpose fix a bounded continuous function $G : \Omega \times \mathcal{R} \to \mathbb{R}$. G is called the *potential* in this context. For any measurable event $A \subset \Omega_+ = \Omega \times \mathcal{R}^{\mathbb{N}}$, define

$$Q_n^\omega(A) = \frac{1}{Z_n^\omega} E_0\big[\mathbb{1}_A(\omega, Z_{1,\infty})\, e^{\sum_{k=0}^{n-1} G(T_{X_k}\omega\,,\, Z_{k+1})}\big] \tag{16.14}$$

with normalization, or *partition function,*

$$Z_n^\omega = E_0\big[e^{\sum_{k=0}^{n-1} G(T_{X_k}\omega\,,\, Z_{k+1})}\big]. \tag{16.15}$$

The measures Q_n^ω are not consistent: that is, Q_n^ω is not the marginal of Q_{n+1}^ω on n-step paths (Exercise 16.7).

The situation is not exactly that of Gibbs measures of Part II, but the spirit is the same. The quantity $-\sum_{k=0}^{n-1} G(T_{x_k}\omega\,,\, z_{k+1})$ represents the "energy" of a path $(x_k)_{k=0}^n$ in environment ω.

We state one result, namely the existence of the limiting free energy. As in Section 8.1, it is a consequence of Varadhan's theorem and the LDP we stated above. Let J denote the rate function of the LDP of Theorem 16.4 for the constant transition probability $\pi_{x,x+e_i}(\omega) = 1/2$ for $i = 1, 2$.

Corollary 16.6. *For $\mathbb{P}$-almost every ω we have the finite, nonrandom limit*

$$\lim_{n\to\infty} n^{-1} \log \mathcal{Z}_n^\omega = \sup_{\nu\in\mathcal{M}_1} \{E^\nu[G] - J(\nu)\}. \tag{16.16}$$

The simplest and most studied potential is $G(\omega, z) = \omega_0$. We assumed G bounded in order to apply Varadhan's theorem to the continuous function $\nu \mapsto E^\nu[G]$, but the a.s. limit (16.16) exists with much weaker assumptions.

Despite superficial similarities, the RWRE model discussed above and the DPRM behave quite differently. Under the averaged measure the RWRE X_n is a standard random walk and consequently its fluctuations at time n are of order $n^{1/2}$. In contrast, the walk X_n under the averaged measure $\mathbb{E}Q_n^\omega(\cdot)$ is expected to have fluctuations of order $n^{2/3}$, in accordance with behavior that falls within the Kardar-Parisi-Zhang universality class, at least under some fairly general assumptions on the potential. Pointers for further study appear in the literature notes at the end of the chapter.

Exercise 16.7. Show that probabilities $Q_n^\omega(X_n = x_n)$ and $Q_{n+1}^\omega(X_n = x_n)$ do not agree in general.

Exercise 16.8. Deduce a large deviation principle for the quenched distributions $Q_n^\omega\{L_n \in \cdot\}$.

16.2. Proofs via the Baxter-Jain theorem

In this section we prove the theorems of Section 16.1 for the dynamical RWRE. We begin by stating the general auxiliary large deviation theorem that we utilize.

Preliminary result: An extension of the Baxter-Jain theorem to a noncompact space.

We extend Theorem 15.5 to a noncompact space. Let $\mathcal{X}$ and $\mathcal{L}$ be two real vector spaces in duality, endowed with their weak topologies, assumed Hausdorff, and let $\mathcal{X}_0$ be a bounded, closed, convex subset of $\mathcal{X}$. Boundedness means that for every open neighborhood B of the origin in $\mathcal{X}$ there exists $0 < c < \infty$ such that $\mathcal{X}_0 \subset cB$. Boundedness implies that each linear function $\varphi \in \mathcal{L}$ is a bounded, continuous function on $\mathcal{X}_0$.

The setting is from Section 15.2. A normalizing sequence $0 < r_n \nearrow \infty$ is given. For each i from an arbitrary index set $\mathcal{I}$ there is a sequence $\{\mu_n^{(i)}\}_{n\in\mathbb{N}}$ of Borel probability measures on $\mathcal{X}_0$ with local rate functions

$$\overline{\kappa}^{(i)}(x) = -\inf_{G\subset\mathcal{X}\,:\,x\in G,\ G\text{ open}} \overline{\lim_{n\to\infty}} \frac{1}{r_n} \log \mu_n^{(i)}(G), \tag{16.17}$$

(16.18) $$\underline{\kappa}^{(i)}(x) = -\inf_{G\subset\mathcal{X}\,:\,x\in G,\,G\text{ open}} \ \varliminf_{n\to\infty} \frac{1}{r_n}\log\mu_n^{(i)}(G),$$

and

(16.19) $$\kappa(x) = \sup_{i\in\mathcal{I}} \underline{\kappa}^{(i)}(x), \quad x\in\mathcal{X}.$$

The sequences $\{\mu_n^{(i)} : n\in\mathbb{N}\}$ determine the same pressure:

(16.20) $$p(\varphi) = \lim_{n\to\infty}\frac{1}{r_n}\log\int_{\mathcal{X}_0} e^{r_n\langle x,\varphi\rangle}\,\mu_n^{(i)}(dx), \qquad \varphi\in\mathcal{L},\ i\in\mathcal{I}.$$

The convex conjugate $J:\mathcal{X}\to[0,\infty]$ of p is

(16.21) $$J(x) = \sup_{\varphi\in\mathcal{L}}\{\langle x,\varphi\rangle - p(\varphi)\}, \quad x\in\mathcal{X}.$$

Theorem 16.9. *Let $\mathcal{X}_0$ be a bounded, closed, convex subset of $\mathcal{X}$. Make the following three assumptions on the setting above.*

(i) *For each $i\in\mathcal{I}$ the sequence $\{\mu_n^{(i)}\}_{n\in\mathbb{N}}$ is exponentially tight with normalization r_n.*

(ii) *The limit $p(\varphi)$ in (16.20) exists for all $\varphi\in\mathcal{L}$ and $i\in\mathcal{I}$, and the sublevel sets $\{x\in\mathcal{X} : J(x)\le c\}$ are compact $\forall c\in\mathbb{R}$.*

(iii) *The function κ in (16.19) is convex.*

Then for each $i\in\mathcal{I}$ the sequence $\{\mu_n^{(i)}\}_{n\in\mathbb{N}}$ satisfies the LDP with rate function $J=\overline{\kappa}^{(i)}=\underline{\kappa}^{(i)}=\kappa$ and normalization $\{r_n\}$.

Note that assumption (i) does not ask for uniformity in $i\in\mathcal{I}$. That is, the choice of compact sets required for exponential tightness can vary with i. The theorem is proved by adapting the arguments of Chapter 14. Exercise 16.15 gives some hints.

Proofs for the RWRE setting.

The proofs are organized around a derivation of Theorem 16.3 from Theorem 16.9. The duality is between the space $\mathcal{X}=\mathcal{M}$ of real-valued Borel measures on $\Omega\times\mathcal{R}$ and the space $\mathcal{L}=\mathscr{U}_b$ of bounded local functions on $\Omega\times\mathcal{R}$ that are uniformly continuous under a totally bounded metric on $\Omega\times\mathcal{R}$. The set $\mathcal{X}_0=\mathcal{M}_1$, the subspace of probability measures. The weak $\mathscr{U}_b$-topology restricted to $\mathcal{M}_1$ is the standard weak topology of probability measures. We leave the boundedness of $\mathcal{M}_1$ as Exercise 16.16.

The key to applying Theorem 16.9 lies in extending the quenched distributions $P_0^\omega\{L_n\in\cdot\,,\,X_n=\hat{x}_n(\xi)\}$ to an appropriate larger indexed class of measures that enables us to prove the convexity of the function κ in (16.19) while retaining the other properties needed for Theorem 16.9.

Recall the set $\mathcal{U} = \{\xi = (\xi_1, 1-\xi_1) : 0 \le \xi_1 \le 1\}$ of admissible velocities and for the moment consider a fixed $\xi \in \mathcal{U}$. Let the index set be $\mathcal{I} = \mathbb{Z}_+^2 \times \{0,1\}$. For $(a,b,\tau) \in \mathcal{I}$ define the following measures on the space $\mathcal{M}_1$:

$$\mu_n^{a,b,\tau,\omega}(B) = P^\omega_{\hat{x}_{an+b}(\xi)}\bigl\{L_{n+\tau} \in B, X_{n+\tau} = \hat{x}_{(a+1)n+b+\tau}(\xi)\bigr\}. \tag{16.22}$$

The relevant local rate functions are defined for $\nu \in \mathcal{M}_1$ by

$$\underline{\kappa}^{a,b,\tau,\omega}(\nu,\xi) = -\inf_{B \ni \nu} \varliminf_{n\to\infty} \frac{1}{n} \log \mu_n^{a,b,\tau,\omega}(B) \tag{16.23}$$

where the infimum is taken over weak neighborhoods B of ν, and by

$$\kappa^\omega(\nu,\xi) = \sup_{(a,b,\tau)\in\mathcal{I}} \underline{\kappa}^{a,b,\tau,\omega}(\nu,\xi).$$

It follows from the definitions that $\kappa^\omega(\nu,\xi)$ is a lower semicontinuous function of ν. At this point it is not evident that $\kappa^\omega(\nu,\xi)$ is almost surely independent of ω. That will be a consequence of the equality $J(\nu,\xi) = \kappa^\omega(\nu,\xi)$ that comes from Theorem 16.9. Obviously, $\mu_n^{a,b,\tau,\omega}$ is not a probability measure. When we apply Theorem 16.9 we will normalize $\mu_n^{a,b,\tau,\omega}$ but for now it is convenient to restrict $P^\omega_{\hat{x}_{an+b}(\xi)}$ to the event $X_{n+\tau} = \hat{x}_{(a+1)n+b+\tau}(\xi)$ as in (16.22), rather than conditioning it.

Lemma 16.10. *The function $\kappa^\omega(\nu,\xi)$ is convex in ν.*

Proof. We begin with a general computation that illustrates how we will use the Markov property many times over. Let $\ell = i + j$, and let B, B', B'' be subsets of $\mathcal{M}_1$ such that $\frac{i}{\ell}B' + \frac{j}{\ell}B'' \subset B$. By the Markov property,

$$\begin{aligned}
&P^\omega_{\hat{x}_k}\{L_\ell \in B,\, X_\ell = \hat{x}_{k+\ell}\} \\
&\quad\ge P^\omega_{\hat{x}_k}\Bigl\{L_i \in B',\, X_i = \hat{x}_{k+i},\, j^{-1}\sum_{h=i}^{\ell-1} \delta_{(T_{X_h}\omega\,,\,Z_{h+1})} \in B'',\, X_\ell = \hat{x}_{k+\ell}\Bigr\} \\
&\quad= P^\omega_{\hat{x}_k}\{L_i \in B',\, X_i = \hat{x}_{k+i}\}\, P^\omega_{\hat{x}_{k+i}}\{L_j \in B'',\, X_j = \hat{x}_{k+\ell}\}.
\end{aligned}$$

With $k = an+b$, $\ell = n+\tau$, and $i,j \in \{m, m+1\}$, the above translates into the following inequalities for $m,n \in \mathbb{N}$:

$$\begin{aligned}
&\mu_n^{a,b,\tau,\omega}(B) \;\ge\; \mu_m^{2a,b,0,\omega}(B')\,\mu_m^{2a+1,b,\tau,\omega}(B'') \\
&\text{if } n = 2m \;\text{ and }\; \tfrac{m}{n+\tau}B' + \tfrac{m+\tau}{n+\tau}B'' \subset B,
\end{aligned} \tag{16.24}$$

while

$$\begin{aligned}
&\mu_n^{a,b,\tau,\omega}(B) \;\ge\; \mu_m^{2a,a+b,\tau,\omega}(B')\,\mu_m^{2a+1,a+b+\tau,1,\omega}(B'') \\
&\text{if } n = 2m+1 \;\text{ and }\; \tfrac{m+\tau}{n+\tau}B' + \tfrac{m+1}{n+\tau}B'' \subset B.
\end{aligned} \tag{16.25}$$

Now let $\nu = (\mu+\rho)/2$ in $\mathcal{M}_1$. Fix ω and let $c < \kappa^\omega(\nu,\xi)$. Pick $(a,b,\tau) \in \mathcal{I}$ such that $\underline{\kappa}^{a,b,\tau,\omega}(\nu,\xi) > c$ and then a neighborhood $B = B_\nu$ of ν such that

$$-c > \varliminf_{n\to\infty} \frac{1}{n} \log \mu_n^{a,b,\tau,\omega}(B). \tag{16.26}$$

Pick neighborhoods B_μ of μ and B_ρ of ρ so that $\frac{1}{2}B_\mu + \frac{1}{2}B_\rho \subset B$. By taking further smaller neighborhoods $\mu \in B' \subset B_\mu$ and $\rho \in B'' \subset B_\rho$, we can guarantee that for large enough n the conditions for (16.24) and (16.25) are satisfied.

Suppose the liminf on the right-hand side of (16.26) happens along a subsequence of even n. Then take logarithms and limits in (16.24). (If the liminf happens along odd n use (16.25) similarly.) We have

$$\begin{aligned}
-c &> \varliminf_{n=2m\to\infty} n^{-1} \log \mu_n^{a,b,\tau,\omega}(B) \\
&\geq \tfrac{1}{2} \varliminf_{m\to\infty} m^{-1} \log \mu_m^{2a,b,0,\omega}(B') + \tfrac{1}{2} \varliminf_{m\to\infty} m^{-1} \log \mu_m^{2a+1,b,\tau,\omega}(B'') \\
&\geq -\tfrac{1}{2}\underline{\kappa}^{2a,b,0,\omega}(\mu,\xi) - \tfrac{1}{2}\underline{\kappa}^{2a+1,b,\tau,\omega}(\rho,\xi) \\
&\geq -\tfrac{1}{2}\kappa^\omega(\mu,\xi) - \tfrac{1}{2}\kappa^\omega(\rho,\xi).
\end{aligned}$$

Letting $c \nearrow \kappa^\omega(\nu,\xi)$ gives

$$\kappa^\omega(\nu,\xi) \leq \tfrac{1}{2}\kappa^\omega(\mu,\xi) + \tfrac{1}{2}\kappa^\omega(\rho,\xi). \tag{16.27}$$

By lower semicontinuity this is enough for convexity (Exercise 4.7). $\square$

Proposition 16.11. *Let g be a bounded measurable local function on $\Omega\times\mathcal{R}$. Then there exists an event Ω_1 such that $\mathbb{P}(\Omega_1) = 1$ and for all $\omega \in \Omega_1$ the finite nonrandom limit*

$$p(g,\xi) = \lim_{n\to\infty} \frac{1}{n} \log E^\omega_{\hat{x}_{an+b}(\xi)}\big[e^{(n+\tau)L_{n+\tau}(g)}, X_{n+\tau} = \hat{x}_{(a+1)n+b+\tau}(\xi)\big] \tag{16.28}$$

exists for all $(a,b,\tau) \in \mathcal{I}$ and $\xi \in \mathcal{U}$. Part of the conclusion is that for a given (g,ξ) pair the limit is the same for all $(a,b,\tau) \in \mathcal{I}$. The limit holds also in $L^1(\mathbb{P})$. Furthermore, $p(g,\xi)$ is continuous and concave as a function of $\xi \in \mathcal{U}$, and it is convex as a function of g.

Proof. The event Ω_1 will be the event on which the limits hold for rational ξ. The proof shows that on this event the limits hold also for irrational ξ and in fact that the function $p(g,\xi)$ on irrational ξ is the unique continuous extension of $p(g,\,\cdot\,)$ from rationals to irrationals on the relative interior $\operatorname{ri}\mathcal{U} = \{\xi = (\xi_1, 1-\xi_1) : 0 < \xi_1 < 1\}$. Convexity in g is the by-now familiar consequence of Hölder's inequality. Once the limits hold a.s., L^1 convergence follows from boundedness, which comes from boundedness of g and uniform ellipticity (16.2).

Step 1. Case $a = b = \tau = 0$ for rational ξ.

Abbreviate $\hat{x}_n = \hat{x}_n(\xi)$. Fix $k \in \mathbb{N}$ so that $k\xi \in \mathbb{Z}^2$ and then $\hat{x}_{nk}(\xi) = nk\xi$. By the Markov property

$$
\begin{aligned}
&E_0^\omega\big[e^{(m+n)kL_{(m+n)k}(g)}, X_{(m+n)k} = (m+n)k\xi\big] \\
&\quad \ge E_0^\omega\big[e^{\sum_{j=0}^{mk-1} g(T_{X_j}\omega, Z_{j+1})}, X_{mk} = mk\xi, \\
&\qquad\qquad e^{\sum_{j=mk}^{(m+n)k-1} g(T_{X_j}\omega, Z_{j+1})}, X_{(m+n)k} = (m+n)k\xi\,\big] \\
&\quad = E_0^\omega\big[e^{mkL_{mk}(g)}, X_{mk} = mk\xi\big] \cdot E_{mk\xi}^\omega\big[e^{nkL_{nk}(g)}, X_{nk} = nk\xi\big] \\
&\quad = E_0^\omega\big[e^{mkL_{mk}(g)}, X_{mk} = mk\xi\big] \cdot E_0^{T_{mk\xi}\omega}\big[e^{nkL_{nk}(g)}, X_{nk} = nk\xi\big].
\end{aligned}
\tag{16.29}
$$

By ellipticity (16.2) we have the bound

$$
\big|\log E_0^\omega[\,e^{nkL_{nk}(g)}, X_{nk} = nk\xi\,]\big| \le nk\,\|g\|_\infty + nk\,|\log\eta|\,. \tag{16.30}
$$

This is sufficient for an application of the subadditive ergodic theorem [**27**, Theorem 7.4.1] to the process

$$
Y_{m,n} = -\log E_0^{T_{mk\xi}\omega}\big[e^{(n-m)kL_{(n-m)k}(g)}, X_{(n-m)k} = (n-m)k\xi\big], \;\; 0 \le m < n.
$$

We get an almost sure and L^1 limit

$$
p(g,\xi) = \lim_{m\to\infty} \frac{1}{mk} \log E_0^\omega[\,e^{mkL_{mk}(g)}, X_{mk} = mk\xi\,]. \tag{16.31}
$$

The underlying i.i.d. process implies that each process $\{Y_{n\ell,n\ell+\ell} : n \ge 1\}$ is ergodic, and this guarantees that the limit $p(g,\xi)$ is nonrandom. The limit is independent of k because if k_1 and k_2 satisfy $k_1\xi, k_2\xi \in \mathbb{Z}^2$ and give two distinct limits, then the limit along the subsequence of multiples of k_1k_2 would not be defined.

Next we show that the limit in (16.31) can be taken along the full sequence, not only along a subsequence. Given n, choose m so that $mk \le n < (m+1)k$. There are steps $u_1, \dots, u_k \in \mathcal{R}$ such that

$$
mk\xi + u_1 + \cdots + u_{n-mk} = \hat{x}_n
$$

and

$$
\hat{x}_n + u_{n-mk+1} + \cdots + u_k = (m+1)k\xi.
$$

We derive bounds using the Markov property and ellipticity:

$$
\begin{aligned}
&\log E_0^\omega\big[e^{nL_n(g)}, X_n = \hat{x}_n\big] \\
&\quad \ge \log E_0^\omega\big[e^{mkL_{mk}(g)}, X_{mk} = mk\xi,\, X_n = \hat{x}_n\big] - k\,\|g\|_\infty \\
&\quad = \log E_0^\omega\big[e^{mkL_{mk}(g)}, X_{mk} = mk\xi\big] \\
&\qquad + \log P_{mk\xi}^\omega\{X_{n-mk} = \hat{x}_n\} - k\,\|g\|_\infty \\
&\quad \ge \log E_0^\omega\big[e^{mkL_{mk}(g)}, X_{mk} = mk\xi\big] + k\log\eta - k\,\|g\|_\infty\,.
\end{aligned}
\tag{16.32}
$$

Similarly the complementary bound follows:

$$\begin{aligned}(16.33)\qquad &\log E_0^\omega\big[e^{(m+1)kL_{(m+1)k}(g)}, X_{(m+1)k} = (m+1)k\xi\big]\\ &\quad\ge \log E_0^\omega\big[e^{nL_n(g)}, X_n = \hat{x}_n\big] + k\log\eta - k\,\|g\|_\infty\,.\end{aligned}$$

Divide by n and take $n \to \infty$ in (16.32) and (16.33). The error terms vanish as $n \to \infty$ and the limit in (16.31) can be taken along the full sequence.

Step 2. Continuity of $p(g, \xi)$ for rational ξ.

Let $0 < \xi_1 < 1$ and $\varepsilon \in (0, 1/2)$. Let $\delta < \frac{1}{2}\varepsilon((1 - \xi_1) \wedge \xi_1)$. Let $\xi_1' \in [0, 1]$ satisfy $|\xi_1 - \xi_1'| < \delta$. Then one can check that for large enough n and $n_1 = \lfloor (1 - \varepsilon)n \rfloor$,

$$0 \le \lfloor n\xi_1 \rfloor - \lfloor n_1\xi_1' \rfloor \le n - n_1.$$

This implies that an admissible path of $n - n_1$ steps can be constructed from $\hat{x}_{n_1}(\xi')$ to $\hat{x}_n(\xi)$. By the Markov property and ellipticity,

$$\begin{aligned}&\log E_0^\omega\big[e^{nL_n(g)}, X_n = \hat{x}_n(\xi)\big]\\ (16.34)\qquad &\ge \log E_0^\omega\big[e^{n_1 L_{n_1}(g)},\ X_{n_1} = \hat{x}_{n_1}(\xi'),\ X_n = \hat{x}_n(\xi)\big]\\ &\quad - (n - n_1)\,\|g\|_\infty\\ &\ge \log E_0^\omega\big[e^{n_1 L_{n_1}(g)},\ X_{n_1} = \hat{x}_{n_1}(\xi')\big] + (n - n_1)(\log\eta - \|g\|_\infty).\end{aligned}$$

Suppose first that $\xi_1, \xi_1' \in (0, 1) \cap \mathbb{Q}$. Dividing by n and letting $n \to \infty$ gives

$$\begin{aligned}p(g, \xi) &\ge (1 - \varepsilon)p(g, \xi') + \varepsilon(\log\eta - \|g\|_\infty)\\ &\ge p(g, \xi') + 2\varepsilon(\log\eta - \|g\|_\infty).\end{aligned}$$

The roles of ξ and ξ' can be reversed. Note also that the size of δ depends on how close ξ_1, ξ_1' are to the endpoints of the interval. This gives

(16.35) $p(g, \xi)$ is a uniformly continuous function of rational ξ_1 in any compact subinterval of $(0, 1)$.

The argument above can be adapted to build paths from $\xi' = e_1$ to nearby rationals, and similarly from $\xi' = e_2$. This will show lower semicontinuity at $\xi' = e_1$ and $\xi' = e_2$. Upper semicontinuity needs a different argument because if $0 < \xi_1 < 1$, it is impossible to construct a path from $\hat{x}_n(\xi)$ to either $\hat{x}_{n'}(e_1)$ or $\hat{x}_{n'}(e_2)$ for any $n' > n$. We leave upper semicontinuity at e_1 and e_2 as Exercise 16.17. The argument sketched in the exercise uses the i.i.d. property of the environment in an essential way.

Step 3. Case $a = b = \tau = 0$ for irrational ξ.

Let $\xi_1 \in (0, 1)$ be irrational, and pick a rational ξ_1' close enough so that the conditions in Step 2 above for δ are satisfied for both ξ_1 and ξ_1'. Then bounds (16.34) applied both ways give

$$
\begin{aligned}
&(1-\varepsilon)p(g,\xi') - \varepsilon(\log\eta - \|g\|_\infty) \\
&\le \varliminf_{n\to\infty} n^{-1}\log E_0^\omega\bigl[e^{nL_n(g)}, X_n = \hat{x}_n(\xi)\bigr] \\
&\le \varlimsup_{n\to\infty} n^{-1}\log E_0^\omega\bigl[e^{nL_n(g)}, X_n = \hat{x}_n(\xi)\bigr] \\
&\le \frac{1}{1-\varepsilon}p(g,\xi') - \frac{\varepsilon}{1-\varepsilon}(\log\eta - \|g\|_\infty).
\end{aligned}
$$

This tells us first that the limit

$$
p(g,\xi) = \lim_{n\to\infty} n^{-1}\log E_0^\omega\bigl[e^{nL_n(g)}, X_n = \hat{x}_n(\xi)\bigr]
$$

exists for all irrational ξ on the full probability event Ω_1 on which the limit exists for rational ξ. Second, it tells us that the extension of $p(g,\xi)$ from rational to irrational ξ in $(0,1)$ is the unique extension by continuity.

The limit (16.28) has been proved for the case $a=b=\tau=0$. The case $(a,b,\tau)=(0,0,1)$ follows immediately by replacing n with $n+1$ and noting that normalizations n^{-1} and $(n+1)^{-1}$ in the front lead to the same limit because $\frac{n+1}{n}\to 1$.

Step 4. The general case $(a,b,\tau)\in\mathcal{I}$.

The development in (16.29) gives

$$
\begin{aligned}
&\log E_0^\omega\bigl[e^{((a+1)n+b+\tau)L_{(a+1)n+b+\tau}(g)}, X_{(a+1)n+b+\tau} = \hat{x}_{(a+1)n+b+\tau}(\xi)\bigr] \\
&\qquad \ge \log E_0^\omega\bigl[e^{(an+b)L_{an+b}(g)}, X_{an+b} = \hat{x}_{an+b}(\xi)\bigr] \\
&\qquad\quad + \log E^\omega_{\hat{x}_{an-b}(\xi)}\bigl[e^{(n+\tau)L_{n+\tau}(g)}, X_{n+\tau} = \hat{x}_{(a+1)n+b+\tau}(\xi)\bigr].
\end{aligned} \tag{16.36}
$$

Dividing by n gives

$$
\begin{aligned}
&\varlimsup_{n\to\infty} n^{-1}\log E^\omega_{\hat{x}_{an+b}(\xi)}\bigl[e^{(n+\tau)L_{n+\tau}(g)}, X_{n+\tau} = \hat{x}_{(a+1)n+b+\tau}(\xi)\bigr] \\
&\qquad \le p(g,\xi).
\end{aligned} \tag{16.37}
$$

This upper bound applies to all ξ, rational and irrational.

It remains to derive the complementary lower bound for the liminf. The i.i.d. property of the measure $\mathbb{P}$ is important for this argument.

Let $\varepsilon>0$. Suppose first that ξ is rational and $k\xi\in\mathbb{Z}^2$. Using the L^1 limit in (16.31) fix a large m_0 so that, for $m\ge m_0$,

$$
\frac{1}{mk}\mathbb{E}\bigl[\log E_0^\omega\bigl(e^{mkL_{mk}(g)}, X_{mk} = mk\xi\bigr)\bigr] \ge p(g,\xi)-\varepsilon. \tag{16.38}
$$

Fix ℓ_0 so that $g(\omega,z)$ depends only on $\{\omega_{i,j} : -\ell_0\le i+j<\ell_0\}$. Let $m\ge m_0$ and $m_1 = m+2\ell_0$. Write a general $n\in\mathbb{N}$ as $n = jm_1k + r$ for a remainder

$0 \le r < m_1 k$. Write the increment of the walk in terms of multiples of $m_1 k\xi$ and a remainder y^n:

$$\hat{x}_{(a+1)n+b+\tau}(\xi) - \hat{x}_{an+b}(\xi) = jm_1 k\xi + y^n$$

where the coordinates satisfy $y_1^n, y_2^n \ge 0$ and $y_1^n + y_2^n = r + \tau$.

To get a lower bound, shift ω so that the walk starts at the origin, and add the intermediate restrictions $X_{im_1k} = im_1 k\xi$ for $i = 1, \dots, j$ inside the expectation. Use the Markov property as was done in (16.29). Drop the $r+\tau$ last terms from $nL_n(g)$. As the last step, restrict each term further by $X_{mk} = mk\xi$ and drop $(m_1 - m)k$ terms from the exponent. Thereby

$$\begin{aligned}
&\frac{n}{jm_1k} \cdot \frac{1}{n} \log E^{\omega}_{\hat{x}_{an+b}(\xi)}\bigl[e^{(n+\tau)L_{n+\tau}(g)}, X_{n+\tau} = \hat{x}_{(a+1)n+b+\tau}(\xi)\bigr] \\
&= \frac{1}{jm_1k} \log E_0^{T_{\hat{x}_{an+b}(\xi)}\omega}\bigl[e^{(n+\tau)L_{n+\tau}(g)}, X_{n+\tau} = \hat{x}_{(a+1)n+b+\tau}(\xi) - \hat{x}_{an+b}(\xi)\bigr] \\
&\ge \frac{1}{j}\sum_{i=0}^{j-1} \frac{1}{m_1k} \log E_0^{T_{\hat{x}_{an+b}(\xi)+im_1k\xi}\omega}\bigl[e^{m_1kL_{m_1k}(g)}, X_{m_1k} = m_1k\xi\bigr] \\
&\quad + \frac{1}{jm_1k} \log P^{\omega}_{\hat{x}_{an+b}(\xi)+jm_1k\xi}\{X_{r+\tau} = \hat{x}_{(a+1)n+b+\tau}(\xi)\} - \frac{1}{j}\,\|g\|_\infty \\
&\ge \frac{1}{j}\sum_{i=0}^{j-1} \frac{1}{m_1k} \log E_0^{T_{\hat{x}_{an+b}(\xi)+im_1k\xi}\omega}\bigl[e^{m_1kL_{m_1k}(g)}, X_{m_1k} = m_1k\xi\bigr] \\
&\quad + \frac{\log\eta - \|g\|_\infty}{j} \\
&\ge \frac{1}{j}\sum_{i=0}^{j-1} \frac{1}{m_1k} \log E_0^{T_{\hat{x}_{an+b}(\xi)+im_1k\xi}\omega}\bigl[e^{mkL_{mk}(g)}, X_{mk} = mk\xi\bigr] \\
&\quad + \left(\frac{2\ell_0}{m_1} + \frac{1}{j}\right)(\log\eta - \|g\|_\infty).
\end{aligned}$$

The terms in the i-sum after the last $\ge$ are bounded i.i.d. random variables as functions of ω. Thus standard large deviation bounds apply. This means that deviations from the mean decay exponentially. The required estimate follows from Example 1.4, or from bounds (2.10) and (2.11) with suitable choices of θ. Consequently, by the Borel-Cantelli lemma, as $n \to \infty$ (equivalently, $j \to \infty$) the average converges almost surely to the expectation. By (16.38) this gives

$$\begin{aligned}
&\varliminf_{n\to\infty} \frac{1}{n} \log E^{\omega}_{\hat{x}_{an+b}(\xi)}\bigl[e^{(n+\tau)L_{n+\tau}(g)}, X_{n+\tau} = \hat{x}_{(a+1)n+b+\tau}(\xi)\bigr] \\
&\qquad \ge \frac{m}{m_1}\bigl(p(g,\xi) - \varepsilon\bigr) + \frac{2\ell_0}{m_1}(\log\eta - \|g\|_\infty).
\end{aligned}$$

Next take $m \to \infty$ so that $m/m_1 \to 1$, and last let $\varepsilon \to 0$. Combining the resulting lower bound with (16.37) concludes the proof of (16.28) for rational ξ.

Suppose ξ is irrational. Again, pick a rational ξ' close enough so that $p(g, \xi') \geq p(g, \xi) - \varepsilon$ and an admissible path from $\hat{x}_{n_1}(\xi')$ to $\hat{x}_{(a+1)n+b+\tau}(\xi) - \hat{x}_{an+b}(\xi)$ is possible for some $n_1 > (1-\varepsilon)n$, for a given ε. The calculation begins now as follows:

$$\begin{aligned}\frac{1}{n}\log E^{\omega}_{\hat{x}_{an+b}(\xi)}\big[e^{(n-\tau)L_{n+\tau}(g)}, X_{n+\tau} = \hat{x}_{(a+1)n+b+\tau}(\xi)\big] \\ \geq \frac{1}{n}\log E_0^{T_{\hat{x}_{an+b}(\xi)}\omega}\big[e^{n_1 L_{n_1}(g)}, X_{n_1} = \hat{x}_{n_1}(\xi')\big] \\ + \frac{n - n_1 + 1}{n}(\log \eta - \|g\|_{\infty}).\end{aligned}$$

Continue as above, choosing m_0 as in (16.38) but this time for ξ', then with a decomposition into a sum of i.i.d. variables. In the end we have

$$\begin{aligned}\varliminf_{n\to\infty} \frac{1}{n}\log E^{\omega}_{\hat{x}_{an+b}(\xi)}\big[e^{(n+\tau)L_{n+\tau}(g)}, X_{n+\tau} = \hat{x}_{(a+1)n+b+\tau}(\xi)\big] \\ \geq (1-\varepsilon)p(g,\xi) - 2\varepsilon + \varepsilon(\log\eta - \|g\|_{\infty}).\end{aligned}$$

Letting $\varepsilon \to 0$ completes the proof of the lower bound for irrational ξ.

Step 5. Concavity of $p(g, \xi)$ in ξ.

Begin with rationals: for $t \in [0,1] \cap \mathbb{Q}$ and $\xi', \xi'' \in \mathcal{U} \cap \mathbb{Q}^2$, fix $k \in \mathbb{N}$ so that $kt \in \mathbb{N}$ and $kt\xi', k(1-t)\xi'' \in \mathbb{Z}^2$. By the Markov property

$$\begin{aligned}\log E_0^{\omega}\big[e^{mkL_{mk}(g)}, X_{mk} = mk(t\xi' + (1-t)\xi'')\big] \\ \geq \log E_0^{\omega}\big[e^{mktL_{mkt}(g)}, X_{mkt} = mkt\xi'\big] \\ + \log E_0^{T_{mkt\xi'}\omega}\big[e^{mk(1-t)L_{mk(1-t)}(g)}, X_{mk(1-t)} = mk(1-t)\xi''\big].\end{aligned}$$

When we divide by mk and let $m \to \infty$, the last term converges to $(1-t)p(g, \xi'')$ in probability, hence a.s. along some subsequence $\{m_j\}$. Now take $m \to \infty$ along this subsequence and in the limit

$$p(g, t\xi' + (1-t)\xi'') \geq tp(g, \xi') + (1-t)p(g, \xi'').$$

The result extends to irrationals by continuity. This completes the proof of Proposition 16.11. □

Lemma 16.12. *Let g be a bounded measurable function on $\Omega \times \mathcal{R}$. Define*

$$p(g) = \sup_{\xi \in \mathbb{Q}} p(g, \xi). \tag{16.39}$$

Then $\mathbb{P}$-almost surely

$$p(g) = \lim_{n\to\infty} n^{-1}\log E_0^{\omega}[e^{nL_n(g)}]. \tag{16.40}$$

Proof. The immediate direction: for any ξ,

$$\varliminf_{n\to\infty} \frac{1}{n} \log E_0^\omega[e^{nL_n(g)}] \geq \varlimsup_{n\to\infty} \frac{1}{n} \log E_0^\omega\big[e^{nL_n(g)}, X_n = \hat{x}_n(\xi)\big] = p(g,\xi)$$

so $\leq$ holds in (16.40) for $\varliminf$ on the right.

Let $\varepsilon \in (0, 1/2)$. Fix $k \in \mathbb{N}$ such that $k > 4/\varepsilon$ and let $\zeta^i = (\frac{i}{k}, 1 - \frac{i}{k}) \in \mathcal{U}$. For $n \in \mathbb{N}$ let $n_1 = \lceil n(1 + \frac{3}{k})\rceil$. Then, if n is large enough, for each $\xi \in \mathcal{U}$ there exists a path of $n_1 - n$ steps from $\hat{x}_n(\xi)$ to $\hat{x}_{n_1}(\zeta^i)$ for some $i = i(n,\xi) \in \{0, \dots, k\}$. This can be checked for example as follows. Pick j so that $\frac{j}{k} \leq \xi_1 \leq \frac{j+1}{k}$. If $j \geq k/3$, then the path is constructed from $\hat{x}_n(\xi)$ to $\hat{x}_{n_1}(\zeta^j)$. That this is possible follows from $0 \leq \lfloor n_1 j/k\rfloor - \lfloor n\xi_1\rfloor \leq n_1 - n$. The path takes $m = \lfloor n_1 j/k\rfloor - \lfloor n\xi_1\rfloor$ e_1-steps and $n_1 - n - m$ e_2-steps. If $j \leq k/2$, the path is constructed from $\hat{x}_n(\xi)$ to $\hat{x}_{n_1}(\zeta^{j+1})$ and similar inequalities are verified.

We derive an upper bound:

$$\begin{aligned}
n^{-1} \log E_0^\omega[e^{nL_n(g)}] &= n^{-1} \log \sum_{\xi\in(n^{-1}\mathbb{Z}^2)\cap\mathcal{U}} E_0^\omega[e^{nL_n(g)}, X_n = \hat{x}_n(\xi)\,] \\
&\leq \max_{\xi\in(n^{-1}\mathbb{Z}^2)\cap\mathcal{U}} n^{-1} \log E_0^\omega\big[e^{nL_n(g)}, X_n = \hat{x}_n(\xi)\big] + \frac{C\log n}{n} \\
&\leq \max_{\xi\in(n^{-1}\mathbb{Z}^2)\cap\mathcal{U}} n^{-1} \log E_0^\omega\big[e^{nL_n(g)},\, X_n = \hat{x}_n(\xi),\, X_{n_1} = \hat{x}_{n_1}(\zeta^{i(n,\xi)})\,\big] \\
&\quad - \frac{n_1 - n}{n} \log\eta + \frac{C\log n}{n} \\
&\leq \max_{0\leq i\leq k} n^{-1} \log E_0^\omega\big[e^{n_1 L_{n_1}(g)}, X_{n_1} = \hat{x}_{n_1}(\zeta^i)\,\big] \\
&\quad + \frac{n_1 - n}{n}(\,\|g\|_\infty - \log\eta) + \frac{C\log n}{n}.
\end{aligned}$$

Letting $n \to \infty$ gives a.s.

$$\begin{aligned}
\varlimsup_{n\to\infty} n^{-1} \log E_0^\omega[e^{nL_n(g)}] &\leq \max_{0\leq i\leq k} (1+\varepsilon)p(g,\zeta^i) + \varepsilon(\,\|g\|_\infty - \log\eta) \\
&\leq (1+\varepsilon)p(g) + \varepsilon(\,\|g\|_\infty - \log\eta).
\end{aligned}$$

Letting $\varepsilon \to 0$ completes the proof. □

For the next lemma, let $\mathfrak{S}_\ell$ denote the σ-algebra generated by the z-coordinate and $\{\omega_{i,j} : -\ell \leq i + j < \ell\}$ on $\Omega \times \mathcal{R}$. $b\mathfrak{S}_\ell$ denotes the space of bounded $\mathfrak{S}_\ell$-measurable functions. $\mathbf{E}$ denotes expectation under the joint distribution $\mathbf{P}(d\omega, dz_{1,\infty}) = P_0^\omega(dz_{1,\infty})\mathbb{P}(d\omega)$.

Lemma 16.13. *The functions $J(\nu)$, $p(g)$, $J(\nu,\xi)$, and $p(g,\xi)$ have the following properties.*

(a) *If $E^\nu[Z_1] \neq \xi$, then $J(\nu,\xi) = \infty$.*

(b) *If g is bounded and $\mathfrak{S}_\ell$-measurable, then*

$$p(g,\xi) \le p(g) \le (2\ell)^{-1} \log \mathbf{E}\big[e^{2\ell g(T_{X_\ell}\omega, Z_{\ell+1})}\big]. \tag{16.41}$$

(c) *The sublevel sets $\{\nu \in \mathcal{M}_1 : J(\nu) \le c\}$ and $\{\nu \in \mathcal{M}_1 : J(\nu,\xi) \le c\}$ are compact for all $\xi \in \mathcal{U}$ and $c \in \mathbb{R}$.*

Proof. Part (a). Take $g(\omega, z) = cz \cdot e_1$ in (16.7).

Part (b). The assumptions are that $\ell \ge 1$ and $g(\omega,z)$ depends only on $\{\omega_{i,j} : -\ell \le i+j < \ell\}$. We develop a bound in several steps. Apply first Jensen's and Hölder's inequalities. Then we have

$$\begin{aligned}
&\mathbb{E}\big[\log E_0^\omega\big(e^{2m\ell L_{2m\ell}(g)}\big)\big] \\
&\quad \le \log \mathbb{E}E_0^\omega\big[e^{\sum_{i=0}^{2\ell-1}\sum_{k=0}^{m-1} g(T_{X_{i+2k\ell}}\omega, Z_{i+2k\ell+1})}\big] \\
&\quad \le (2\ell)^{-1}\sum_{i=0}^{2\ell-1} \log \mathbb{E}E_0^\omega\big[e^{2\ell\sum_{k=0}^{m-1} g(T_{X_{i+2k\ell}}\omega, Z_{i+2k\ell+1})}\big] \\
&\quad \le (2\ell)^{-1}\sum_{i=0}^{2\ell-1} \log \sum_x \mathbb{E}E_0^\omega\big[e^{2\ell\sum_{k=0}^{m-2} g(T_{X_{i+2k\ell}}\omega, Z_{i+2k\ell+1})}, \\
&\qquad\qquad\qquad\qquad X_{i+2(m-2)\ell+\ell} = x\big] \cdot E_x^\omega\big[e^{2\ell g(T_{X_\ell}\omega, Z_{\ell+1})}\big]
\end{aligned}$$

In the last step we conditioned on $X_{i+2(m-2)\ell+\ell}$ and used the Markov property. The two quenched expectations inside $\mathbb{E}$ above are independent under $\mathbb{P}$ because the first one depends on $\{\omega_{y_1,y_2} : y_1+y_2 < i+2(m-2)\ell+\ell\}$ while the second one depends on $\{\omega_{y_1,y_2} : y_1+y_2 \ge i+2(m-2)\ell+\ell\}$. The expectation of the second one is, by a shift of ω,

$$\mathbb{E}E_x^\omega\big[e^{2\ell g(T_{X_\ell}\omega, Z_{\ell+1})}\big] = \mathbf{E}\big[e^{2\ell g(T_{X_\ell}\omega, Z_{\ell+1})}\big].$$

The bound above develops as follows, upon repeating the step $m-1$ times:

$$\begin{aligned}
&\mathbb{E}\big[\log E_0^\omega\big(e^{2m\ell L_{2m\ell}(g)}\big)\big] \\
&\quad \le (2\ell)^{-1}\sum_{i=0}^{2\ell-1} \log \mathbb{E}E_0^\omega\big[e^{2\ell\sum_{k=0}^{m-2} g(T_{X_{i+2k\ell}}\omega, Z_{i+2k\ell+1})}\big] + \log \mathbf{E}\big[e^{2\ell g(T_{X_\ell}\omega, Z_{\ell+1})}\big] \\
&\quad \cdots \le (2\ell)^{-1}\sum_{i=0}^{2\ell-1} \log \mathbb{E}E_0^\omega\big[e^{2\ell g(T_{X_i}\omega, Z_{i+1})}\big] + (m-1)\log \mathbf{E}\big[e^{2\ell g(T_{X_\ell}\omega, Z_{\ell+1})}\big].
\end{aligned}$$

Divide through by $2m\ell$ and let $m \to \infty$ to get (16.41).

Part (c). The sublevel sets are closed because $J(\nu,\xi)$ and $J(\nu)$ are lower semicontinuous in ν as suprema of continuous functions. Since $J(\nu,\xi) \ge J(\nu)$ it suffices to prove compactness for $J(\nu)$. We bound $J(\nu)$ below by an entropy. Let Q_ℓ denote the distribution of $\big(\{(T_{X_\ell}\omega)_{i,j} : -\ell \le i+j < \ell\}, Z_{\ell+1}\big)$ under the joint measure $\mathbf{P}$. Then the right-hand side of (16.41) equals $(2\ell)^{-1}\log E^{Q_\ell}[e^{2\ell g}]$. We have

$$
\begin{aligned}
J(\nu) &= \sup_{g\in\mathscr{U}_b}\{E^\nu[g]-p(g)\} = \sup_{\ell\in\mathbb{N}}\ \sup_{g\in\mathscr{U}_b\cap b\mathfrak{S}_\ell}\{E^\nu[g]-p(g)\}\\
&\ge \sup_{\ell\in\mathbb{N}}\ \sup_{g\in\mathscr{U}_b\cap b\mathfrak{S}_\ell}\big\{E^\nu[g]-(2\ell)^{-1}\log E^{Q_\ell}[e^{2\ell g}]\big\}\\
&\ge \sup_{\ell\in\mathbb{N}}\ \sup_{h\in\mathscr{U}_b\cap b\mathfrak{S}_\ell}(2\ell)^{-1}\big\{E^\nu[h]-\log E^{Q_\ell}[e^{h}]\big\}\\
&= \sup_{\ell\in\mathbb{N}}(2\ell)^{-1}H\big(\nu_{\mathfrak{S}_\ell}\,\big|\,Q_\ell\big).
\end{aligned}
\tag{16.42}
$$

This bound suffices for the conclusion. For, given a sequence $\{\nu^j\}$ such that $J(\nu^j)\le c$, the $\mathfrak{S}_\ell$-marginals satisfy $H\big(\nu^j_{\mathfrak{S}_\ell}\,\big|\,Q_\ell\big)\le 2\ell c$ and hence lie in a compact set (Proposition 5.7). A diagonal argument produces a convergent subsequence $\{\nu^{j_k}\}$. □

Proof of Theorem 16.2. Part (a) is contained in Proposition 16.11 and Lemma 16.12.

Part (b). Lower semicontinuity and compact sublevel sets of $J(\nu,\xi)$ and $J(\nu)$ are in Lemma 16.13(c). That $J(\nu,\xi)\ge 0$ comes from (16.7) with $g=0$. Convexity of $J(\nu)$ and $J(\nu,\xi)$ comes from their definitions and the concavity of $p(g,\xi)$ in ξ. The statement for $E^\nu[Z_1]\ne\xi$ is Lemma 16.13(a).

In the calculation below apply the minimax theorem from Section A.5 to the compact set $\mathcal{U}$ and the space $\mathscr{U}_b$. Then

$$
\begin{aligned}
J(\nu) &= \sup_{g\in\mathscr{U}_b}\{E^\nu[g]-p(g)\}\\
&= \sup_{g\in\mathscr{U}_b}\inf_{\xi\in\mathcal{U}}\{E^\nu[g]-p(g,\xi)\}\\
&= \inf_{\xi\in\mathcal{U}}\sup_{g\in\mathscr{U}_b}\{E^\nu[g]-p(g,\xi)\} = \inf_{\xi\in\mathcal{U}}J(\nu,\xi) = J(\nu,E^\nu[Z_1]).
\end{aligned}
$$

Part (c). What is needed here is that p is its own double conjugate, both in the point-to-point and point-to-line version. The argument is the same for both so we discuss the point-to-line version.

In the supremum norm topology of $\mathscr{U}_b$, $p(g)$ is continuous and convex in g. Hence $\{g\in\mathscr{U}_b : p(g)\le s\}$ is a norm-closed and convex set. By the Hahn-Banach separation theorem this set is closed also in the weak topology generated by the dual space $\mathscr{U}_b^*$. Thus $p(g)$ is lower semicontinuous in the weak topology $\sigma(\mathscr{U}_b, \mathscr{U}_b^*)$. However, the dual space $\mathscr{U}_b^*$ is not necessarily $\mathcal{M}$ unless $\Omega \times \mathcal{R}$ is compact. Consequently we move the duality to a compactification.

Let us be specific about the compactification. Pick a totally bounded metric d_0 on $\mathcal{S}$ and define a totally bounded metric d on $\Omega \times \mathcal{R}$ by

$$d((\omega, z), (\tilde{\omega}, \tilde{z})) = \sum_{x \in \mathbb{Z}^2} 2^{-|x|_\infty} d_0(\omega_x, \tilde{\omega}_x) + |z - \tilde{z}| .$$

Let $\bar{\mathcal{S}}$ be the (compact) completion of $\mathcal{S}$ under d_0. Then $\mathcal{T} = \bar{\mathcal{S}}^{\mathbb{Z}^2} \times \mathcal{R}$ is the completion of $\Omega \times \mathcal{R}$ under d. $\mathcal{T}$ is compact, $\mathscr{U}_b$ is in bijective, isometric and isomorphic, correspondence with $\mathscr{C}(\mathcal{T})$, and the dual of $\mathscr{U}_b \cong \mathscr{C}(\mathcal{T})$ is the measure space $\mathcal{M}(\mathcal{T})$. (Recall that similar considerations arose earlier on pages 190 and 227.) Define $J(\nu)$ for $\nu \in \mathcal{M}(\mathcal{T})$ by (16.5). As before, if ν is not a probability measure, this definition gives $J(\nu) = \infty$. Now, by lower semicontinuity and convexity,

$$p(g) = p^{**}(g) = \sup_{\nu \in \mathcal{M}_1(\mathcal{T})} \{E^\nu[g] - J(\nu)\} \quad \text{for } g \in \mathscr{U}_b. \tag{16.43}$$

It remains to argue that $\mathcal{M}_1(\mathcal{T})$ can be replaced by the smaller space $\mathcal{M}_1 = \mathcal{M}_1(\Omega \times \mathcal{R})$ in the supremum above.

The kind of inequality developed in (16.42) shows that for $\nu \in \mathcal{M}_1(\mathcal{T})$, $J(\nu) = \infty$ unless ν is supported on $\Omega \times \mathcal{R}$ (Exercise 16.18). Consequently the identity above becomes

$$p(g) = \sup_{\nu \in \mathcal{M}_1} \{E^\nu[g] - J(\nu)\} = J^*(g). \tag{16.44}$$

□

Lemma 16.14. *Let* $(a, b, \tau) \in \mathcal{I}$ *and* $\xi \in \mathcal{U}$. *For* $\mathbb{P}$*-a.e.* ω *the distributions* $P^\omega_{\hat{x}_{an+b}(\xi)}\{L_{n+\tau} \in \cdot\}$ *are exponentially tight.*

Proof. Recall that $\mathbf{P}(dz_{1,\infty}, d\omega) = P^\omega(dz_{1,\infty})\mathbb{P}(d\omega)$ denotes the joint distribution of the random walk path and the environment. Suppose we had exponential tightness for the distributions $\mathbf{P}\{L_{n+\tau} \in \cdot\}$. Given $s < \infty$, pick a compact set $A \subset \mathcal{M}_1$ such that $\mathbf{P}\{L_{n+\tau} \in A^c\} \le e^{-2sn}$ for large enough n. Then by Chebyshev's inequality and shift invariance, for large n,

$$\begin{aligned}\mathbb{P}\{\omega : P^\omega_{\hat{x}_{an+b}(\xi)}(L_{n+\tau} \in A^c) \ge e^{-sn}\} &\le e^{sn}\mathbb{E}\big[P^\omega_0(L_{n+\tau} \in A^c)\big] \\ &= e^{sn}\mathbf{P}\{L_{n+\tau} \in A^c\} \le e^{-sn}.\end{aligned}$$

By the Borel-Cantelli lemma, for $\mathbb{P}$-a.e. ω

$$P^\omega_{\hat{x}_{an+b}(\xi)}(L_{n+\tau} \in A^c) \le e^{-sn} \quad \text{for large enough } n.$$

This suffices for quenched exponential tightness.

It remains to show exponential tightness for $\mathbf{P}\{L_n \in \cdot\}$ (the "$+\tau$" makes no difference in the large n limit; hence it is dropped). This proof provides an excellent technical exercise so we indicate the main steps of the argument. Since $\mathcal{R}$ is compact, it suffices to show exponential tightness of the distributions $\mathbf{P}\{n^{-1}\sum_{k=0}^{n-1}\delta_{T_{X_k}\omega} \in \cdot\,\}$ on $\mathcal{M}_1(\Omega)$. For this purpose note that for any $x \in \mathbb{Z}^2$ the variables $\{\omega_{x+X_k}\}_{k\in\mathbb{Z}_+}$ are i.i.d. under $\mathbf{P}$ because the walk never visits the same point twice. Then adapt the construction in the proof of Lemma 6.18. $\square$

Proof of Theorem 16.3. Fix $\xi \in \mathcal{U}$. The event Ω_0 is defined as follows. Let $\{g_j\}$ be a countable subset of $\mathscr{U}_b$ that includes the zero function and is dense in the supremum norm topology. ($\mathscr{U}_b$ is separable because it is isometric to the separable space $\mathscr{C}(\mathcal{T})$ defined in the proof of Theorem 16.2.) Let Ω_0 be the event on which limit (16.28) holds for all g_j and all $(a,b,\tau) \in \mathcal{I}$, and the exponential tightness of Lemma 16.14 holds for all $(a,b,\tau) \in \mathcal{I}$. As a countable intersection of full probability events, $\mathbb{P}(\Omega_0) = 1$. Next note that by the density of $\{g_j\}$, limit (16.28) holds simultaneously for all $g \in \mathscr{U}_b$ and all $(a,b,\tau) \in \mathcal{I}$ if $\omega \in \Omega_0$.

Fix $\omega \in \Omega_0$. Apply the version of the Baxter-Jain theorem given in Theorem 16.9 to the probability measures

$$\begin{aligned}\tilde{\mu}_n^{a,b,\tau,\omega}(B) &= P^{\omega}_{\hat{x}_{an+b}(\xi)}\{L_{n+\tau} \in B \,|\, X_{n+\tau} = \hat{x}_{(a+1)n+b+\tau}(\xi)\}\\ &= \frac{\mu_n^{a,b,\tau,\omega}(B)}{P^{\omega}_{\hat{x}_{an+b}(\xi)}\{X_{n+\tau} = \hat{x}_{(a+1)n+b+\tau}(\xi)\}}, \qquad (a,b,\tau) \in \mathcal{I}.\end{aligned}$$

Proposition 16.11 gives the limit

$$p(0,\xi) = \lim_{n\to\infty}\frac{1}{n}\log P^{\omega}_{\hat{x}_{an+b}(\xi)}\{X_{n+\tau} = \hat{x}_{(a+1)n+b+\tau}(\xi)\}.$$

The hypotheses of Theorem 16.9 for $\{\tilde{\mu}_n^{a,b,\tau,\omega}\}$ are established above in the first paragraph of the proof and in Lemmas 16.10 and 16.13(c). The LDP for $\{\tilde{\mu}_n^{0,0,0,\omega}\}$ gives the bounds claimed in the theorem. $\square$

Proof of Theorem 16.4. (i) Upper bound. Let Ω_1 be the event on which the point-to-point upper bound (16.10) holds for all rational $\xi \in \mathcal{U}$. We show that for $\omega \in \Omega_1$ we also have the upper bound

$$\varlimsup_{n\to\infty}\frac{1}{n}\log P_0^{\omega}\{L_n \in A\} \le -\inf_{\nu\in A} J(\nu) \tag{16.45}$$

for all closed $A \subset \mathcal{M}_1$.

Let $\varepsilon > 0$. Pick a smaller $\varepsilon_1 > 0$ such that if $L_n \in A$, n is large enough, and $n_1 = \lceil n(1+\varepsilon_1)\rceil$, then

$$L_{n_1} \in A^{\varepsilon} = \{\mu \in \mathcal{M}_1 : \exists \nu \in A \text{ such that } d(\mu,\nu) < \varepsilon\}.$$

Here d is some metric of weak convergence, for example

$$d(\mu,\nu)=\sum_{j=1}^{\infty}2^{-j}\,\|g_j\|_\infty^{-1}\,|\langle g_j,\mu-\nu\rangle|$$

where $\{g_j\}$ is a countable norm-dense subset of $\mathscr{U}_b$. Imitate the proof of Lemma 16.12 to deduce the bound

$$\begin{aligned}&n^{-1}\log P_0^\omega\{L_n\in A\}\\&\quad\le\max_{0\le i\le k}n^{-1}\log P_0^\omega\{L_{n_1}\in\overline{A^\varepsilon},X_{n_1}=\hat{x}_{n_1}(\zeta^i)\}-(n_1-n)\log\eta.\end{aligned}$$

Since $J(\nu,\zeta^i)\ge J(\nu)$, we get the asymptotic bound

$$\begin{aligned}\overline{\lim_{n\to\infty}}\,n^{-1}\log P_0^\omega\{L_n\in A\}&\le-\min_{0\le i\le k}\inf_{\nu\in\overline{A^\varepsilon}}J(\nu,\zeta^i)-\varepsilon_1\log\eta\\&\le-\inf_{\nu\in\overline{A^\varepsilon}}J(\nu)-\varepsilon_1\log\eta.\end{aligned}$$

Let $\varepsilon\to0$ and apply Exercise 2.15. This proves (16.45).

(ii) Lower bound. To construct an event Ω_2 on which the lower bound

$$\underline{\lim_{n\to\infty}}\frac{1}{n}\log P_0^\omega\{L_n\in G\}\ge-\inf_{\nu\in G}J(\nu) \tag{16.46}$$

holds for all open sets, start by fixing a countable base $\{B_j\}$ of open neighborhoods for the topology of $\mathcal{M}_1$. By picking countably many measures in each B_j, we can collect a countable set $\{\nu_\ell\}$ in $\mathcal{M}_1$ such that for each B_j, $\inf_{\nu\in B_j}J(\nu)=\inf_{\ell:\nu_\ell\in B_j}J(\nu_\ell)$. This implies $\inf_{\nu\in G}J(\nu)=\inf_{\ell:\nu_\ell\in G}J(\nu_\ell)$ for every open set G. Now let Ω_2 be the event on which the point-to-point lower bound (16.11) holds for all $\xi^\ell=E^{\nu_\ell}[Z_1]$.

Given open G, let ℓ be such that $\nu_\ell\in G$. Then

$$\begin{aligned}\underline{\lim_{n\to\infty}}\,n^{-1}\log P_0^\omega\{L_n\in G\}&\ge\underline{\lim_{n\to\infty}}\,n^{-1}\log P_0^\omega\{L_n\in G,X_n=\hat{x}_n(\xi^\ell)\}\\&\ge-\inf_{\nu\in G}J(\nu,\xi^\ell)\ge-J(\nu_\ell,\xi^\ell)=-J(\nu_\ell).\end{aligned}$$

The supremum over $\nu_\ell\in G$ on the right establishes (16.46). □

***Exercise 16.15.** Adapt the arguments of Chapter 14 to prove Theorem 16.9.

Hint: (a) In Lemma 14.1, $\overline{\kappa}^{(i)}$ gives a large deviation upper bound for all closed sets. (b) Lemma 14.2 is proved for each $i\in\mathcal{I}$ separately, for the limits

$$\overline{p}^{(i)}(g)=\overline{\lim_{n\to\infty}}\frac{1}{r_n}\log\int_{\mathcal{X}_0}e^{r_ng(x)}\,\mu_n^{(i)}(dx)$$

assuming g bounded above. Use exponential tightness. (c) For Theorem 14.5 assume f has compact sublevel sets. Then A is compact by virtue of the boundedness of g on $\mathcal{X}_0$.

***Exercise 16.16.** Show that $\mathcal{M}_1$ is a bounded subset in the weak topology of $\mathcal{M}$ generated by $\mathscr{U}_b$.

Hint: A neighborhood base at the origin of $\mathcal{M}$ is given by sets of the type

$$\{\mu \in \mathcal{M} : |\langle g_i, \mu\rangle| < \varepsilon \text{ for } i = 1, \dots, m\}$$

for $\varepsilon > 0$, $m \in \mathbb{N}$, and $g_i \in \mathscr{U}_b$.

***Exercise 16.17.** Complete Step 2 of the proof of Proposition 16.11 by showing upper semicontinuity of $p(g, \xi)$ at $\xi = (1,0)$ and $(0,1)$.

Hint: This exercise is a little tricky. It is enough to show that, given $\varepsilon > 0$,

$$\mathbb{P}\Big\{E_0^\omega[e^{nL_n(g)}, X_n = \hat{x}_n(\xi)] \geq e^{np(g,(1,0))+n\varepsilon}\Big\} \longrightarrow 0 \tag{16.47}$$

along some subsequence if $\xi \in \mathcal{U}$ is rational and close enough to $(1,0)$. The idea is to take a union bound over all paths to $\hat{x}_n(\xi)$ and for each path cut and paste the environment so that the path becomes a straight path to $(n,0)$.

Consider a particular path $(x_k)_{k=0}^n$ from 0 to $\hat{x}_n(\xi) = (n-\ell, \ell)$. This path is of the form $x_k = (k-j, j)$ for $k_j \leq k < k_{j+1}$ where $k_j = \min\{k : x_k \cdot e_2 = j\}$ marks the time when the path takes its jth e_2-step, $0 \leq j \leq \ell$, with $k_{\ell+1} = n+1$. Define a new environment $\breve{\omega}$ by

$$\breve{\omega}_{a,b} = \begin{cases} \omega_{a,b}, & a+b < k_1, \\ \omega_{a-j,b+j}, & k_j \leq a+b < k_{j+1},\ 1 \leq j \leq \ell-1, \\ \omega_{a-\ell,b+\ell}, & a+b \geq k_\ell. \end{cases}$$

The point of the definition is that $\breve{\omega}_{k,0} = \omega_{x_k}$ for $0 \leq k \leq n$. Similarly $g(T_{k,0}\breve{\omega}, z) = g(T_{x_k}\omega, z)$ as long as the support of $g(T_{k,0}\breve{\omega}, z)$ is fully contained in one of the cases in the definition of $\breve{\omega}$. Thus except for some boundary effects between the cases in the definition of $\breve{\omega}$, the $\breve{\omega}$-environment seen by the path $y_k = (k,0)$ from 0 to $(n,0)$ is the same as the original ω-environment seen by the path x_k. By virtue of the i.i.d. distribution $\mathbb{P}$, $\breve{\omega}$ and ω are equal in distribution. If ξ is close to $(1,0)$, ℓ is very small compared to n, the number of different paths x_k from 0 to $(n-\ell,\ell)$ is small on an exponential scale, and the construction above leads to a good enough estimate for (16.47). A result for more general paths and higher dimensions can be found in Theorem 3.2 of [**64**].

***Exercise 16.18.** Complete the proof of part (c) of Theorem 16.2. Namely, use the reasoning in (16.42) to show that for $\nu \in \mathcal{M}_1(\mathcal{T})$, $J(\nu) = \infty$ unless $\nu(\Omega \times \mathcal{R}) = 1$.

Hint: If $\{I_\ell\}$ is a sequence of finite sets that increase up to $\mathbb{Z}^2$, then $\Omega \times \mathcal{R} = \bigcap_\ell \mathcal{S}^{I_\ell} \times \bar{\mathcal{S}}^{\mathbb{Z}^2 \smallsetminus I_\ell} \times \mathcal{R}$.

***Exercise 16.19.** Complete the proof of Lemma 16.14.

Literature notes. The results of this chapter are special cases of the results in [**64**] and [**65**]. The approach to prove large deviations for RWRE via the Baxter-Jain theorem is original. The approach in [**64**] and [**65**] generalizes the treatment of Section 13.2 and gives a formula for the rate function (16.5) as the convex conjugate of a functional much like K in (13.11), instead of the duality with the limiting pressure (16.4). A general introduction to RWRE can be found in lectures [**81**], and [**16**] contains RWRE as an example of large deviations. The directed polymer model was introduced in 1985 by Huse and Henley [**49**]. Lectures [**17**] provide a general introduction to this model and its relatives, while an introduction to the Kardar-Parisi-Zhang (KPZ) universality class appears in the review paper [**13**].

Appendixes

Appendix A

Analysis

This appendix and Appendix B collect some basic definitions and results without proof. The purpose is not to teach the subjects but to help the reader remember things encountered in the past and to point to fuller treatments elsewhere. Proofs can be found in the references mentioned in the literature notes at the end of the appendixes and in myriad other textbooks. Some more specialized results used in the present text are proved.

A.1. Metric spaces and topology

We begin with the definition of a metric space and then discuss some features of metric spaces and more general topological spaces. In general a space is simply a set, but the term *space* is preferred when the set carries additional structure.

Definition A.1. Let $\mathcal{X}$ be a space. A function $d : \mathcal{X} \times \mathcal{X} \to [0, \infty)$ is a *metric* if for all $x, y, z \in \mathcal{X}$,

(i) $d(x, y) = 0$ if and only if $x = y$,

(ii) $d(x, y) = d(y, x)$, and

(iii) $d(x, y) \le d(x, z) + d(z, y)$.

The pair $(\mathcal{X}, d)$, or $\mathcal{X}$ alone, is called a *metric space*.

Examples of metrics are the Euclidean metric

$$d(x, y) = |x - y| = \Big(\sum_{i=1}^{d} (x_i - y_i)^2\Big)^{1/2} \tag{A.1}$$

for points $x = (x_1, \ldots, x_d)$ and $y = (y_1, \ldots, y_d)$ in $\mathbb{R}^d$, and the sup-norm metric

$$d(f, g) = \|f - g\|_\infty = \sup_{s \in \mathcal{S}} |f(s) - g(s)| \tag{A.2}$$

on any space of bounded real-valued functions on an underlying space $\mathcal{S}$. A function $f : \mathcal{S} \to \mathbb{R}$ is bounded if there exists a constant C such that $|f(s)| \le C$ for all $s \in \mathcal{S}$.

A basic role for a metric is the definition of limits. Convergence of a sequence $\{x_n\}$ to a point x in $\mathcal{X}$ means that the distance vanishes in the limit: $x_n \to x$ if $d(x_n, x) \to 0$. The sequence $\{x_n\}$ is a *Cauchy sequence* if $\sup_{m:m>n} d(x_m, x_n) \to 0$ as $n \to \infty$. *Completeness* of a metric space means that every Cauchy sequence in the space has a limit in the space. A subset $D \subset \mathcal{X}$ is *dense* in $\mathcal{X}$ if for every $x \in \mathcal{X}$ and every $\varepsilon > 0$, there exists $y \in D$ such that $d(x, y) < \varepsilon$. $\mathcal{X}$ is a *separable* metric space if it has a countable dense set. A complete, separable metric space is called a *Polish space.*

Two metric spaces (S, d) and $(\tilde{S}, \tilde{d})$ are *isometric* if there exists a bijection $f : S \to \tilde{S}$ such that $d(x, y) = \tilde{d}(f(x), f(y))$. Every metric space (S, d) can be completed in an abstract way: there exists a complete metric space $(\mathcal{T}, \rho)$ such that (S, d) is isometric to a dense subset of $\mathcal{T}$. $(\mathcal{T}, \rho)$ is the *completion* of (S, d).

A norm is a function $x \mapsto \|x\| \in \mathbb{R}_+$ on a real vector space for which $\|\alpha x\| = |\alpha| \, \|x\|$ for real α, $\|x + y\| \le \|x\| + \|y\|$, and $\|x\| = 0$ implies $x = 0$. A norm defines a metric $d(x, y) = \|x - y\|$. A space with a norm is called a *normed space.* A complete normed space is called a *Banach space.* Both examples (A.1) and (A.2) are metrics defined by norms on Banach spaces.

For $x \in \mathcal{X}$ and $r > 0$ the *open ball* centered at x with radius $r > 0$ is $B(x, r) = \{y \in \mathcal{X} : d(x, y) < r\}$. A set $G \subset \mathcal{X}$ is *open* if for every $x \in G$ there exists an open ball $B(x, r) \subset G$. A set is *closed* if its complement is open. Equivalently, $F \subset \mathcal{X}$ is closed if $F \ni x_n \to x$ implies $x \in F$.

The open subsets of a metric space $\mathcal{X}$ form a *topology.* A metric is not the only way to define a topology. Hence it is necessary to have an axiomatic approach to topology.

Definition A.2. A collection $\mathscr{T}$ of subsets of a space $\mathcal{X}$ is a *topology* if it has the following properties.

(i) $\varnothing \in \mathscr{T}$ and $\mathcal{X} \in \mathscr{T}$.

(ii) Any union of sets in $\mathscr{T}$ is again a set in $\mathscr{T}$.

(iii) If $A, B \in \mathscr{T}$, then $A \cap B \in \mathscr{T}$.

A space with a topology is a *topological space.* Members of a topology are called open sets. The important point about the definition is that any union of open sets is open, but only finite intersections of open sets are required to be open. It is now a simple exercise to check that the open sets of a metric space satisfy the axioms of Definition A.2.

Any open set that contains a point x is an *open neighborhood* of x. In terms of neighborhoods, $x_n \to x$ if every open neighborhood of x contains all but finitely many of the terms x_n. The *interior* A° of a set A is the union of all open sets contained in A The *closure* $\overline{A}$ of A is the intersection of all closed sets that contain A, or equivalently $\overline{A} = \mathcal{X} \smallsetminus (\mathcal{X} \smallsetminus A)^\circ$.

A topology $\mathscr{T}$ is said to be *metrizable* if there exists a metric whose topology is $\mathscr{T}$. A space is *topologically complete* if it has at least one complete metric that induces the given topology. A metric space is topologically complete if and only if it is a countable intersection of open sets in its completion [**26**, Theorem 2.5.4].

Topologies can be defined in terms of bases, as was done for the weak topology in Proposition 4.6.

Definition A.3. A collection $\mathscr{B}$ of subsets of a space $\mathcal{X}$ is a *base for a topology* if it has the following properties.

(i) $\forall x \in \mathcal{X}\ \exists B \in \mathscr{B}$ such that $x \in B$.

(ii) If $x \in A \cap B$ for $A, B \in \mathscr{B}$, then $\exists C \in \mathscr{B}$ such that $x \in C \subset A \cap B$.

The topology $\mathscr{T}$ generated by a base $\mathscr{B}$ is the collection of all unions of sets in $\mathscr{B}$ (including the empty union to get $\varnothing$). A collection $\mathscr{B}_x$ of open neighborhoods of x is a *base of neighborhoods at* x if every open neighborhood of x contains a member of $\mathscr{B}_x$.

Topological spaces are characterized by *separation axioms* that come in a hierarchy. We list here the two that we use in the text. Metric spaces satisfy both axioms.

• A *Hausdorff space* $\mathcal{X}$ satisfies this condition: if x and y are two distinct points of $\mathcal{X}$, then there exist disjoint open sets G and H such that $x \in G$ and $y \in H$.

• A *regular topological space* $\mathcal{X}$ satisfies this condition: if $x \in \mathcal{X}$, $F \subset \mathcal{X}$ is closed, and $x \notin F$, then there exist disjoint open sets G and H such that $x \in G$ and $F \subset H$.

Compactness is an extremely important property of a space or a set but somewhat nontrivial to work with in infinite-dimensional spaces.

Definition A.4. A subset K of a topological space $\mathcal{X}$ is *compact* if it satisfies the following condition: if $\mathscr{G}$ is a collection of open sets such that $K \subset \bigcup_{G \in \mathscr{G}} G$, then there exists a finite subcollection $\{G_1, \dots, G_m\} \subset \mathscr{G}$ such that $K \subset \bigcup_{i=1}^m G_i$. Succinctly stated, every open cover of K has a finite subcover.

Compactness is the quintessential mathematical property that is seldom checked by appeal to the definition. A student of mathematics must master the characterizations and properties of compact sets in the spaces that he or she works in. A subset of $\mathbb{R}^d$ is compact if and only if it is closed and bounded. (A set is *bounded* if it lies inside some ball of finite radius.) A subset of a general metric space is compact if and only if it is complete and totally bounded. (A set is *totally bounded* if for every $\varepsilon > 0$ it can be covered by finitely many ε-balls, that is, balls of radius ε.) A metric space $(\mathcal{X}, d)$ is separable if and only if it has a totally bounded metric $\tilde{d}$ that generates the same topology as d [**26**, Theorem 2.8.2]. (A metric is totally bounded if the space with this metric is totally bounded.) In a Hausdorff space compact sets are closed.

We end this section with some properties of compact sets. A subset K of a topological space is *sequentially compact* if every sequence in K has a convergent subsequence whose limit is in K. A family $\{F_i : i \in \mathcal{I}\}$ of sets has the *finite intersection property* if $\bigcap_{i\in J} F_i \neq \varnothing$ for all finite $J \subset \mathcal{I}$.

Theorem A.5. (i) *A topological space is compact if and only if every family $\{F_i : i \in \mathcal{I}\}$ of closed sets with the finite intersection property satisfies $\bigcap_{i\in\mathcal{I}} F_i \neq \varnothing$.*

(ii) *Compactness implies sequential compactness.*

(iii) *A subset of a metric space is compact if and only if it is sequentially compact.*

Exercise A.6. Observe that part (i) of Theorem A.5 is a restatement of the definition of compactness because $\bigcap_{i\in\mathcal{I}} F_i = \varnothing$ if and only if $\{F_i^c : i \in \mathcal{I}\}$ covers the space.

A.2. Measure and integral

For integration, a space $\mathcal{X}$ needs a σ-algebra and a measure.

Definition A.7. A collection $\mathscr{A}$ of subsets of $\mathcal{X}$ is a *σ-algebra* (also called a *σ-field*) if it has the following properties.

(i) $\mathcal{X} \in \mathscr{A}$ and $\varnothing \in \mathscr{A}$.

(ii) If $A \in \mathscr{A}$, then also $A^c \in \mathscr{A}$.

(iii) If $\{A_i\}$ is a sequence of sets in $\mathscr{A}$, then their union $\bigcup_i A_i \in \mathscr{A}$.

Members of $\mathscr{A}$ are called *measurable sets.* A pair $(\mathcal{X}, \mathscr{A})$ where $\mathscr{A}$ is a σ-algebra on the space $\mathcal{X}$ is a *measurable space.*

The restriction to countable unions in part (iii) is important. Unions over arbitrarily large collections of sets are not permitted. If we change part

(iii) to permit only finite unions, then $\mathscr{A}$ is called an *algebra* of sets. This structure is not rich enough for developing a useful theory of integration.

The *Borel σ-algebra* $\mathscr{B}_{\mathcal{X}}$ of a metric space $\mathcal{X}$ is the smallest σ-algebra on $\mathcal{X}$ that contains all open sets. The members of $\mathscr{B}_{\mathcal{X}}$ are called *Borel sets*. These are the most important σ-algebras.

Let $(\mathcal{X}, \mathscr{A})$ be a measurable space and let $f : \mathcal{X} \to \mathbb{R}$ be a real-valued function on $\mathcal{X}$. Then f is a *measurable function* if for each Borel subset B of $\mathbb{R}$, the inverse image

$$f^{-1}(B) = \{x \in X : f(x) \in B\} = \{f \in B\}$$

is a member of $\mathscr{A}$. These are the functions that can be integrated. This notion is very flexible. All the usual operations that create new functions from given functions (algebraic operations, composition, supremum and infimum, limits) preserve measurability. The reader not interested in the finer points of mathematical theory need not worry about whether sets and functions encountered along the way are measurable. The ones that arise in practice always are.

More generally, measure theory and probability need functions $f : \mathcal{X} \to [-\infty, \infty]$ whose values are extended reals, that is, real numbers and $\pm\infty$. A Borel subset of $[-\infty, \infty]$ is a set that either coincides with a Borel subset of $\mathbb{R}$ or differs from one by ∞, $-\infty$, or both. A function $f : \mathcal{X} \to [-\infty, \infty]$ is measurable if $f^{-1}(B) \in \mathscr{A}$ for all Borel $B \subset [-\infty, \infty]$.

Definition A.8. Let $(\mathcal{X}, \mathscr{A})$ be a measurable space. A *measure* is a function $\mu : \mathscr{A} \to [0, \infty]$ that satisfies the following properties.

(i) $\mu(\varnothing) = 0$.

(ii) If $\{A_i\}$ is a sequence of sets in $\mathscr{A}$ such that $A_i \cap A_j = \varnothing$ for all $i \neq j$ (pairwise disjoint), then

$$\mu\Big(\bigcup_i A_i\Big) = \sum_i \mu(A_i).$$

Property (ii) is called *countable additivity*. The triple $(\mathcal{X}, \mathscr{A}, \mu)$ is called a *measure space*. If $\mu(\mathcal{X}) = 1$, then μ is a *probability measure*. If $\mathcal{X} = \bigcup_i A_i$ for countably many measurable sets A_i such that $\mu(A_i) < \infty$, then μ is *σ-finite*. A *signed measure* is a function from $\mathscr{A}$ into $(-\infty, \infty]$ or $[-\infty, \infty)$ that satisfies (i) and (ii) above.

The most important measure is the *Lebesgue measure* m on Euclidean spaces. On $\mathbb{R}$ it gives the length of intervals: $m[a, b] = b - a$. More generally on $\mathbb{R}^d$, $m(B)$ is the natural volume of the set B.

Now we can define the Lebesgue integral. It is done in three steps, beginning with simple functions.

Definition A.9. Fix a measure space $(\mathcal{X}, \mathscr{A}, \mu)$.

Step 1. Integral of a nonnegative simple function. A nonnegative measurable simple function is a function f that has finitely many distinct values $\alpha_1, \dots, \alpha_n \in [0, \infty)$ and each $A_i = \{f = \alpha_i\} \in \mathscr{A}$. In other words,

$$f(x) = \sum_{i=1}^{n} \alpha_i \mathbb{1}_{A_i}(x) \qquad \text{where} \qquad \mathbb{1}_A(x) = \begin{cases} 1, & x \in A \\ 0, & x \notin A \end{cases}$$

is the *indicator function* (also called *characteristic function*) of the set A. The integral $\int f \, d\mu$ is defined by

$$\int f \, d\mu = \sum_{i=1}^{n} \alpha_i \mu(A_i).$$

The convention is $0 \cdot \infty = 0$.

Step 2. Integral of a $[0, \infty]$-valued function. Let $f : \mathcal{X} \to [0, \infty]$ be measurable. Then we define

$$\int f \, d\mu = \sup \biggl\{ \int g \, d\mu : \ g \text{ is a simple function such that } 0 \le g \le f \biggr\}.$$

This integral is a well-defined number in $[0, \infty]$.

Step 3. Integral of a general measurable function. Let $f : \mathcal{X} \to [-\infty, \infty]$ be measurable. The positive and negative parts of f are $f^+(x) = f(x) \vee 0 = \max\{f(x), 0\}$ and $f^-(x) = -(f(x) \wedge 0) = \max\{-f(x), 0\}$. $f^{\pm}$ are nonnegative functions and satisfy $f = f^+ - f^-$ and $|f| = f^+ + f^-$. The integral of f is defined by

$$\int f \, d\mu = \int f^+ \, d\mu - \int f^- \, d\mu$$

provided at least one of the integrals on the right is finite.

Notation for integrals is flexible. If the measure μ is understood, $\int f$ is enough. If it is desirable to indicate the space, then write $\int_{\mathcal{X}} f \, d\mu$. An integral restricted to a measurable subset $A \subset \mathcal{X}$ is $\int_A f \, d\mu = \int f \mathbb{1}_A \, d\mu$. The integration variable is included in the notation by writing $\int f(x) \, \mu(dx)$ or $\int f(x) \, d\mu(x)$.

All the integrals the reader may have encountered in mathematics or other disciplines are covered by this integral, more or less. A Riemann integral $\int_a^b f(x) \, dx$ from calculus is an integral with respect to Lebesgue measure. A series $\sum_{i=1}^{\infty} \alpha_i$ is also an integral with underlying space $\mathcal{X} = \mathbb{N}$, counting measure $\mu(i) = 1$, and integrand $f(i) = \alpha_i$.

A terminological quirk is that a function f is *integrable* if $\int |f| \, d\mu < \infty$, in other words, if $\int f \, d\mu$ exists and is finite. This generalizes to the definition of the L^p spaces

$$L^p(\mu) = \left\{ f : \mathcal{X} \to \mathbb{R} \text{ measurable} : \int |f|^p \, d\mu < \infty \right\}.$$

For $1 \le p < \infty$, $L^p(\mu)$ is a Banach space with norm $\|f\|_p = \left(\int |f|^p \, d\mu\right)^{1/p}$.

Properties of functions on sets of measure zero are often irrelevant in measure theory. If a property holds outside some set of measure zero, it is said to hold *almost everywhere*, *almost surely*, or *for almost every x*, abbreviated a.e., a.s., or for a.e. x. For example, if $f_n(x) \to f(x)$ for all $x \in A$ and $\mu(A^c) = 0$, then $f_n \to f$ a.e.

Analogously, the usual supremum and infimum of a function are replaced by the essential supremum and the essential infimum, defined by

$$\text{ess inf}\, f = \sup\{a : \mu(f < a) = 0\} \quad \text{and} \quad \text{ess sup}\, f = \inf\{b : \mu(f > b) = 0\}.$$

The L^∞ norm of a function is $\|f\|_\infty = \text{ess sup}\, |f|$.

One main source of the power of the Lebesgue integral lies in the next three convergence theorems.

Fatou's lemma. *Let $0 \le f_n \le \infty$ be measurable functions. Then*

$$\int \big(\varliminf_{n\to\infty} f_n \big) \, d\mu \le \varliminf_{n\to\infty} \int f_n \, d\mu.$$

Monotone convergence theorem. *Let f_n be a sequence of nonnegative measurable functions, and assume $f_n \le f_{n+1}$ almost everywhere for each n. Let $f = \lim_{n\to\infty} f_n$. This limit exists at least almost everywhere. Then*

$$\int f \, d\mu = \lim_{n\to\infty} \int f_n \, d\mu.$$

Dominated convergence theorem. *Let f_n be measurable functions and assume the limit $f = \lim_{n\to\infty} f_n$ exists almost everywhere. Assume there exists a function $g \ge 0$ such that $|f_n| \le g$ almost everywhere for each n and $\int g \, d\mu < \infty$. Then*

$$\int f \, d\mu = \lim_{n\to\infty} \int f_n \, d\mu.$$

There are many inequalities that involve integrals. The next one is among the most basic.

Hölder's inequality. *Let $1 < p, q < \infty$ satisfy $p^{-1} + q^{-1} = 1$. Then for measurable functions f and g*

$$\int |fg| \, d\mu \le \left(\int |f|^p \, d\mu \right)^{1/p} \left(\int |g|^q \, d\mu \right)^{1/q}.$$

The left- and right-hand sides are finite and equal if and only if there are constants a and b not both zero such that $a\,|f|^p = b\,|g|^q$.

The condition on p and q above is expressed by saying that p and q are *conjugate exponents*. The case $p = q = 2$ is the *Cauchy-Schwarz inequality*.

We turn to some technical generation results. The most useful σ-algebras are complicated and typically impossible to describe except in terms of simpler classes of sets that generate them. The σ-algebra $\sigma(\mathscr{E})$ generated by a collection $\mathscr{E}$ of subsets of $\mathcal{X}$ is by definition the intersection of all σ-algebras on $\mathcal{X}$ that contain $\mathscr{E}$. This definition is sensible because, as one can check readily, an intersection of σ-algebras is again a σ-algebra. This is the precise way in which a Borel σ-algebra is generated by open sets.

Definition A.10. A class of sets $\mathscr{C}$ is a *monotone class* if it is closed under monotone limits of sets. That is, if $A_1 \subset A_2 \subset \cdots \subset A_i \subset \cdots$ are members of $\mathscr{C}$, then so is $\bigcup_i A_i$. Also, if $A_1 \supset A_2 \supset \cdots \supset A_i \supset \cdots$ are members of $\mathscr{C}$, then so is $\bigcap_i A_i$.

An algebra $\mathscr{C}$ is a σ-algebra if and only if it is a monotone class. Any intersection of monotone classes is a monotone class. Thus, given a collection of sets $\mathscr{E}$ we can define the monotone class generated by $\mathscr{E}$ as the intersection of all monotone classes containing $\mathscr{E}$. The next theorem shows how a monotone class can be useful in describing a σ-algebra.

Monotone class theorem. *The monotone class generated by an algebra is the same as the σ-algebra generated by that algebra.*

After this preliminary discussion we get to the result we need. If $(\mathcal{X}, \mathscr{A})$ is a measurable space, then $b\mathscr{A}$ denotes the space of bounded measurable functions $f : \mathcal{X} \to \mathbb{R}$.

Lemma A.11. *Let $\mathcal{X}$ be a metric space with Borel σ-algebra $\mathscr{B}_{\mathcal{X}}$. Let $\mathcal{H}$ be a class of bounded functions that contains the space $\mathscr{U}_b(\mathcal{X})$ of bounded uniformly continuous functions and is closed under uniformly bounded pointwise convergence (that is, if $f_n \in \mathcal{H}$ for all n, $\max_n \sup_x |f_n(x)| < \infty$, and $f_n(x) \to f(x)$ for all $x \in \mathcal{X}$, then $f \in \mathcal{H}$). Then $b\mathscr{B}_{\mathcal{X}} \subset \mathcal{H}$.*

For this we need another technical lemma.

Lemma A.12. *Let $\mathcal{X}$ be a metric space and let $\mathcal{H}$ be a class of bounded functions that contains the space $\mathscr{U}_b(\mathcal{X})$ of bounded uniformly continuous functions and is closed under uniformly bounded pointwise convergence. Fix an arbitrary function g. Suppose $g + f \in \mathcal{H}$ for all $f \in \mathscr{U}_b(\mathcal{X})$. Then $g + \alpha \mathbb{1}_A + f \in \mathcal{H}$ for all real α, all $f \in \mathscr{U}_b(\mathcal{X})$, and all Borel sets $A \in \mathscr{B}_{\mathcal{X}}$.*

Proof. Let

$$\mathscr{C} = \{A \subset \mathcal{X} : g + \alpha \mathbb{1}_A + f \in \mathcal{H}, \forall \alpha \in \mathbb{R}, \forall f \in \mathscr{U}_b(\mathcal{X})\}.$$

$\mathscr{C}$ contains the algebra

$$\mathscr{A} = \{A \subset \mathcal{X} : \exists f_n \in \mathscr{U}_b(\mathcal{X}) \text{ uniformly bounded} : \mathbb{1}_A = \lim_{n\to\infty} f_n\}$$

because if $A \in \mathscr{A}$, then

$$g + \alpha \mathbb{1}_A + f = \lim_{n\to\infty} (g + \alpha f_n + f) \in \mathcal{H}$$

by the hypothesis and closedness under uniformly bounded pointwise limits. $\mathscr{C}$ is also a monotone class because if $A_k \in \mathscr{C}$ and $\mathbb{1}_{A_k} \to \mathbb{1}_A$, then

$$g + \alpha \mathbb{1}_A + f = \lim_{k\to\infty} (g + \alpha \mathbb{1}_{A_k} + f) \in \mathcal{H}.$$

By the monotone class theorem $\mathscr{C}$ contains the σ-algebra generated by $\mathscr{A}$. This is all of $\mathscr{B}_{\mathcal{X}}$ because $\mathscr{A}$ contains all open and closed sets. This follows from the fact that if A is closed, $\mathbb{1}_A$ is approximated by $f_n(x) = (1 + nd(x, A))^{-1}$, where d is the metric on $\mathcal{X}$. □

Proof of Lemma A.11. The hypothesis of the lemma above is true for $g = 0$. Consequently $\alpha \mathbb{1}_A - f \in \mathcal{H}$ for all $f \in \mathscr{U}_b(\mathcal{X})$, $\alpha \in \mathbb{R}$, and A measurable.

Suppose we have shown that $\mathcal{H}$ contains all functions of the form $h + f$ where h is a simple function with at most n terms and $f \in \mathscr{U}_b(\mathcal{X})$. (The case $n = 1$ has been verified in the previous paragraph.) Then the lemma above shows again that $\mathcal{H}$ contains also all $h + f$ where h is a simple function with at most $n + 1$ terms and $f \in \mathscr{U}_b(\mathcal{X})$. Since $f = 0$ is a member of $\mathscr{U}_b(\mathcal{X})$, all simple functions are in $\mathcal{H}$. Since bounded measurable functions are uniform limits of simple ones, $\mathcal{H}$ contains all bounded measurable functions. □

A.3. Product spaces

Cartesian products of topological spaces and metric spaces arise naturally. If $\mathcal{X}_1, \dots, \mathcal{X}_n$ are topological spaces, then the product topology on the space $\mathcal{X} = \mathcal{X}_1 \times \cdots \times \mathcal{X}_n$ is generated by the basis $\{B_1 \times \cdots \times B_n : B_i \subset \mathcal{X}_i \text{ is open}\}$. If d_i is a metric on $\mathcal{X}_i$, then $\mathcal{X}$ can be metrized by

$$d(x, y) = \sum_{i=1}^{n} d_i(x_i, y_i)$$

for points $x = (x_1, \dots, x_n), y = (y_1, \dots, y_n) \in \mathcal{X}$. On $\mathbb{R}^n$ this is not the usual Euclidean metric, but it is equivalent to it in the sense that these metrics are bounded by constant multiples of each other. To define a metric d on the Cartesian product $\mathcal{X} = \prod_{i\in\mathcal{I}} \mathcal{X}_i$ of metric spaces $(\mathcal{X}_i, d_i)$ with a countably infinite index set $\mathcal{I}$, pick weights $\alpha_i > 0$ that are summable ($\sum_{i\in\mathcal{I}} \alpha_i < \infty$) and let

$$d(x, y) = \sum_{i\in\mathcal{I}} \alpha_i \big(1 \wedge d_i(x_i, y_i)\big) \quad \text{for } x = (x_i)_{i\in\mathcal{I}},\ y = (y_i)_{i\in\mathcal{I}} \in \mathcal{X}. \tag{A.3}$$

Several desirable properties of the metric spaces $\mathcal{X}_i$ are inherited by the countable product space $\mathcal{X}$. Examples include compactness, completeness, and separability.

Let $(\mathcal{X}_i, \mathscr{A}_i)$, $i \in \mathcal{I}$, be a family of measurable spaces indexed by an arbitrary, possibly uncountable, index set $\mathcal{I}$. The *product σ-algebra* $\mathscr{A} = \bigotimes_i \mathscr{A}_i$ on $\mathcal{X} = \prod_{i\in\mathcal{I}} \mathcal{X}_i$ is by definition the σ-algebra generated by the *cylinder sets* $\prod_{i\in\mathcal{I}} A_i$ where $A_i \in \mathscr{A}_i$ for all $i \in \mathcal{I}$ and $A_i = \mathcal{X}_i$ except for finitely many i. Equivalently, $\bigotimes_i \mathscr{A}_i$ is the smallest σ-algebra on $\mathcal{X}$ that makes the coordinate projections $\pi_i : \mathcal{X} \to \mathcal{X}_i$ measurable. Coordinate projections are defined by $\pi_i(x) = x_i$ for $x = (x_i)_{i\in\mathcal{I}} \in \mathcal{X}$.

In general, the product measure space $(\mathcal{X}, \mathscr{A}, \mu) = \left(\prod_i \mathcal{X}_i, \bigotimes_i \mathscr{A}_i, \bigotimes_i \mu_i\right)$ can be defined for finite $\mathcal{I}$. The *product measure* $\mu = \bigotimes_{i\in\mathcal{I}} \mu_i$ is constructed by setting $\mu(A) = \prod_{i\in\mathcal{I}} \mu(A_i)$ for product sets $A = \prod_{i\in\mathcal{I}} A_i$ with $A_i \in \mathscr{A}_i$ and by extending this to a measure on all of $\mathscr{A}$ by using general measure extension machinery, for example Carathéodory's extension theorem [**34**, Theorem 1.11].

For practical computations the key point about product measures is the Fubini-Tonelli theorem, valid for σ-finite spaces: if $f_i : \mathcal{X}_i \to \mathbb{R}$ are given measurable functions, then $f : \mathcal{X} \to \mathbb{R}$ defined by $f(x) = \prod_{i\in\mathcal{I}} f_i(x_i)$ for $x = (x_i)_{i\in\mathcal{I}}$ is measurable and satisfies

$$\int_{\mathcal{X}} f\, d\mu = \prod_{i\in\mathcal{I}} \int_{\mathcal{X}_i} f_i\, d\mu_i \tag{A.4}$$

provided all the integrals are well-defined.

Infinite products exist for probability measures and are constructed by the Kolmogorov extension theorem (Section B.2).

A.4. Separation theorem

A subset A of a real vector space $\mathcal{Z}$ is *convex* if for all $x, y \in A$ and $t \in [0,1]$, $tx + (1-t)y \in A$. A topological vector space is a vector space equipped with a topology under which vector addition and multiplication by a scalar are continuous operations. A topological vector space is *locally convex* if every neighborhood of $x \in \mathcal{Z}$ contains a convex neighborhood of x. The *dual space* $\mathcal{Z}^*$ of a topological vector space $\mathcal{Z}$ is the vector space of continuous linear functionals $z^* : \mathcal{Z} \to \mathbb{R}$. Common notation is $z \mapsto \langle z, z^* \rangle$.

The following theorem is foundational for functional analysis. It is used a number of times in the book. For a proof see Theorem 3.4 of [**69**].

Hahn-Banach separation theorem. *Suppose A and B are nonempty, disjoint, convex subsets of a real topological vector space $\mathcal{Z}$.*

(a) *If A is open, there exist $z^* \in \mathcal{Z}^*$ and $\gamma \in \mathbb{R}$ such that*

$$\langle z', z^* \rangle < \gamma \le \langle z'', z^* \rangle \quad \forall z' \in A, \forall z'' \in B.$$

(b) *If $\mathcal{Z}$ is locally convex, A is compact, and B is closed, there exist $z^* \in \mathcal{Z}^*$ and $\gamma_1, \gamma_2 \in \mathbb{R}$ such that*

$$\langle z', z^* \rangle < \gamma_1 < \gamma_2 < \langle z'', z^* \rangle \quad \forall z' \in A, \forall z'' \in B.$$

A.5. Minimax theorem

In this section we prove a theorem of König [**55**] following the proof given by Kassay [**53**].

Let $\mathcal{X}$ and $\mathcal{Y}$ be nonempty sets and let $f : \mathcal{X} \times \mathcal{Y} \to \mathbb{R}$ be a given function. We say f is *uniformly Jensen-concave-convex-like* if

$$\forall y_0, y_1 \in \mathcal{Y},\ \exists y \in \mathcal{Y} : \forall x \in \mathcal{X},\ f(x,y) \le (f(x,y_0) + f(x,y_1))/2 \tag{A.5}$$

and

$$\forall x_0, x_1 \in \mathcal{X},\ \exists x \in \mathcal{X} : \forall y \in \mathcal{Y},\ f(x,y) \ge (f(x_0,y) + f(x_1,y))/2. \tag{A.6}$$

Of course, if $\mathcal{X}$ and $\mathcal{Y}$ are convex sets and $f(x,y)$ is concave in x for each fixed y and convex in y for each fixed x, then f is also uniformly Jensen-concave-convex-like.

***Exercise A.13.** Fix an integer $k \ge 2$. Show that there exists a dense set

$$M_k \subset \Big\{(t_1, \dots, t_k) \in [0,1]^k : \sum_{i=1}^k t_i = 1\Big\}$$

such that for $t = (t_1, \dots, t_k) \in M_k$ the following hold:

(a) For every $y_1, \dots, y_k \in \mathcal{Y}$, there exists $y_t \in \mathcal{Y}$ such that

$$f(x, y_t) \le \sum_{j=1}^k t_j f(x, y_j) \quad \forall x \in \mathcal{X}.$$

(b) For every $x_1, \dots, x_k \in \mathcal{X}$, there exists $x_t \in \mathcal{X}$ such that

$$f(x_t, y) \ge \sum_{i=1}^k t_i f(x_i, y) \quad \forall y \in \mathcal{Y}.$$

Hint: Let M_k consist of vectors of dyadic rationals $j2^{-n}$, $n \in \mathbb{N}$, $j \in \{0, 1, \dots, 2^n\}$. Use induction on k. For the case $k = 2$ use induction on the exponent n of the denominator.

König's minimax theorem. *Suppose $\mathcal{X}$ is a compact Hausdorff space and $f : \mathcal{X} \times \mathcal{Y} \to \mathbb{R}$ is uniformly Jensen-concave-convex-like. Assume $f(\cdot, y) : \mathcal{X} \to \mathbb{R}$ is upper semicontinuous for every fixed $y \in \mathcal{Y}$. Then*

$$\sup_{x\in\mathcal{X}} \inf_{y\in\mathcal{Y}} f(x,y) = \inf_{y\in\mathcal{Y}} \sup_{x\in\mathcal{X}} f(x,y).$$

Proof. One direction is easy. Indeed, it is always true that

$$\sup_{x\in\mathcal{X}} \inf_{y\in\mathcal{Y}} f(x,y) \leq \inf_{y\in\mathcal{Y}} \sup_{x\in\mathcal{X}} f(x,y).$$

Hence we can assume that $c^+ = \inf_y \sup_x f(x,y) > -\infty$. For $c < c^+$ and $y \in \mathcal{Y}$ define the nonempty closed set

$$H_{c,y} = \{x \in \mathcal{X} : f(x,y) \geq c\}.$$

Lemma A.14. *Under the assumptions of König's minimax theorem, for each $c < c^+$ and $y_1, \ldots, y_n \in \mathcal{Y}$, we have $\bigcap_{j=1}^n H_{c,y_j} \neq \varnothing$.*

Proof. Fix $c < c^+$ and assume $\bigcap_{j=1}^n H_{c,y_j} = \varnothing$. Define the function $h : \mathcal{X} \to \mathbb{R}^n$ by

$$h(x) = (f(x,y_1) - c, \ldots, f(x,y_n) - c)$$

and set $K = [0,\infty)^n$. Let $\overline{\mathrm{co}}(h(\mathcal{X}))$ be the closed convex hull of $h(\mathcal{X})$, that is, the intersection of all closed convex sets containing $h(\mathcal{X})$. Let K° be the interior of K.

If $\overline{\mathrm{co}}(h(\mathcal{X})) \bigcap K^\circ \neq \varnothing$, then there exist $s_1, \ldots, s_k \in [0,1]$ with $\sum_{i=1}^k s_i = 1$ and $x_1, \ldots, x_k \in \mathcal{X}$ such that $\sum_{i=1}^k s_i h(x_i) \in K^\circ$. By Exercise A.13 we can choose $(t_1, \ldots, t_k) \in M_k$ such that $\sum_{i=1}^k t_i h(x_i) \in K$. By the choice of x_t, $f(x_t, y_j) \geq \sum_{i=1}^k t_i f(x_i, y_j) \geq c$ for each j. Hence $x_t \in \bigcap_{j=1}^n H_{c,y_j}$, contradicting the assumption $\bigcap_{j=1}^n H_{c,y_j} = \varnothing$. Thus, $\overline{\mathrm{co}}(h(\mathcal{X})) \cap K^\circ = \varnothing$.

By the Hahn-Banach separation theorem, there exists a hyperplane that separates $\overline{\mathrm{co}}(h(\mathcal{X}))$ and K°: there exist $\gamma \in \mathbb{R}$ and $\alpha \in \mathbb{R}^n$ such that for all $x \in \mathcal{X}$ and $u \in K^\circ$,

$$\sum_{j=1}^n \alpha_j u_j < \gamma \leq \sum_{j=1}^n \alpha_j (f(x,y_j) - c).$$

Taking $u_j \to \infty$ shows that $\alpha_j \leq 0$. Taking $u \to 0$ shows that $\gamma \geq 0$. Clearly, there exists a $j \leq n$ such that $\alpha_j \neq 0$. Setting $\delta = \alpha / \sum_{j=1}^n \alpha_j \in [0,1]^n$ one has $\sum_{j=1}^n \delta_j = 1$ and $\sum_{j=1}^n \delta_j f(x,y_j) \leq c$.

Now take $c < c_1 < c^+$ and set $d = c_1 - c$. Then, for all $x \in \mathcal{X}$,

$$\sum_{j=1}^n \delta_j [f(x,y_j) - c_1] \leq -d.$$

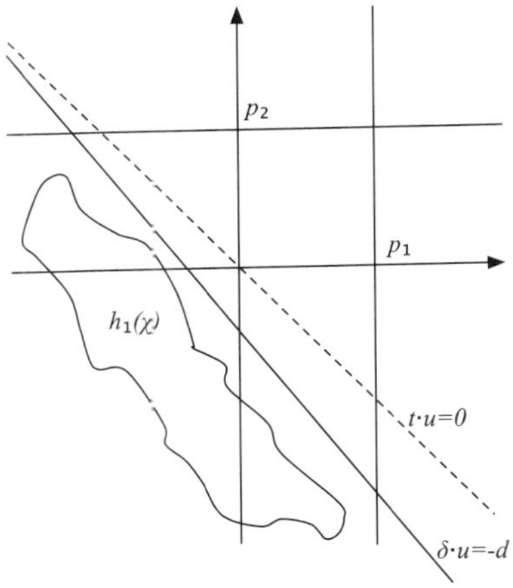

Figure A.1. The hyperplane separating $h_1(\mathcal{X})$ and K and passing through the origin.

Hence, if $h_1(x) = (f(x, y_1) - c_1, \ldots, f(x, y_n) - c_1)$, then $h_1(\mathcal{X})$ is separated from K by the hyperplane $\sum_{j=1}^n \delta_j u_j = -d$.

For each fixed y the closed sets $\{x : f(x, y) \geq m\}$ must be empty for large enough m because otherwise, by compactness and Theorem A.5(i), there exists x such that $f(x, y) = \infty$. Hence there exists $p \in K^\circ$ such that $h_1(\mathcal{X}) \subset \prod_{j=1}^n (-\infty, p_j]$. By Exercise A.13 there exists $t \in M_n$ close enough to δ such that the hyperplane $\sum_{j=1}^n t_j u_j = 0$ separates $h_1(\mathcal{X})$ and K; see Figure A.1. This means that

$$\sum_{j=1}^{n} t_j f(x, y_j) \leq c_1$$

for all $x \in \mathcal{X}$. Consider now $y_t \in \mathcal{Y}$ such that $f(x, y_t) \leq \sum_{j=1}^n t_j f(x, y_j)$ for all $x \in \mathcal{X}$. Then, $c^+ \leq \sup_{x \in \mathcal{X}} f(x, y_t) \leq c_1$. This contradicts $c_1 < c^+$ and proves that $\bigcap_{j=1}^n H_{c,y_j} \neq \varnothing$. ☐

König's theorem now follows. By Lemma A.14 the closed sets $\{H_{y,c} : y \in \mathcal{Y}\}$ have the finite intersection property. By Theorem A.5(i) $\bigcap_y H_{c,y} \neq \varnothing$ which implies that $\sup_x \inf_y f(x, y) \geq c$. Let $c \nearrow c^+$. ☐

Literature notes. For general topology and undergraduate analysis there are many good references, such as the books by Munkres [**60**] and Rudin [**69**]. Excellent graduate-level textbooks include Dudley [**26**] and Folland [**34**].

Appendix B

Probability

The framework for probability theory is taken directly from measure theory, with notation and terminology adapted to probabilistic conventions. A random experiment or collection of experiments is modeled by a *probability space* $(\Omega, \mathscr{F}, P)$, which is a measure space with total mass $P(\Omega) = 1$. A real-valued *random variable* is a measurable function $X : \Omega \to \mathbb{R}$. Its *expectation* is the Lebesgue integral $E[X] = \int_\Omega X(\omega) P(d\omega)$ under the *probability measure* P.

More generally, if $(\mathcal{S}, \mathscr{S})$ is a measurable space, an $\mathcal{S}$-valued random variable is a measurable function $X : \Omega \to \mathcal{S}$. This means that $\{X \in A\} \in \mathscr{F}$ for each $A \in \mathscr{S}$. The *law*, or *distribution*, of a random variable X is the probability measure μ on $\mathcal{S}$ defined by $\mu(A) = P\{X \in A\}$ for $A \in \mathscr{S}$.

Here are the most important probability distributions.

- Degenerate random variable: $P(X = x) = 1$ for some point x. The distribution is denoted by δ_x and is called a *point mass at x*.
- Bernoulli(p): $P(X = 1) = 1 - P(X = 0) = p$, $p \in (0, 1)$.
- Binomial(n, p): $P(X = k) = \binom{n}{k} p^k (1-p)^{n-k}$, $k \in \{0, 1, \ldots, n\}$, $n \in \mathbb{N}$.
- Geometric(p): $P(X = k) = p(1-p)^{k-1}$, $k \in \mathbb{N}$.
- Poisson(λ): $P(X = k) = e^{-\lambda} \lambda^k / k!$, $k \in \mathbb{Z}_+$, $0 < \lambda < \infty$.
- Exponential(λ): $P(X \le x) = \max\{0, 1 - e^{-\lambda x}\}$, $x \in \mathbb{R}$.
- Normal(μ, σ^2): $P(X \le x) = \int_{-\infty}^x \frac{1}{\sqrt{2\pi\sigma^2}} e^{-(y-\mu)^2/(2\sigma^2)}\, dy$, $x \in \mathbb{R}$, $\mu \in \mathbb{R}$, and $\sigma > 0$. The *standard normal* is the case $(\mu, \sigma^2) = (0, 1)$.

• Multivariate normal: the $\mathbb{R}^d$-valued random vector $\mathbf{X} = (X_1, \dots, X_d)$ has the normal distribution with mean $\mu \in \mathbb{R}^d$ and covariance matrix Γ if the $d \times d$ matrix Γ is symmetric and nonnegative definite ($\langle \mathbf{x}, \Gamma\mathbf{x}\rangle \ge 0 \ \forall \mathbf{x} \in \mathbb{R}^d$) and the characteristic function of $\mathbf{X}$ satisfies

$$E(e^{i\langle t, \mathbf{X}\rangle}) = e^{-\frac{1}{2}\langle t, \Gamma t\rangle} \quad \text{for } t \in \mathbb{R}^d.$$

If Γ is positive definite ($\langle \mathbf{x}, \Gamma\mathbf{x}\rangle > 0$ for all nonzero $\mathbf{x} \in \mathbb{R}^d$), then $\mathbf{X}$ has a density on $\mathbb{R}^d$ given by

$$P\{\mathbf{X} \in A\} = \int \cdots \int_A \frac{1}{(2\pi)^{d/2}\sqrt{\det \Gamma}} e^{-\frac{1}{2}\langle \mathbf{x}-\mu, \Gamma^{-1}(\mathbf{x}-\mu)\rangle}\, d\mathbf{x}$$

where $\mathbf{x} = (x_1, \dots, x_d)$ is the integration variable on $\mathbb{R}^d$, A is a Borel subset of $\mathbb{R}^d$, and $\det \Gamma$ is the determinant of Γ.

The next two inequalities are ubiquitous in probability theory. Recall from Definition 2.26 what it means for a function ϕ to be convex. ϕ is *strictly convex* if $\phi(tx + (1-t)y) < t\phi(x) + (1-t)\phi(y)$ for all $x \ne y$ in its domain and all $0 < t < 1$.

Jensen's inequality. *Let $-\infty \le a < b \le \infty$, $P(a < X < b) = 1$, and let ϕ be a convex function on (a, b). Assume that EX and $E[\phi(X)]$ are well-defined. Then $\phi(EX) \le E[\phi(X)]$. If ϕ is strictly convex, then $\phi(EX) = E[\phi(X)]$ if and only if $P(X = EX) = 1$.*

Chebyshev's inequality. *Let ϕ be a nondecreasing function on $\mathbb{R}$ and let $\phi(t) > 0$. Then $P(X \ge t) \le \phi(t)^{-1}E[\phi(X)]$.*

Commonly used instances of Chebyshev's inequality are

$$P\{|X| \ge t\} \le t^{-p}E[|X|^p] \quad \text{and} \quad P\{X \ge t\} \le e^{-\lambda t}E[e^{\lambda X}]$$

for positive t, p, and λ.

Definition B.1. A sequence of random variables X_n converges to X

(a) almost surely if $P\{\omega : \lim_{n\to\infty} X_n(\omega) = X(\omega)\} = 1$ and

(b) in probability if, for each $\varepsilon > 0$, $P\{|X_n - X| > \varepsilon\} \to 0$ as $n \to \infty$.

B.1. Independence

Fix a probability space $(\Omega, \mathscr{F}, P)$. Full information about the outcome of the experiment would mean knowing, for each event $A \in \mathscr{F}$, whether A occurred or not. Typically $\mathscr{F}$ contains all singletons $\{\omega\}$, and then this means knowing the exact sample point $\omega \in \Omega$ that occurred. Partial information about the outcome is represented by a sub-σ-algebra $\mathscr{A} \subset \mathscr{F}$. The notions of independence and conditioning naturally involve information, and hence their mathematical definitions are expressed in terms of σ-algebras.

Let $\{\mathscr{A}_i : i \in \mathcal{I}\}$ be a collection of sub-σ-algebras of $\mathscr{F}$. They are *independent* if $P(\bigcap_{k=1}^n A_{i_k}) = \prod_{k=1}^n P(A_{i_k})$ for any choice of finitely many distinct indices $i_1, \dots, i_n \in \mathcal{I}$ and events $A_{i_k} \in \mathscr{A}_{i_k}$. From this definition flow all the familiar elementary definitions of independence. For example, random variables $\{X_i\}_{i\in\mathcal{I}}$ are independent if the σ-algebras $\sigma(X_i) = \{\{X_i \in B\} : B \in \mathcal{B}_{\mathbb{R}}\}$ are independent. Events $\{A_i\}_{i\in\mathcal{I}}$ are independent if their indicator random variables $\{\mathbb{1}_{A_i}\}_{i\in\mathcal{I}}$ are independent.

For a sequence of events $\{A_k\}_{k\in\mathbb{N}}$, the event that A_k occurs infinitely often (i.o.) is defined by

$$\{A_k \text{ i.o.}\} = \bigcap_{n\in\mathbb{N}} \bigcup_{k\geq n} A_k.$$

Borel-Cantelli lemma. *Let $\{A_k\}$ be a sequence of events in $(\Omega, \mathscr{F}, P)$.*

If $\sum_k P(A_k) < \infty$, then $P\{A_k \text{ i.o.}\} = 0$.

If $\{A_k\}$ are independent and $\sum_k P(A_k) = \infty$, then $P\{A_k \text{ i.o.}\} = 1$.

B.2. Existence of stochastic processes

A *stochastic process* $\{X_i : i \in \mathcal{I}\}$ is a collection of random variables X_i on a probability space $(\Omega, \mathscr{F}, P)$, where $\mathcal{I}$ is an arbitrary index set and $X_i : \Omega \to \mathcal{S}_i$ takes its values in some measurable space $(\mathcal{S}_i, \mathscr{S}_i)$. The existence question is the following: given a probability distribution μ_J on the product space $\mathcal{S}^J = \prod_{i\in J} \mathcal{S}_i$ for each finite index set $J \subset \mathcal{I}$, does there exist a process $\{X_i : i \in \mathcal{I}\}$ with these marginal distributions, that is, a process that satisfies $P\{(X_i)_{i\in J} \in B\} = \mu_J(B)$ for all product measurable sets $B \subset \mathcal{S}^J$.

The *canonical construction* takes the product space $\Omega = \mathcal{S}^{\mathcal{I}} = \prod_{i\in\mathcal{I}} \mathcal{S}_i$ with coordinate variables $X_i(\omega) = \omega_i$ for $\omega = (\omega_i)_{i\in\mathcal{I}} \in \Omega$ and the product σ-algebra $\mathscr{F} = \sigma\{X_i : i \in \mathcal{I}\}$ generated by the variables $\{X_i : i \in \mathcal{I}\}$. The question then becomes one of existence of a probability measure on an infinite-dimensional product space with prescribed finite-dimensional marginals. If the underlying spaces are reasonable, this problem can be solved under minimal assumptions.

Kolmogorov's extension theorem ([**26**, Theorem 12.1.2]). *Assume each $(\mathcal{S}_i, \mathscr{S}_i)$ is a Polish space with its Borel σ-algebra. Assume that the measures μ_J are consistent in the sense that, for finite $I \subset J$,*

$$\mu_J\{(\omega_i)_{i\in I} \in B\} = \mu_I(B) \quad \forall \text{ measurable set } B \subset \mathcal{S}^I.$$

Then there exists a unique probability measure P on the product space $(\Omega, \mathscr{F})$ with finite-dimensional marginal distributions μ_J.

A special case of the construction is a product probability measure which is the analytical counterpart of independence. (Note the product form in (A.4).) Let μ_i be a given distribution of X_i on $(\mathcal{S}_i, \mathscr{S}_i)$. The finite-dimensional marginals $\mu_J = \bigotimes_{i \in J} \mu_i$ are immediately consistent. Kolmogorov's theorem gives the existence of a process $\{X_i : i \in \mathcal{I}\}$ that satisfies

$$P\Big(\bigcap_{i \in J} \{X_i \in B_i\}\Big) = \prod_{i \in J} \mu_i(B_i) = \prod_{i \in J} P\{X_i \in B_i\}$$

for any finite $J \subset \mathcal{I}$ and Borel sets $B_i \subset \mathcal{S}_i$. In other words, these variables are independent. The random variables $\{X_i\}_{i \in \mathcal{I}}$ are *independent and identically distributed* (i.i.d.) if they are independent and they all have the same probability distribution.

B.3. Conditional expectation

The elementary definition of the conditional probability of an event A, given event B, is

$$P(A|B) = \frac{P(A \cap B)}{P(B)}$$

under the assumption that $P(B) > 0$. This extends naturally to an elementary conditional expectation defined by

$$E(X|B) = \frac{1}{P(B)} E[X \mathbb{1}_B] \tag{B.1}$$

for any integrable random variable X and event B with $P(B) > 0$. Advanced probability needs a more refined conditioning machinery, which we now discuss briefly.

Radon-Nikodym theorem. *Let λ be a signed measure and μ a nonnegative measure on a measurable space $(\mathcal{X}, \mathscr{A})$. Then there exists an $\mathscr{A}$-measurable function $f : \mathcal{X} \to \mathbb{R}$ such that*

$$\lambda(A) = \int_A f \, d\mu$$

for all $A \in \mathscr{A}$ if and only if λ is absolutely continuous with respect to μ. Absolute continuity, abbreviated $\lambda \ll \mu$, means that for all $A \in \mathscr{A}$, $\mu(A) = 0$ implies $\lambda(A) = 0$. The function $f = \frac{d\lambda}{d\mu}$ is the Radon-Nikodym derivative of λ with respect to μ, uniquely determined μ-almost everywhere.

Fix a probability space $(\Omega, \mathscr{F}, P)$ and a sub-σ-algebra $\mathscr{D}$ of $\mathscr{F}$. Let X be an integrable real-valued random variable on $(\Omega, \mathscr{F}, P)$ (that is, $E\,|X| <$

∞). To define the conditional expectation $E[X|\mathscr{D}](\omega)$ as a random variable, define the signed measure λ on $(\Omega, \mathscr{D})$ by

$$\lambda(D) = \int_D X\, dP = E[X\mathbb{1}_D] \qquad \text{for } D \in \mathscr{D}.$$

Since $\lambda \ll P$, the Radon-Nikodym theorem gives an integrable $\mathscr{D}$-measurable function (random variable), which we denote by $E[X \mid \mathscr{D}](\omega)$, that satisfies for all $D \in \mathscr{D}$

$$E[X\mathbb{1}_D] = \lambda(D) = \int_D E[X \mid \mathscr{D}]\, dP = E\big[E[X \mid \mathscr{D}]\mathbb{1}_D\big]\,. \tag{B.2}$$

This random variable $E[X \mid \mathscr{D}]$ is the *conditional expectation of X, given $\mathscr{D}$.* Property (B.2) determines it P-almost surely.

If the conditioning σ-algebra $\mathscr{D}$ is generated by an $\mathcal{S}$-valued random variable Y, then $\mathscr{D} = \sigma(Y) = \{\{Y \in A\} : A \in \mathscr{S}\}$ and an alternative notation is $E[X|\mathscr{D}] = E[X|Y]$.

The next two exercises relate the conditional expectation defined above to the ones from elementary probability.

Exercise B.2. Let $\{D_k\} \subset \mathscr{F}$ be a finite or countable partition of Ω, which means that the sets D_k are pairwise disjoint ($k \neq \ell$ implies $D_k \cap D_\ell = \varnothing$) and $\Omega = \bigcup D_k$. Let $\mathscr{D}$ be the σ-algebra generated by $\{D_k\}$, which now is concretely the collection of all unions of D_k's. Show that

$$E(X|\mathscr{D})(\omega) = \sum_{k:P(D_k)>0} \mathbb{1}_{D_k}(\omega) E(X|D_k).$$

Exercise B.3. Suppose (X, Y) has joint density $f(x, y)$ on $\mathbb{R}^2$. Define the marginal density of Y and the conditional density of X, given $Y = y$, by

$$f_Y(y) = \int_{\mathbb{R}} f(x, y)\, dx \quad \text{and} \quad f_{X|Y}(x|y) = \frac{f(x, y)}{f_Y(y)} \cdot \mathbb{1}\{f_Y(y) > 0\}.$$

Let h be a bounded Borel measurable function on $\mathbb{R}$. Show that

$$E[h(X)|Y](\omega) = \int_{\mathbb{R}} h(x) f_{X|Y}(x \mid Y(\omega))\, dx.$$

The next theorem lists basic properties of conditional expectations that are used in calculations.

Theorem B.4. *Assume X and Y are random variables such that $E[|X|]$ and $E[|Y|]$ are finite. The following hold almost surely.*

(a) $E[E[X \mid \mathscr{D}]] = E[X]$ *and* $E[1 \mid \mathscr{D}] = 1$.

(b) *If* $P(X \geq 0) = 1$, *then* $E[X \mid \mathscr{D}] \geq 0$.

(c) $E[aX + bY \mid \mathscr{D}] = aE[X \mid \mathscr{D}] + bE[Y \mid \mathscr{D}]$ *for all* $a, b \in \mathbb{R}$.

(d) *If* Y *is a* $\mathscr{D}$*-measurable random variable such that* $E[|XY|] < \infty$, *then* $E[XY \,|\, \mathscr{D}] = YE[X \,|\, \mathscr{D}]$.

(e) *If* $\mathscr{G} \subset \mathscr{D}$ *is a smaller* σ*-algebra, then* $E[E[X \,|\, \mathscr{D}] \,|\, \mathscr{G}] = E[X \,|\, \mathscr{G}]$.

(f) *If* ϕ *is a convex function, then* $\phi\big(E[X \,|\, \mathscr{D}]\big) \le E[\phi(X) \,|\, \mathscr{D}]$.

The conditional expectation of an indicator variable is a conditional probability, with an alternative notation $P(A \,|\, \mathscr{D})(\omega) = E[\mathbb{1}_A \,|\, \mathscr{D}](\omega)$. This does not necessarily define a genuine probability measure for any fixed ω because conditional expectations are defined only up to sets of measure zero. So while we know that $P(\bigcup_i A_i \,|\, \mathscr{D})(\omega) = \sum_i P(A_i \,|\, \mathscr{D})(\omega)$ almost surely for a pairwise disjoint sequence $\{A_i\}$, we do not know if there exists some fixed ω for which this holds for *all* pairwise disjoint sequences $\{A_i\}$.

This difficulty can be overcome with additional assumptions.

Theorem B.5. *Let* $(\mathcal{S}, \mathscr{B}_{\mathcal{S}})$ *be a Polish space with its Borel* σ*-algebra, let* $\mathscr{D}$ *be a sub-*σ*-algebra of* $\mathscr{F}$ *on* Ω, *and let* $Y : \Omega \to \mathcal{S}$ *be an* $\mathcal{S}$*-valued random variable. Then there exists a function* $p : \Omega \times \mathscr{B}_{\mathcal{S}} \to [0,1]$ *with the following properties.*

(a) *For each* $\omega \in \Omega$, $p(\omega, \cdot\,)$ *is a probability measure on* $(\mathcal{S}, \mathscr{B}_{\mathcal{S}})$.

(b) *For each* $B \in \mathscr{B}_{\mathcal{S}}$, $p(\omega, B) = P(Y \in B \,|\, \mathscr{D})(\omega)$ *almost surely.*

See Theorem 10.2.2 in [**26**] for a proof. The function $p(\omega, B)$ is the *conditional distribution of* Y, *given* $\mathscr{D}$. In case $(\Omega, \mathscr{F}) = (\mathcal{S}, \mathscr{B}_{\mathcal{S}})$ is itself a Polish space and we take Y to be the identity mapping, we get the *conditional probability measure* P^ω, which is a genuine probability measure on $(\Omega, \mathscr{F})$ for each ω and satisfies $P^\omega(A) = P(A|\mathscr{D})(\omega)$ a.s. for each $A \in \mathscr{F}$. When a conditional probability measure exists, conditional expectations can be calculated as expectations:

$$E[X \,|\, \mathscr{D}](\omega) = \int_\Omega X(\widetilde{\omega})\, P^\omega(d\widetilde{\omega}).$$

B.4. Weak topology of probability measures

This section goes over the basics of the weak topology of probability measures. Parts of this material can be found in the books by Billingsley [**9**], Dudley [**26**], and Parthasarathy [**62**].

Let $(\mathcal{S}, d)$ be a metric space. Let $\mathcal{M}(\mathcal{S})$ be the space of bounded signed Borel measures on $\mathcal{S}$ and let $\mathcal{M}_1(\mathcal{S})$ be the subspace of probability measures. Let $\mathscr{C}_b(\mathcal{S})$ be the space of bounded continuous real-valued functions on $\mathcal{S}$.

Definition B.6. A sequence of probability measures $\mu_n \in \mathcal{M}_1(\mathcal{S})$ *converges weakly* to $\mu \in \mathcal{M}_1(\mathcal{S})$ if $\int_{\mathcal{S}} f\, d\mu_n \to \int_{\mathcal{S}} f\, d\mu$ for all $f \in \mathscr{C}_b(\mathcal{S})$.

Since knowing $\int_{\mathcal{S}} f \, d\mu$ for all $f \in \mathscr{C}_b(\mathcal{S})$ uniquely determines μ, μ_n cannot converge weakly to two different measures.

A duality between $\mathcal{M}(\mathcal{S})$ and $\mathscr{C}_b(\mathcal{S})$ in the sense of Section 4.1 is defined by $\langle \mu, f \rangle = \int_{\mathcal{S}} f \, d\mu$ for $f \in \mathscr{C}_b(\mathcal{S})$ and $\mu \in \mathcal{M}(\mathcal{S})$. Convergence of probability measures in the weak topology $\sigma(\mathcal{M}(\mathcal{S}), \mathscr{C}_b(\mathcal{S}))$ is the same as the weak convergence of Definition B.6. By Exercise 5.1 this weak topology is Hausdorff. Weak convergence of probability measures has several useful characterizations. We drop the subscript from $\int_{\mathcal{S}}$ when it is clear that integration is over the space $\mathcal{S}$.

Portmanteau theorem. *Let $\{\mu_n\}_{n \in \mathbb{N}}$ and μ be Borel probability measures on a metric space $(\mathcal{S}, d)$. The following are equivalent.*

(a) *μ_n converges weakly to μ.*

(b) *$\int f \, d\mu_n \to \int f \, d\mu$ for all $f \in \mathscr{U}_{b,\widetilde{d}}(\mathcal{S})$, where $\mathscr{U}_{b,\widetilde{d}}(\mathcal{S})$ is the space of bounded uniformly continuous functions on $(\mathcal{S}, \widetilde{d})$ and $\widetilde{d}$ is any metric on $\mathcal{S}$ that defines the same topology as d.*

(c) *$\overline{\lim}_{n \to \infty} \mu_n(F) \le \mu(F)$ for closed sets $F \subset \mathcal{S}$.*

(d) *$\underline{\lim}_{n \to \infty} \mu_n(G) \ge \mu(G)$ for open sets $G \subset \mathcal{S}$.*

(e) *$\lim_{n \to \infty} \mu_n(A) = \mu(A)$ for continuity sets of μ which by definition are Borel sets $A \subset \mathcal{S}$ such that $\mu(\overline{A} \smallsetminus A^\circ) = 0$.*

The next exercise shows that the weak topology $\sigma(\mathcal{M}(\mathcal{S}), \mathscr{C}_b(\mathcal{S}))$ cannot be metrized unless $\mathcal{S}$ is finite.

Exercise B.7. Let $\mathcal{X}$ and $\mathcal{Y}$ be two vector spaces in duality and let them induce weak topologies on each other.

(a) Prove that if $\mathcal{X}$ is metrizable, $\mathcal{Y}$ must have a countable algebraic basis.

Hint: If $\mathcal{X}$ is metrizable, there is a countable collection $\{B_k\}$ of neighborhoods around 0, of the form in part (a) of Proposition 4.6, such that every open neighborhood of 0 contains some B_k. Let $\{v_k\}$ be the entire, at most countable, collection of vectors in $\mathcal{Y}$ appearing in the expressions of the B_k's. Show that each $v \in \mathcal{Y}$ is a linear combination of finitely many v_k's. You may need to use Lemma 3.9 of [**69**].

(b) Prove that $\mathcal{M}(\mathbb{R})$ and $\mathcal{M}(\mathbb{N})$ are not metrizable.

It is of critical importance that, despite the exercise above, the space $\mathcal{M}_1(\mathcal{S})$ of probability measures is a metric space if $\mathcal{S}$ is separable, and even Polish if $\mathcal{S}$ is Polish. We discuss these facts in the remainder of this section. We first develop a metric based on a rich enough countable collection of functions.

Definition B.8. A sequence of bounded measurable functions $\{g_k\}$ *determines weak convergence* on $\mathcal{M}_1(\mathcal{S})$ if weak convergence $\mu_n \to \mu$ is equivalent to having $\int g_k \, d\mu_n \to \int g_k \, d\mu$ for all k.

If $\{g_k\}$ determines weak convergence, then $\mu_n \to \mu$ in $\mathcal{M}_1(\mathcal{S})$ is equivalent to convergence $\delta(\mu_n, \mu) \to 0$ under the metric

$$\delta(\mu, \nu) = \sum_{k=1}^{\infty} \frac{1}{2^k \, \|g_k\|_\infty} \left| \int g_k \, d\mu - \int g_k \, d\nu \right|. \tag{B.3}$$

The next item is that the weak topology $\sigma(\mathcal{M}(\mathcal{S}), \mathscr{C}_b(\mathcal{S}))$ restricted to $\mathcal{M}_1(\mathcal{S})$ coincides with the topology defined by the δ-metric.

***Exercise B.9.** Let $\mathcal{S}$ be a metric space. Let $\{g_k\}$ be any countable collection of $\mathscr{C}_b(\mathcal{S})$-functions that determines weak convergence on $\mathcal{M}_1(\mathcal{S})$. Prove that on $\mathcal{M}_1(\mathcal{S})$ the weak topology generated by $\mathscr{C}_b(\mathcal{S})$ coincides with the topology given by the metric δ in (B.3).

Hint: A weak neighborhood of $\mu \in \mathcal{M}_1(\mathcal{S})$ restricted to $\mathcal{M}_1(\mathcal{S})$ is a set of the type

$$U = \left\{ \nu \in \mathcal{M}_1(\mathcal{S}) : \left| \int f_i \, d\nu - \int f_i \, d\mu \right| < \varepsilon \text{ for } i = 1, \dots, N \right\}$$

where $\varepsilon > 0$, $N \in \mathbb{N}$, and $f_i \in \mathscr{C}_b(\mathcal{S})$. (See part (a) of Proposition 4.6.) What needs to be shown is that each weak neighborhood of type U contains an open ball $B_\delta(\mu, r) = \{\nu \in \mathcal{M}_1(\mathcal{S}) : \delta(\mu, \nu) < r\}$ in the δ-metric with $r > 0$, and vice versa.

So the weak topology of $\mathcal{M}_1(\mathcal{S})$ is metrizable if there exists a countable weak convergence determining collection of $\mathscr{C}_b(\mathcal{S})$-functions. Such a collection of functions exists if $\mathcal{S}$ is separable because a separable metric space is isometric to a dense subset of a compact metric space. This last claim is true because $\mathcal{S}$ has a totally bounded metric $\bar{d}$ and then the completion $(\bar{\mathcal{S}}, \bar{d})$ of $(\mathcal{S}, \bar{d})$ is compact (Theorem 2.8.2 in [**26**]). Uniformly continuous functions on $(\mathcal{S}, \bar{d})$ extend uniquely to continuous functions on $(\bar{\mathcal{S}}, \bar{d})$, and consequently the function spaces $\mathscr{U}_{b,\bar{d}}(\mathcal{S})$ and $\mathscr{C}(\bar{\mathcal{S}})$ are isometric. $\mathscr{C}(\bar{\mathcal{S}})$ is separable (Theorem 11.2.5 in [**26**]) and hence so is $\mathscr{U}_{b,\bar{d}}(\mathcal{S})$. Thus there exists a sup-norm-dense countable set $\{g_k\} \subset \mathscr{U}_{b,\bar{d}}(\mathcal{S})$. By (b) in the portmanteau theorem, weak convergence in $\mathcal{M}_1(\mathcal{S})$ is determined by $\{g_k\}$. By Exercise B.9, metric δ of (B.3) with this sequence $\{g_k\}$ metrizes the weak topology on $\mathcal{M}_1(\mathcal{S})$ generated by $\mathscr{C}_b(\mathcal{S})$.

Exercise B.10. Show that if $\mathcal{S}$ is separable, then $\mathcal{M}_1(\mathcal{S})$ is a separable metric space.

Hint: Think of convex combinations of point masses.

A metric of type (B.3) does not have to be complete. On $\mathbb{R}$ we can choose a weak convergence determining collection $\{g_k\}$ of compactly supported functions. Then the sequence $\mu_n = \delta_n$ of point masses is a Cauchy sequence but cannot converge to a probability measure on $\mathbb{R}$.

Another metric on $\mathcal{M}_1(\mathcal{S})$ is the *Prohorov metric*

$$\rho(\mu, \nu) = \inf\{\varepsilon > 0 : \mu(F) \le \nu(F^\varepsilon) + \varepsilon \text{ for all closed } F \subset \mathcal{S}\} \tag{B.4}$$

with $F^\varepsilon = \{x : d(x, y) < \varepsilon \text{ for some } y \in F\}$.

Exercise B.11. Show that ρ is a metric.

Exercise B.12. Assuming that $\mathcal{S}$ is separable, show that ρ metrizes the topology on $\mathcal{M}_1(\mathcal{S})$ induced by the weak topology $\sigma(\mathcal{M}(\mathcal{S}), \mathscr{C}_b(\mathcal{S}))$.

Hint: Since the weak topology is metrizable, it is enough to prove that $\rho(\mu_n, \mu) \to 0$ is equivalent to $\mu_n \to \mu$.

Exercise B.13. Let $\mathcal{S} = \mathbb{Z}$. Show that the Prohorov metric is given by

$$\rho(\mu, \nu) = \tfrac{1}{2} \sum_{k \in \mathbb{Z}} |\mu(k) - \nu(k)|.$$

The importance of the Prohorov metric stems from the next fact (Theorem 11.5.5 in [**26**]).

Theorem B.14. *Suppose* $(\mathcal{S}, d)$ *is a Polish space. Then so is* $(\mathcal{M}_1(\mathcal{S}), \rho)$, *the space of Borel probability measures on* $\mathcal{S}$ *with the Prohorov metric defined by* (B.4).

Last we address compactness in $\mathcal{M}_1(\mathcal{S})$. To get a feeling for the situation, consider first the case $\mathcal{S} = \mathbb{R}$.

Exercise B.15. Let μ_n be a sequence of probability measures on $\mathbb{R}$. Consider statements (i) and (ii).

(i) Every subsequence μ_{n_k} has a further subsequence $\mu_{n_{k_j}}$ that converges weakly to a probability measure on $\mathbb{R}$.

(ii) For every $\varepsilon > 0$ there exist $a < b$ such that $\inf_n \mu_n\{[a, b]\} \ge 1 - \varepsilon$.

(a) Show that (i) implies (ii). The statements are actually equivalent but proving that (ii) implies (i) requires Helly's selection theorem (Theorem 3.2.6. in [**27**]).

(b) Show that (i) and (ii) are equivalent for probability measures supported on $\mathbb{Z}$.

We now give a general definition and then the theorem.

Definition B.16. Let $\mathcal{S}$ be a topological space. A set of Borel probability measures $A \subset \mathcal{M}_1(\mathcal{S})$ is *tight* if for every $\varepsilon > 0$ there exists a compact subset $K \subset \mathcal{S}$ such that $\sup_{\nu \in A} \nu(K^c) \leq \varepsilon$.

The following is a characterization of compactness in the weak topology of probability measures (Theorems 6.1 and 6.2 in [**9**], Theorem 11.5.4 in [**26**]).

Prohorov's theorem. *Let $\mathcal{S}$ be a metric space and let $A \subset \mathcal{M}_1(\mathcal{S})$. If A is tight and μ_n is a sequence of probability measures in A, then there exist $\mu \in \mathcal{M}_1(\mathcal{S})$ and a subsequence $\mu_{n_k} \to \mu$. If $\mathcal{S}$ is Polish and A is closed, then A is tight if and only if A is compact.*

Note that unless A is closed, the limit point μ provided by the theorem above does not necessarily lie in A. The converse statement of Prohorov's theorem for a single measure is also worth noting.

Ulam's theorem. *Let $\mathcal{S}$ be Polish. For any probability measure $\lambda \in \mathcal{M}_1(\mathcal{S})$ and for any $\varepsilon > 0$ there exists a compact set $K \subset \mathcal{S}$ such that $\lambda(K^c) \leq \varepsilon$.*

***Exercise B.17.** Let $\mathcal{S}$ be Polish. Prove that a weakly converging sequence of probability measures on $\mathcal{S}$ is tight.

B.5. First limit theorems

We record here the two most fundamental limit theorems of probability theory. Let $\{X_n\}$ be a sequence of independent and identically distributed (i.i.d.) random variables. Let $\mu = E[X_i]$ and $\sigma^2 = E[X_i^2] - \mu^2$ denote the common mean and variance, when they exist.

Law of large numbers. *If $E\,|X_1| < \infty$, then*

$$\frac{X_1 + \cdots + X_n}{n} \underset{n\to\infty}{\longrightarrow} \mu \quad \textit{almost surely.}$$

Central limit theorem. *If $E[X_1^2] < \infty$, then the distribution of*

$$\frac{X_1 + \cdots + X_n - n\mu}{\sigma\sqrt{n}}$$

converges weakly to a standard normal distribution.

B.6. Ergodic theory

Let $(\Omega, \mathscr{F})$ be a measurable space, let $d \in \mathbb{N}$, and let $(\theta_i)_{i\in\mathbb{Z}^d}$ be a group of invertible measurable transformations on Ω such that θ_0 is the identity and $\theta_i \circ \theta_j = \theta_{i+j}$ for all $i, j \in \mathbb{Z}^d$. The σ-algebra of invariant events is $\mathcal{I} = \{A \in \mathscr{F} : \theta_i A = A \; \forall i \in \mathbb{Z}^d\}$.

Exercise B.18. Prove that $f : \Omega \to \mathbb{R}$ is $\mathcal{I}$-measurable if and only if $f \circ \theta_i = f$ for all $i \in \mathbb{Z}^d$.

The space of invariant probability measures is $\mathcal{M}_\theta(\Omega) = \{P \in \mathcal{M}_1(\Omega) : P(\theta_i A) = P(A)\ \forall i \in \mathbb{Z}^d,\ A \in \mathscr{F}\}$. The example $\theta_k(\omega) = \omega + k$ on $\Omega = \mathbb{Z}$ shows that this space can be empty.

Definition B.19. $P \in \mathcal{M}_\theta(\Omega)$ is *ergodic* if $P(A) \in \{0, 1\}$ for all $A \in \mathcal{I}$.

In particular, P is ergodic if and only if $E[f \,|\, \mathcal{I}] = E[f]$ P-almost surely for any $f \in L^1(P)$.

Let $(V_n)_{n\ge1}$ be an increasing sequence of cubes in $\mathbb{Z}^d$ such that $|V_n| \to \infty$ as $n \to \infty$. For $\omega \in \Omega$ let

$$R_n(\omega) = |V_n|^{-1} \sum_{i \in V_n} \delta_{\theta_i \omega}.$$

For the expectation of a function f under the empirical measure, we use the equivalent notation $R_n(\omega, f) = E^{R_n(\omega)}[f] = |V_n|^{-1} \sum_{i \in V_n} f(\theta_i \omega)$.

Multidimensional ergodic theorem. *Let $P \in \mathcal{M}_\theta(\Omega)$. Then, for all $p \ge 1$ and $f \in L^p(P)$, $R_n(\omega, f) \to E[f \,|\, \mathcal{I}](\omega)$ in $L^p(P)$ and for P-a.e. ω.*

To illuminate the theorem somewhat, note that for $f \in L^2(P, \mathscr{F})$ the conditional expectation $E[f \,|\, \mathcal{I}]$ is the orthogonal projection of f onto the space $L^2(P, \mathcal{I})$. Thus the L^2 version of the theorem says that the limit of $|V_n|^{-1} \sum_{i \in V_n} f \circ \theta_i$ is the orthogonal projection of f onto the space of fixed points of the group (θ_i). The L^p version is obtained from this by a density argument. The almost sure convergence is the hardest part. See Appendix 14.A of [**41**] for the proof of the theorem.

A corollary of the ergodic theorem is that distinct ergodic measures are supported on disjoint sets.

***Exercise B.20.** Prove that if P_1 and P_2 are two different ergodic measures, then they are mutually singular, that is, there exists a measurable set A such that $P_1(A) = 0$ and $P_2(A) = 1$. Furthermore, we can take $A \in \mathcal{I}$.

The extreme points of the convex set $\mathcal{M}_\theta(\Omega)$ are the measures $P \in \mathcal{M}_\theta(\Omega)$ that cannot be written as $P = tP_1 + (1-t)P_2$ for $t \in (0, 1)$ and $P_1, P_2 \in \mathcal{M}_\theta(\Omega)$ except when $P_1 = P_2 = P$. The extreme points turn out to be precisely the ergodic measures. This is another corollary of the ergodic theorem.

***Exercise B.21.** A probability measure $P \in \mathcal{M}_\theta(\Omega)$ is ergodic if and only if it is an extreme point of $\mathcal{M}_\theta(\Omega)$.

Hint: Use the exercise above.

A natural question about a convex set is whether the entire set consists of weighted averages of its extreme points. This is the case with $\mathcal{M}_\theta(\Omega)$ for a Polish Ω.

Ergodic decomposition theorem. *Assume that $(\Omega, \mathscr{F})$ is a Polish space with its Borel σ-algebra. Then for each $P \in \mathcal{M}_\theta(\Omega)$ there is a unique probability measure μ_P supported on the set $\mathcal{M}_e(\Omega)$ of ergodic probability measures such that*

$$P = \int_{\mathcal{M}_e(\Omega)} Q \, \mu_P(dQ).$$

This theorem can be seen as a consequence of the existence of a conditional probability measure P^ω of P, given $\mathcal{I}$. If $P \in \mathcal{M}_\theta(\Omega)$, then P^ω is ergodic for P-a.e. ω, and μ_P is the distribution on $\mathcal{M}_e(\Omega)$ induced by the map $\omega \mapsto P^\omega$. Exercises B.30 and B.31 sketch a derivation.

In Exercises B.22–B.29 we specialize to a product space $\Omega = \mathcal{S}^{\mathbb{Z}^d}$ where $(\mathcal{S}, \mathscr{B})$ is a measurable space, the coordinate random variables $X_i : \Omega \to \mathcal{S}$ are defined by $X_i(\omega) = \omega_i$ for $\omega = (\omega_i)_{i \in \mathbb{Z}^d} \in \Omega$, $\mathscr{F} = \sigma\{X_i : i \in \mathbb{Z}^d\}$ is the product σ-algebra on Ω, and $(\theta_i \omega)_j = \omega_{i+j}$ is the shift map. Let $\mathscr{F}_V = \sigma\{X_i : i \in V\}$. Let $\{V_n\}$ be an increasing sequence of cubes that exhausts $\mathbb{Z}^d$. The *tail σ-algebra* is $\mathscr{T} = \bigcap_n \mathscr{F}_{V_n^c}$. As always, when $\mathcal{S}$ is a metric space, $\mathscr{B} = \mathscr{B}_{\mathcal{S}}$ is the Borel σ-algebra of $\mathcal{S}$.

***Exercise B.22** (Kolmogorov's 0-1 law)**.** Let $\lambda_i \in \mathcal{M}_1(\mathcal{S})$ and let $P = \bigotimes_{i \in \mathbb{Z}^d} \lambda_i$ be the product probability measure on $(\Omega, \mathscr{F})$ under which the variables $\{X_i\}$ are independent and X_i is λ_i-distributed. Show that $P(A) \in \{0, 1\}$ for any $\mathscr{T}$-measurable set A.

Hint: Observe that A is independent of $\mathscr{F}_{V_n}$, for all $n \geq 1$. Now prove that the class of sets that are independent of A is a monotone class. Conclude that A is independent of itself.

***Exercise B.23.** Let P be a shift-invariant probability measure on Ω. Show that if A is $\mathcal{I}$-measurable, then there exists $B \in \mathscr{T}$ such that $P(A \triangle B) = 0$.

Hint: For each $k \in \mathbb{N}$ there exists $B_k \in \mathscr{F}_{V_{m(k)}}$ such that $P(A \triangle B_k) < 2^{-k}$. This is true because sets of a σ-algebra can be approximated by sets from a generating algebra. Try $B = \bigcup_{m \geq n} \bigcap_{k \geq m} \theta_{-k-m(k)} B_k$. Note that B is independent of n.

***Exercise B.24.** Let $\lambda \in \mathcal{M}_1(\mathcal{S})$. Show that the i.i.d. product measure $P = \lambda^{\otimes \mathbb{Z}^d}$ is ergodic for the shifts $(\theta_j)_{j \in \mathbb{Z}^d}$.

Exercise B.25. Suppose $\mathcal{S}$, and therefore also Ω, is a separable metric space. Suppose $f : \Omega \to \mathbb{R}$ is continuous and $\mathcal{I}$-measurable. Show that f is constant.

Hint: Let $\{x_k\}$ be a dense set in $\mathcal{S}$. Find $\tilde{\omega} \in \Omega$ such that for any n and any $\omega \in \Omega$ such that $\omega_i \in \{x_k\}$ for all $i \in V_n$, there exists a j such that $\tilde{\omega}_{j+V_n} = \omega_{V_n}$. Prove that $\{\theta_i \tilde{\omega} : i \in \mathbb{Z}^d\}$ is dense.

In the separable metric case the product σ-algebra $\mathscr{F}$ is also the Borel σ-algebra of Ω. It is countably generated. For example, balls of rational radius centered at the points of a countable dense set generate $\mathscr{F}$.

Exercise B.26. Show that $\mathcal{I}$ is not countably generated.

Hint: Pick $x \neq y \in \mathcal{S}$ and let $P = \lambda^{\otimes \mathbb{Z}^d}$, with $\lambda = (\delta_x + \delta_y)/2$. If $\mathcal{I}$ is generated by $(A_k)_{k \in \mathbb{N}}$, then let $B_k = A_k$ if $P(A_k) = 1$ and $B_k = A_k^c$ if $P(A_k) = 0$. Check that $C = \bigcap B_k = \{\theta_i \omega : i \in \mathbb{Z}^d\}$ for some $\omega \in \Omega$. Show that $P(C)$ should then be 0 and 1 at the same time.

Since distinct ergodic measures have disjoint supports, one can actually define a universal version of the conditional probability measure P^ω (given $\mathcal{I}$) *simultaneously* for all measures $P \in \mathcal{M}_\theta(\Omega)$.

Exercise B.27. Let $\mathcal{S}$ be a Polish space and define the set $\Omega_0 = \{\omega : \kappa(\omega) = \lim_{n\to\infty} R_n(\omega) \text{ exists in } \mathcal{M}_1(\Omega)\}$. For $\omega \notin \Omega_0$ define $\kappa(\omega) = Q$ for some fixed $Q \in \mathcal{M}_\theta(\Omega)$. Abbreviate $\kappa(\omega, f) = E^{\kappa(\omega)}[f]$. Prove that Ω_0 is a Borel set and that $\kappa : \Omega \to \mathcal{M}_\theta(\Omega)$ is measurable. Moreover, $\kappa \circ \theta_i = \kappa$ for all $i \in \mathbb{Z}^d$ and, for any bounded measurable function f and bounded $\mathcal{I}$-measurable function g,

(a) $\kappa(\omega, f)$ is $\mathcal{I}$-measurable and

(b) $E^P[gf] = \int g(\omega)\kappa(\omega, f)P(d\omega)$, for any $P \in \mathcal{M}_\theta(\Omega)$.

In other words, for any shift-invariant P, $\kappa(\omega)$ gives a version of P^ω.

Exercise B.28. Prove that the universal kernel κ from Exercise B.27 is $\mathscr{T}$-measurable.

Exercise B.29. Let $P = \lambda^{\otimes \mathbb{Z}^d}$ for $\lambda \in \mathcal{M}_1(\mathcal{S})$. Let $\tilde{h}(\omega) = h(\kappa(\omega) \,|\, P)$, where h is the specific relative entropy introduced in Section 6.2. Prove that $h(\nu \,|\, P) = \int \tilde{h}\, d\nu$, for all $\nu \in \mathcal{M}_\theta(\Omega)$.

Hint: Use the affinity and lower semicontinuity of h together with the ergodic decomposition $\nu = \int \kappa\, d\nu$. If this is too difficult, Lemma 5.4.24 in [**18**] gives a more general result.

The next exercise develops an abstract approach to the ergodic decomposition from article [**28**]. The virtue of this approach is that it applies to several examples, as illustrated by the exercises. The key hypothesis is that a collection of measures has a common kernel for conditional probabilities relative to some σ-algebra. Exercise B.27 showed that such a kernel can be defined for shift-invariant measures relative to the shift-invariant σ-algebra

$\mathcal{I}$. Exercise B.32 below shows the same for exchangeable measures, and Exercises 7.25 and 7.26 apply Exercise B.30 to Gibbs measures.

***Exercise B.30.** Let $(\Omega, \mathscr{F})$ be a measurable space and let $\mathcal{M}_1$ be the space of probability measures on $(\Omega, \mathscr{F})$. Equip $\mathcal{M}_1$ with the smallest σ-algebra $\mathscr{H}$ under which all functions $\mu \mapsto \int g\, d\mu$ are measurable for $g \in b\mathscr{F}$. Fix a nonempty measurable subset $\mathcal{P}$ of $\mathcal{M}_1$ and a sub-σ-algebra $\mathscr{A}$ of $\mathscr{F}$. Define the subset of $\mathscr{A}$-ergodic measures by

$$\mathcal{P}_e = \{\mu \in \mathcal{P} : \mu(A) = 0 \text{ or } 1 \quad \forall A \in \mathscr{A}\}.$$

Make the following two assumptions.

(i) There exists a countable family $\mathscr{W} \subset b\mathscr{F}$ of bounded measurable functions that distinguishes elements of $\mathcal{P}$: if $\mu, \nu \in \mathcal{P}$ and $\int g\, d\mu = \int g\, d\nu$ for all $g \in \mathscr{W}$, then $\mu = \nu$.

(ii) There exists a stochastic kernel κ from $(\Omega, \mathscr{A})$ into $(\Omega, \mathscr{F})$ with the following properties:

$$\text{(B.5)} \qquad \begin{aligned} &\kappa(\omega, \cdot\,) \in \mathcal{P} \text{ for all } \omega \in \Omega. \\ &\text{For every } \mu \in \mathcal{P} \text{ and } B \in \mathscr{F},\ \kappa(\omega, B) \text{ is a version of } \mu(B \,|\, \mathscr{A})(\omega). \end{aligned}$$

(a) Let $\mu \in \mathcal{P}$. Show that statements (i)–(iv) are equivalent.

(i) $\mu \in \mathcal{P}_e$.

(ii) $\kappa(\omega, g) = E^\mu[g]$ for μ-a.e. ω, for all $g \in \mathscr{W}$.

(iii) $\Phi_g(\mu) = 0$ for all $g \in \mathscr{W}$, where

$$\Phi_g(\mu) = \int |\kappa(\omega, g) - E^\mu[g]|^2\, \mu(d\omega) = \int \kappa(\omega, g)^2\, \mu(d\omega) - E^\mu[g]^2.$$

(iv) $\mu\{\omega : \kappa(\omega) = \mu\} = 1$.

Hint: (i)$\Longrightarrow$(ii)$\Longrightarrow$(iii)$\Longrightarrow$(iv)$\Longrightarrow$(i).

(b) Establish the following statements.

(i) $\mathcal{P}_e$ is a measurable subset of $\mathcal{M}_1$ and a subset of the image $\{\kappa(\omega) : \omega \in \Omega\}$.

(ii) For any $\mu \in \mathcal{P}$, $\mu\{\omega : \kappa(\omega) \in \mathcal{P}_e\} = 1$.

(iii) $\mathcal{P}_e$ is not empty.

Hint: $\mathcal{P}_e$ is the intersection of the sets $\{\Phi_g = 0\}$. By conditioning on $\mathscr{A}$ derive $\int \Phi_g(\kappa(\omega))\, \mu(d\omega) = 0$.

(c) Show that for every $\mu \in \mathcal{P}$ there exists a unique probability measure Q_μ on $\mathcal{P}_e$ such that

$$\mu(B) = \int_{\mathcal{P}_e} \nu(B)\, Q_\mu(d\nu) \qquad \forall B \in \mathscr{F}.$$

Hint: Let $Q_\mu(D) = \mu\{\omega : \kappa(\omega) \in D\}$, after adjusting $\kappa(\omega)$ so that all its values are in $\mathcal{P}_e$. For uniqueness, use point (iv) of part (a) to show that if Q_μ gives the decomposition, then it must be the distribution of $\kappa(\omega)$ under μ.

***Exercise B.31** (Ergodic decomposition). Apply Exercise B.30 to deduce the ergodic decomposition. Define the kernel $\kappa(\omega)$ as in Exercise B.27.

***Exercise B.32** (de Finetti's theorem). Let $\mathcal{S}$ be a Polish space and let $\Omega = \mathcal{S}^{\mathbb{N}}$, the space of $\mathcal{S}$-valued sequences with its product σ-algebra $\mathscr{F}$. Let $\mathbf{S}_n$ denote the group of permutations (bijective mappings) on $\{1, 2, \dots, n\}$. Permutations $\pi \in \mathbf{S}_n$ act on Ω in the obvious way: $(\pi\omega)_i = \omega_{\pi(i)}$ where we take $\pi(i) = i$ for $i > n$. Define σ-algebras

$$\mathscr{E}_n = \{A \in \mathscr{F} : \pi^{-1}A = A \ \forall \pi \in \mathbf{S}_n\}$$

and the *exchangeable σ-algebra*

$$\mathscr{E} = \bigcap_{n \geq 1} \mathscr{E}_n.$$

A probability measure $\mu \in \mathcal{M}_1(\Omega)$ is *exchangeable* if it is invariant under all finite permutations.

The goal here is to apply Exercise B.30 to prove the following.

De Finetti's theorem. *Fix an exchangeable probability measure μ. Then, conditional on $\mathscr{E}$, the coordinates $X_i(\omega) = \omega_i$ are i.i.d. under μ, and μ can be decomposed into a mixture of i.i.d. measures.*

Define the empirical measures

$$A_n(\omega) = \frac{1}{n!} \sum_{\pi \in \mathbf{S}_n} \delta_{\pi\omega}.$$

Averages of functions are denoted by $A_n(\omega, f) = \frac{1}{n!} \sum_{\pi \in \mathbf{S}_n} f(\pi\omega)$.

(a) Show that $A_n(\omega, f) = E^\mu[f \,|\, \mathscr{E}_n](\omega)$. Consequently, by the backward-martingale convergence theorem (Theorem 5.6.3 of [**27**] or page 155 of [**54**]), $A_n(f) \to E^\mu[f \,|\, \mathscr{E}]$ μ-a.s. Show that out of these limits we can construct a kernel that satisfies (B.5).

(b) Let f be a bounded function on $\mathcal{S}^k$ and g a bounded function on $\mathcal{S}$. Show that

$$A_n\big(f(X_1, \dots, X_k) g(X_{k+1})\big) - A_n\big(f(X_1, \dots, X_k)\big)\, A_n\big(g(X_1)\big) = \mathcal{O}(n^{-1}).$$

Show that any limit point of A_n is an i.i.d. product measure.

(c) Show that $\mathscr{E}$ is trivial under an exchangeable measure μ if and only if μ is an i.i.d. product measure. The "if" direction is known as the Hewitt-Savage 0-1 law. Complete the proof of de Finetti's theorem.

B.7. Stochastic ordering

In this section we work in the configuration spaces $\Omega = \{-1,1\}^{\mathbb{Z}^d}$ and $\Omega_\Lambda = \{-1,1\}^\Lambda$. Pointwise minima and maxima of configurations are defined by $(\eta \wedge \omega)_i = \eta_i \wedge \omega_i$ and $(\eta \vee \omega)_i = \eta_i \vee \omega_i$. Recall the definition of stochastic ordering (or domination) from page 142. An event A is *increasing* if $\omega \in A$ and $\omega \leq \sigma$ imply $\sigma \in A$. This is the same as saying that the indicator function of A is increasing.

Strassen's lemma. *Let $\mu, \nu \in \mathcal{M}_1(\Omega)$ satisfy $\mu \leq \nu$. Assume $\mu\{\sigma_i = 1\} = \nu\{\sigma_i = 1\}$ for all $i \in \mathbb{Z}^d$. Then $\mu = \nu$.*

Proof. We prove the following claim for all $N \geq 1$.

Claim B.33. *For any $\Lambda \subset \mathbb{Z}^d$ with $|\Lambda| = N$ and $\mu, \nu \in \mathcal{M}_1(\Omega_\Lambda)$, if $\mu \leq \nu$ and $\mu(\sigma_i = 1) = \nu(\sigma_i = 1)$ for all $i \in \Lambda$, then $\mu = \nu$.*

This proves that the finite-dimensional marginals of μ and ν in Strassen's lemma coincide and thus $\mu = \nu$.

The claim is proved by induction over N. The case $N = 1$ is obvious. The case $N = 2$ is a direct computation. Indeed, if $\Lambda = \{i, j\}$, then observe that $\mathbb{1}\{\sigma_i = 1 \text{ or } \sigma_j = 1\}$ is increasing. Thus,

$$\begin{aligned}
&\mu\{\sigma_i = -1, \sigma_j = 1\} + \mu\{\sigma_i = 1, \sigma_j = 1\} + \mu\{\sigma_i = 1, \sigma_j = -1\} \\
&\qquad \leq \nu\{\sigma_i = -1, \sigma_j = 1\} + \nu\{\sigma_i = 1, \sigma_j = 1\} + \nu\{\sigma_i = 1, \sigma_j = -1\}.
\end{aligned}$$

But the sums of the first two terms on each side are equal. Thus,

$$\mu\{\sigma_i = 1, \sigma_j = -1\} \leq \nu\{\sigma_i = 1, \sigma_j = -1\}.$$

Since $\mathbb{1}\{\sigma_i = 1, \sigma_j = 1\}$ is increasing we also have

$$\mu\{\sigma_i = 1, \sigma_j = 1\} \leq \nu\{\sigma_i = 1, \sigma_j = 1\}.$$

The sum of the left-hand sides of the last two inequalities equals the sum of the right-hand sides. Thus these must actually be equalities. Interchanging the roles of i and j also proves that $\mu\{\sigma_i = -1, \sigma_j = 1\} = \nu\{\sigma_i = -1, \sigma_j = 1\}$. Then $\mu\{\sigma_i = -1, \sigma_j = -1\} = \nu\{\sigma_i = -1, \sigma_j = -1\}$ follows, proving the claim when $N = 2$.

Now assume the claim is true for N spins, with $N \geq 2$. Consider measures on Ω_Λ with $|\Lambda| = N + 1$. By Exercise 9.8, if μ and ν agree on all events $K(A) = \{\sigma_i = 1 \ \forall i \in A\}$, $A \subset \Lambda$, then $\mu = \nu$. So suppose they do not agree on all such events. Since their N-spin marginals agree, the only possible case of disagreement is $A = \Lambda$.

So suppose $\exists \varepsilon > 0$ such that $\nu\{K(\Lambda)\} = \mu\{K(\Lambda)\} + \varepsilon$. (Since $\mu \leq \nu$ the difference must go this way.) Again since the marginals on N spins agree, it must be that for any $i \in \Lambda$,

$$\mu\{K(\Lambda \smallsetminus \{i\}), \sigma_i = -1\} = \nu\{K(\Lambda \smallsetminus \{i\}), \sigma_i = -1\} + \varepsilon.$$

Take $i, j \in \Lambda$, with $i \neq j$, and define the increasing event

$$B = K(\Lambda \smallsetminus \{i, j\}) \cap (\{\sigma_i = 1\} \cup \{\sigma_j = 1\}).$$

From the disjoint union

$$B = \big\{K(\Lambda \smallsetminus \{j\}), \sigma_j = -1\big\} \cup \big\{K(\Lambda \smallsetminus \{i\}), \sigma_i = -1\big\} \cup K(\Lambda)$$

and the observations above, $\mu(B) = \nu(B) + 2\varepsilon - \varepsilon = \nu(B) + \varepsilon$. This contradicts $\mu \leq \nu$. □

In the next theorem we prove a stochastic domination $\mu \leq \nu$ in $\mathcal{M}_1(\Omega_\Lambda)$ by proving the existence of a *coupling*: namely, a probability measure q on $\Omega_\Lambda \times \Omega_\Lambda$ with marginals μ and ν and such that $q\{\eta \leq \xi\} = 1$. In fact more is true. Strassen proved that if $\mathcal{X}$ is a Polish space with a partial order relation $\leq$ such that $\{(x, y) : x \leq y\}$ is a closed subset of $\mathcal{X}^2$, then $\mu \leq \nu$ is equivalent to the existence of such a coupling measure. See Strassen's Theorem 11 in [**78**] and for the special case when $\mathcal{X}$ is compact see Theorem 2.4 of [**58**]. For this book such a general result is not needed.

Holley's theorem. *Let $\Lambda \subset \mathbb{Z}^d$ be finite and let μ and ν be strictly positive probability measures on Ω_Λ. Assume*

$$\mu(\eta \wedge \xi)\nu(\eta \vee \xi) \geq \mu(\eta)\nu(\xi) \quad \forall \eta, \xi \in \Omega_\Lambda. \tag{B.6}$$

Then there exists a probability measure q on $\Omega_\Lambda \times \Omega_\Lambda$ such that $q\{(\eta, \xi) : \eta \leq \xi\} = 1$ and q has marginals μ and ν, that is,

$$q(\{\eta\} \times \Omega_\Lambda) = \mu(\eta) \quad \text{and} \quad q(\Omega_\Lambda \times \{\xi\}) = \nu(\xi) \quad \forall \eta, \xi \in \Omega_\Lambda.$$

In particular $\mu \leq \nu$.

Proof. The idea is to couple two Markov chains on $\Omega_\Lambda = \{-1, 1\}^\Lambda$ with stationary distributions μ and ν so that the coupled process is a Markov chain on the space $\mathcal{X} = \{(\eta, \xi) \in \Omega_\Lambda \times \Omega_\Lambda : \eta \leq \xi\}$.

Let η^j denote the configuration after flipping the j-th spin in η. That is, $\eta^j_j = -\eta_j$ and $\eta^j_i = \eta_i$ for $i \neq j$. At each step the joint Markov chain will either do nothing or choose a site in Λ and flip the η-spin and/or the ξ-spin at that site. To define the transition probabilities, first set for each $j \in \Lambda$

$$\begin{array}{ll} r((\eta,\xi),(\eta^j,\xi)) = 1 & \text{if } \eta_j = -1 \text{ and } \xi_j = 1, \\ r((\eta,\xi),(\eta,\xi^j)) = \frac{\nu(\xi^j)}{\nu(\xi)} & \text{if } \eta_j = -1 \text{ and } \xi_j = 1, \\ r((\eta,\xi),(\eta^j,\xi^j)) = 1 & \text{if } \eta_j = \xi_j = -1, \\ r((\eta,\xi),(\eta^j,\xi^j)) = \frac{\nu(\xi^j)}{\nu(\xi)} & \text{if } \eta_j = \xi_j = 1, \\ r((\eta,\xi),(\eta^j,\xi)) = \frac{\mu(\eta^j)}{\mu(\eta)} - \frac{\nu(\xi^j)}{\nu(\xi)} & \text{if } \eta_j = \xi_j = 1, \text{ and} \\ r((\eta,\xi),(\eta',\xi')) = 0 & \text{otherwise.} \end{array}$$

The next-to-last definition above requires $\mu(\eta^j)\nu(\xi) \geq \mu(\eta)\nu(\xi^j)$. This is ensured by condition (B.6) because if $\eta \leq \xi$ and $\eta_j = \xi_j = 1$, then $\eta \wedge \xi^j = \eta^j$ and $\eta \vee \xi^j = \xi$. The list above describes the possible transitions at site j. Each transition preserves the inequality $\eta \leq \xi$.

Now we can either define a continuous-time Markov chain with the rates given above or a discrete-time chain. We do the latter.

Pick

$$C > \max_{(\eta,\xi)} \sum_{(\eta',\xi')} r((\eta,\xi),(\eta',\xi')),$$

and define the transition probabilities

$$p((\eta,\xi),(\eta',\xi')) = r((\eta,\xi),(\eta',\xi'))/C,$$

when $(\eta,\xi) \neq (\eta',\xi')$ and

$$p((\eta,\xi),(\eta,\xi)) = 1 - C^{-1} \sum_{(\eta',\xi')} r((\eta,\xi),(\eta',\xi')) > 0.$$

This transition probability defines a Markov chain on the finite space $\mathcal{X}$. Fix an arbitrary state $(\eta_0,\xi_0) \in \mathcal{X}$. By compactness there exists a probability measure q that is a subsequential limit of the averages of transition probabilities:

$$q(\eta,\xi) = \lim_{\ell\to\infty} n_\ell^{-1} \sum_{k=1}^{n_\ell} p^k((\eta_0,\xi_0),(\eta,\xi)) \quad \text{for } (\eta,\xi) \in \mathcal{X}.$$

It follows that q is invariant for transition p on $\mathcal{X}$. We argue that q has marginals μ and ν and thereby furnishes the desired coupling.

***Exercise B.34.** Prove that each one of the two coordinates of the Markov chain above is itself a Markov chain. Show that the transition probabilities of the η-chain are

$$\begin{array}{ll} p(\eta,\eta^j) = 1/C & \text{if } \eta_j = -1, \\ p(\eta,\eta^j) = \frac{\mu(\eta^j)}{C\mu(\eta)} & \text{if } \eta_j = 1, \text{ and} \\ p(\eta,\eta) = 1 - \sum_j p(\eta,\eta^j) & \text{for all } \eta. \end{array}$$

Also show that the transition probabilities of the ξ-chain are

$$\begin{aligned} &p(\xi, \xi^j) = 1/C && \text{if } \xi_j = -1, \\ &p(\xi, \xi^j) = \tfrac{\nu(\xi^j)}{C\nu(\xi)} && \text{if } \xi_j = 1, \text{ and} \\ &p(\xi, \xi) = 1 - \textstyle\sum_j p(\xi, \xi^j) && \text{for all } \xi. \end{aligned}$$

These marginal chains are irreducible and aperiodic on Ω_Λ.

***Exercise B.35.** Show that μ is the invariant measure for the η-chain and that ν is the invariant measure for the ξ-chain. In fact, these measures are reversible.

Since the marginal chains obey the Markov chain convergence theorem (Exercise 5.16 or [**27**, Theorem 6.6.4]), for arbitrary states $\eta \in \Omega_\Lambda$ and $(\eta_0, \xi_0) \in \mathcal{X}$,

$$\begin{aligned} q(\{\eta\} \times \Omega_\Lambda) &= \lim_{\ell\to\infty} n_\ell^{-1} \sum_{k=1}^{n_\ell} p^k\big((\eta_0, \xi_0), \{\eta\} \times \Omega_\Lambda\big) = \lim_{\ell\to\infty} n_\ell^{-1} \sum_{k=1}^{n_\ell} p^k(\eta_0, \eta) \\ &= \mu(\eta). \end{aligned}$$

The same argument gives $q(\Omega_\Lambda \times \{\xi\}) = \nu(\xi)$. Stochastic domination $\mu \le \nu$ follows: if f is an increasing function on Ω, then

$$\int_{\Omega_\Lambda} f\, d\mu = \int_{\eta \le \xi} f(\eta)\, q(d\eta, d\xi) \le \int_{\eta \le \xi} f(\xi)\, q(d\eta, d\xi) = \int_{\Omega_\Lambda} f\, d\nu. \qquad \square$$

Exercise B.36 (Harris–Fortuin-Kasteleyn-Ginibre inequality)**.** Let $\Lambda \subset \mathbb{Z}^d$ be finite and let μ be a strictly positive probability measure on Ω_Λ. Assume (B.6) holds with $\nu = \mu$. Prove that for any increasing functions f and g we have

$$E^\mu[fg] \ge E^\mu[f] E^\mu[g].$$

Note in particular that a product measure satisfies the hypotheses.

Literature notes. There are many excellent textbooks on probability theory from which the reader can fill in missing background or look up proofs of results mentioned in this appendix. Examples include Dudley [**26**], Durrett [**27**], Khoshnevisan [**54**], and Varadhan [**80**].

Inequalities from statistical mechanics

C.1. Griffiths's inequality

Griffiths's inequality. *Fix a finite volume* $\Lambda \subset \mathbb{Z}^d$. *Let* $\mathcal{E} = \{\{i,j\} : |i-j|_1 = 1, i \in \Lambda\}$ *be the set of nearest-neighbor edges of* Λ, *including edges from* Λ *to its complement. Define the function* $F : [0,\infty)^{\mathcal{E}} \to [-1,1]$ *by*

$$F(\mathbf{J}) = \frac{\sum_{\sigma_\Lambda} \sigma_0 \exp\left(\sum_{\{i,j\}\in\mathcal{E}} J_{i,j}\, \sigma_i \sigma_j\right)}{\sum_{\sigma_\Lambda} \exp\left(\sum_{\{i,j\}\in\mathcal{E}} J_{i,j}\, \sigma_i \sigma_j\right)},$$

where $\mathbf{J} = (J_{i,j})_{\{i,j\}\in\mathcal{E}}$, *the sum runs over* $\sigma_\Lambda \in \{-1,1\}^\Lambda$, *and for* $j \notin \Lambda$ *we set* $\sigma_j = 1$ *(that is, the + boundary condition). Then,*

$$\frac{\partial F}{\partial J_{i,j}}(\mathbf{J}) \geq 0 \ \forall \{i,j\} \in \mathcal{E} \text{ and } \forall \mathbf{J} \in [0,\infty)^{\mathcal{E}}.$$

We start by proving the following lemma.

Lemma C.1. *Take the assumptions as above. For any* $\mathbf{J} \in [0,\infty)^{\mathcal{E}}$ *and* $\{k,\ell\} \in \mathcal{E}$,

$$\sum_{\sigma_\Lambda} \sigma_0 \sigma_k \sigma_\ell \exp\Big\{ \sum_{\{i,j\}\in\mathcal{E}} J_{i,j}\, \sigma_i \sigma_j \Big\} \geq 0.$$

Proof. The proof goes by a direct computation, expanding the exponential:

$$\sum_{\sigma_\Lambda} \sigma_0\sigma_k\sigma_\ell \exp\Big(\sum_{\{i,j\}\in\mathcal{E}} J_{i,j}\,\sigma_i\sigma_j\Big)$$

$$= \sum_{\sigma_\Lambda} \sigma_0\sigma_k\sigma_\ell \sum_{n\geq 0} \frac{1}{n!}\Big(\sum_{\{i,j\}\in\mathcal{E}} J_{i,j}\,\sigma_i\sigma_j\Big)^n$$

$$= \sum_{n\geq 0} \frac{1}{n!} \sum_{\{i_p,j_p\}\in\mathcal{E},\, p=1,\dots,n} \Big(\prod_{p=1}^n J_{i_p,j_p}\Big)\Big[\sum_{\sigma_\Lambda} \sigma_0\sigma_k\sigma_\ell\Big(\prod_{p=1}^n \sigma_{i_p}\sigma_{j_p}\Big)\Big].$$

The expression in square brackets vanishes if some spin inside Λ appears in the product an odd number of times. Otherwise the product contains only squares and $(+1)$-valued boundary spins and hence is positive. □

Proof of Griffiths's inequality. A direct computation shows that

$$\frac{\partial F}{\partial J_{k,\ell}} = Z^{-2} \sum_{\sigma_\Lambda,\tau_\Lambda} \sigma_0(\sigma_k\sigma_\ell - \tau_k\tau_\ell) \exp\Big\{ \sum_{\{i,j\}\in\mathcal{E}} J_{i,j}(\sigma_i\sigma_j + \tau_i\tau_j)\Big\},$$

where $Z = \sum_{\sigma_\Lambda} \exp\big\{\sum_{\{i,j\}\in\mathcal{E}} J_{i,j}\,\sigma_i\sigma_j\big\}$ and, just as for σ_Λ, τ_Λ ranges over $\{-1,1\}^\Lambda$ and $\tau_i = 1$ when $i \notin \Lambda$. Define the configuration η such that $\eta_i = 1$ if $\sigma_i = \tau_i$ and -1 if $\sigma_i \neq \tau_i$. Noting that $\tau_i = \eta_i\sigma_i$ for all i (in particular, $\eta_i = 1$ when $i \notin \Lambda$), a change of variables gives

$$\frac{\partial F}{\partial J_{k,\ell}} = Z^{-2} \sum_{\eta_\Lambda} (1 - \eta_k\eta_\ell) \sum_{\sigma_\Lambda} \sigma_0\sigma_k\sigma_\ell \exp\Big\{ \sum_{\{i,j\}\in\mathcal{E}} J_{i,j}(1+\eta_i\eta_j)\sigma_i\sigma_j\Big\}.$$

The inequality now follows from the lemma above using the couplings $J_{i,j}(1+\eta_i\eta_i) \geq 0$ instead of $J_{i,j}$. □

C.2. Griffiths-Hurst-Sherman inequality

Griffiths-Hurst-Sherman inequality. *Consider the Ising model in* $d \geq 1$. *Fix* $\beta > 0$, $J > 0$, *and the volume* V_n. *Define*

$$M_n(h) = E^{\mu^+_{V_n,\beta,h,J}}\Big[\frac{1}{|V_n|}\sum_{i\in V_n} \sigma_i\Big].$$

Then $\frac{\partial^2 M_n}{\partial h^2} \leq 0$ *for* $h > 0$.

Abbreviate $\mu^+_{n,h} = \mu^+_{V_n,\beta,h,J}$ and let $(\mu^+_{n,h})^{\otimes 4}$ be the product of 4 copies of $\mu^+_{n,h}$. Denote the elements of $\Omega^4_{V_n}$ by $(\omega_i, \sigma_i, \omega'_i, \sigma'_i)_{i\in V_n}$.

Lemma C.2. *Assume the same setting as for the Griffiths-Hurst-Sherman inequality. For $i \in V_n$, let*

$$\begin{bmatrix}\alpha_i\\ \beta_i\\ \gamma_i\\ \delta_i\end{bmatrix} = \frac{1}{2}\begin{bmatrix}1 & 1 & 1 & 1\\ 1 & 1 & -1 & -1\\ 1 & -1 & 1 & -1\\ -1 & 1 & 1 & -1\end{bmatrix}\begin{bmatrix}\omega_i\\ \sigma_i\\ \omega'_i\\ \sigma'_i\end{bmatrix}.$$

Let A, B, C, and D be arbitrary subsets of V_n. If $h \geq 0$, then

$$E^{(\mu^+_{n,h})^{\otimes 4}}\Big[\Big(\prod_{i\in A}\alpha_i\Big)\Big(\prod_{i\in B}\beta_i\Big)\Big(\prod_{i\in C}\gamma_i\Big)\Big(\prod_{i\in D}\delta_i\Big)\Big] \geq 0. \tag{C.1}$$

Proof. Since the matrix is orthogonal, it preserves Euclidean scalar products and thus

$$\omega_i\omega_j + \sigma_i\sigma_j + \omega'_i\omega'_j + \sigma'_i\sigma'_j = \alpha_i\alpha_j + \beta_i\beta_j + \gamma_i\gamma_j + \delta_i\delta_j,$$

for any $i, j \in V_n$. Therefore,

$$\begin{aligned}&\mathcal{H}^+_n(\omega) + \mathcal{H}^+_n(\sigma) + \mathcal{H}^+_n(\omega') + \mathcal{H}^+_n(\sigma')\\ &\quad = -J \sum_{\substack{\{i,j\}\subset V_n:\\ |i-j|_1=1}} (\alpha_i\alpha_j + \beta_i\beta_j + \gamma_i\gamma_j + \delta_i\delta_j) - 2J \sum_{\substack{i\in V_n,\, j\notin V_n:\\ |i-j|_1=1}} \alpha_i - 2h\sum_{i\in V_n}\alpha_i.\end{aligned}$$

Expand $e^{-\beta(\mathcal{H}^+_n(\omega)+\mathcal{H}^+_n(\sigma)+\mathcal{H}^+_n(\omega')+\mathcal{H}^+_n(\sigma'))}$ into its Taylor series in the numerator of the expectation in (C.1). This leads to sums of products of integrals of the form

$$\int_{\Omega^4_{V_n}} \alpha_i^k\beta_i^\ell\gamma_i^m\delta_i^p\,\lambda(d\omega_i)\lambda(d\sigma_i)\lambda(d\omega'_i)\lambda(d\sigma'_i),$$

multiplied by nonnegative coefficients.

***Exercise C.3.** Check that $\alpha_i\beta_i = \gamma_i\delta_i = \frac{1}{2}(\omega_i\sigma_i - \omega'_i\sigma'_i)$ and then conclude that $\alpha_i\beta_i\gamma_i\delta_i = \frac{1}{4}(\omega_i\sigma_i - \omega'_i\sigma'_i)^2$.

***Exercise C.4.** Prove that the integral above is 0 if k, ℓ, m, and p are not all of the same parity (that is, all even or all odd).

By the exercise above we can assume that the powers are all of the same parity. If the powers are all even, the integral is obviously nonnegative. If the powers are all odd, use Exercise C.3 to show that the integral is nonnegative. This completes the proof of Lemma C.2. □

Lemma C.5. *Assume the same setting as for the Griffiths-Hurst-Sherman inequality. Let $i, j, k \in V_n$ be arbitrary. Abbreviate $\langle\sigma_\ell\rangle = E^{\mu^+_{n,h}}[\sigma_\ell]$. If $h \geq 0$, then*

$$E^{\mu^+_{n,h}}[(\sigma_i - \langle\sigma_i\rangle)(\sigma_j - \langle\sigma_j\rangle)(\sigma_k - \langle\sigma_k\rangle)] \leq 0. \tag{C.2}$$

Proof. Let

$$\begin{bmatrix} t_i \\ q_i \\ t_i' \\ q_i' \end{bmatrix} = \frac{1}{\sqrt{2}} \begin{bmatrix} 1 & 1 & 0 & 0 \\ 1 & -1 & 0 & 0 \\ 0 & 0 & 1 & 1 \\ 0 & 0 & 1 & -1 \end{bmatrix} \begin{bmatrix} \omega_i \\ \sigma_i \\ \omega_i' \\ \sigma_i' \end{bmatrix}.$$

***Exercise C.6.** Check that the left-hand side of (C.2) is equal to

$$-\sqrt{2}\, E^{(\mu^+_{n,h})^{\otimes 4}}[t_k q_i' q_j' - t_k q_i q_j] = -E^{(\mu^+_{n,h})^{\otimes 4}}[(\alpha_k + \beta_k)(\gamma_i \delta_j + \gamma_j \delta_i)]$$

where the functions α_i, β_i, γ_i, and δ_i are as in the previous lemma.

Hint: The first equality is a direct computation. For the second equality, write $t_k = (\alpha_k + \beta_k)/\sqrt{2}$, $q_i' = (\gamma_i + \delta_i)/\sqrt{2}$, and $q_i = (\gamma_i - \delta_i)/\sqrt{2}$.

The claim then follows from the previous lemma. □

Proof of the Griffiths-Hurst-Sherman inequality. Set $S_n = \sum_{i \in V_n} \sigma_i$ and $\langle S_n \rangle = \sum_{i \in V_n} \langle \sigma_i \rangle$. The lemma above shows that if $h \geq 0$, then

$$E^{\mu^+_{n,h}}[S_n^3] - 3E^{\mu^+_{n,h}}[S_n]E^{\mu^+_{n,h}}[S_n^2] + 2E^{\mu^+_{n,h}}[S_n]^3 = E^{\mu^+_{n,h}}\left[(S_n - \langle S_n \rangle)^3\right] \leq 0.$$

The proof is completed with the next computation.

***Exercise C.7.** Show that

$$\beta^{-2}|V_n| \frac{\partial^2 M_n}{\partial h^2} = E^{\mu^+_{n,h}}[S_n^3] - 3E^{\mu^+_{n,h}}[S_n]E^{\mu^+_{n,h}}[S_n^2] + 2E^{\mu^+_{n,h}}[S_n]^3. \qquad \square$$

Appendix D

Nonnegative matrices

This appendix collects facts from the Perron-Frobenius theory of nonnegative matrices for the large deviation theory of finite Markov chains.

Let $A = (a_{i,j})$ be a square matrix with nonnegative entries. Matrix A is *irreducible* if for each i, j there exists a positive integer $k = k(i,j)$ such that $(A^k)_{i,j} > 0$. The *period* d of index i is the greatest common divisor of the set $\{k \in \mathbb{N} : (A^k)_{i,i} > 0\}$. If A is irreducible, then all indices have the same period, which is then called the period of the matrix. An irreducible matrix is aperiodic if its period is 1.

In general, the *spectral radius* of a matrix A is

$$\rho(A) = \max\{|\lambda| : \lambda \text{ is an eigenvalue of } A\}.$$

If A is a nonnegative irreducible matrix, then $\rho(A)$ is a strictly positive eigenvalue of A, called the Perron-Frobenius eigenvalue. There exist strictly positive left and right eigenvectors u and v of $\rho(A)$, called the Perron-Frobenius eigenvectors. The eigenvectors of $\rho(A)$ are unique up to scalar multiples (Lemma 8.4.1 and Theorem 8.4.4 in [**47**]).

We have the following variational characterization (Corollary 8.1.31 of [**47**]):

$$\rho(A) = \sup_{x>0} \min_i \frac{\sum_j a_{i,j} x_j}{x_i} = \inf_{x>0} \max_i \frac{\sum_j a_{i,j} x_j}{x_i}. \tag{D.1}$$

The supremum (or the infimum) is achieved at x if and only if x is a scalar multiple of the Perron-Frobenius eigenvector v, and for such x the ratios are equal for all i (Theorem 8.3.4 of [**47**] and Theorem 1.6 of [**71**]).

A is aperiodic if and only if $|\lambda| < \rho(A)$ for all eigenvalues $\lambda \neq \rho(A)$ (Corollary 8.4.8 of [**47**]). In this case,

$$\lim_{n\to\infty} \rho(A)^{-n}A^n = (v(i)u(j))_{i,j}, \tag{D.2}$$

where v and u are the right and left Perron-Frobenius eigenvectors, normalized so that $\sum_i v(i)u(i) = 1$ (Theorem 8.5.1 of [**47**]).

Lemma D.1. *Let z and w be the right Perron-Frobenius eigenvectors for B and C, respectively. Let $0 < s < 1$ and $a_{i,j} = b_{i,j}^s c_{i,j}^{1-s}$. Then*

$$\rho(A) \le \rho(B)^s \rho(C)^{1-s}. \tag{D.3}$$

Equality holds if and only if $\rho(B)^{-1}b_{i,j}z_j/z_i = \rho(C)^{-1}c_{i,j}w_j/w_i$ for all i, j. In this case the right Perron-Frobenius eigenvector of A is $v_i = z_i^s w_i^{1-s}$.

Proof. By Hölder's inequality, for each i,

$$\begin{aligned}\frac{\sum_j a_{i,j}v_j}{v_i} &= \frac{\sum_j b_{i,j}^s z_j^s c_{i,j}^{1-s} w_j^{1-s}}{z_i^s w_i^{1-s}} \le \frac{\left(\sum_j b_{i,j}z_j\right)^s \left(\sum_j c_{i,j}w_j\right)^{1-s}}{z_i^s w_i^{1-s}} \\ &= \rho(B)^s\rho(C)^{1-s}.\end{aligned}$$

This says that $Av \le \rho(B)^s\rho(C)^{1-s}v$. Inequality (D.3) now follows from (D.1) and equality holds if and only if v is the Perron-Frobenius eigenvector of A. This holds if and only if equality holds in the Hölder inequality above for all i. This in turn holds if and only if $b_{i,j}z_j/z_i = \beta_i c_{i,j}w_j/w_i\ \forall\, i, j$ for some β_i. Adding over j shows that $\beta_i = \rho(B)/\rho(C)$ for all i. $\square$

Bibliography

1. Michael Aizenman, *Instability of phase coexistence and translation invariance in two dimensions*, Mathematical problems in theoretical physics (Proc. Internat. Conf. Math. Phys., Lausanne, 1979), Lecture Notes in Phys., vol. 116, Springer, Berlin, 1980, pp. 143–147. MR582616

2. ______, *Translation invariance and instability of phase coexistence in the two-dimensional Ising system*, Comm. Math. Phys. **73** (1980), no. 1, 83–94. MR573615

3. Michael Aizenman, Hugo Duminil-Copin, and Vladas Sidoravicius, *Random currents and continuity of Ising model's spontaneous magnetization*, Communications in Mathematical Physics (2014), 1–24.

4. Robert Ash, *Information theory*, Interscience Tracts in Pure and Applied Mathematics, No. 19, Interscience Publishers John Wiley & Sons, New York-London-Sydney, 1965. MR0229475

5. R. R. Bahadur and R. Ranga Rao, *On deviations of the sample mean*, Ann. Math. Statist. **31** (1960), 1015–1027. MR0117775

6. J. R. Baxter, N. C. Jain, and T. O. Seppäläinen, *Large deviations for nonstationary arrays and sequences*, Illinois J. Math. **37** (1993), no. 2, 302–328. MR1208824

7. J. R. Baxter, N. C. Jain, and S. R. S. Varadhan, *Some familiar examples for which the large deviation principle does not hold*, Comm. Pure Appl. Math. **44** (1991), no. 8-9, 911–923. MR1127039

8. John Baxter and Naresh Jain, *Convexity and compactness in large deviation theory*, Unpublished manuscript (1991).

9. Patrick Billingsley, *Convergence of probability measures*, John Wiley & Sons Inc., New York, 1968. MR0233396

10. D. W. Boyd, *Losing runs in Bernoulli trials*, (1972), unpublished manuscript.

11. J. Bricmont, J. R. Fontaine, and L. J. Landau, *On the uniqueness of the equilibrium state for plane rotators*, Comm. Math. Phys. **56** (1977), no. 3, 281–296. MR0489629

12. S. R. Broadbent and J. M. Hammersley, *Percolation processes. I. Crystals and mazes*, Proc. Cambridge Philos. Soc. **53** (1957), 629–641. MR0091567

13. Ivan Corwin, *The Kardar-Parisi-Zhang equation and universality class*, Random Matrices Theory Appl. **1** (2012), no. 1, 1130001, 76. MR2930377

14. A. de Acosta, *Large deviations for empirical measures of Markov chains*, J. Theoret. Probab. **3** (1990), no. 3, 395–431. MR1057523

15. Amir Dembo and Ofer Zeitouni, *Large deviations techniques and applications*, second ed., Applications of Mathematics, vol. 38, Springer-Verlag, New York, 1998. MR1619036

16. Frank den Hollander, *Large deviations*, Fields Institute Monographs, vol. 14, American Mathematical Society, Providence, RI, 2000. MR1739680

17. ______, *Random polymers*, Lecture Notes in Mathematics, vol. 1974, Springer-Verlag, Berlin, 2009, Lectures from the 37th Probability Summer School held in Saint-Flour, 2007. MR2504175

18. Jean-Dominique Deuschel and Daniel W. Stroock, *Large deviations*, Pure and Applied Mathematics, vol. 137, Academic Press Inc., Boston, MA, 1989. MR997938

19. Jean-Dominique Deuschel, Daniel W. Stroock, and Hans Zessin, *Microcanonical distributions for lattice gases*, Comm. Math. Phys. **139** (1991), no. 1, 83–101. MR1116411

20. I. H. Dinwoodie, *A note on the upper bound for i.i.d. large deviations*, Ann. Probab. **19** (1991), no. 4, 1732–1736. MR1127723

21. R. L. Dobrushin, *Existence of a phase transition in the two-dimensional and three-dimensional Ising models*, Soviet Physics Dokl. **10** (1965), 111–113. MR0182405

22. ______, *Description of a random field by means of conditional probabilities and conditions for its regularity*, Teor. Verojatnost. i Primenen **13** (1968), 201–229. MR0231434

23. ______, *Gibbsian random fields for lattice systems with pairwise interactions.*, Funkcional. Anal. i Priložen. **2** (1968), no. 4, 31–43. MR0250630

24. ______, *The Gibbs state that describes the coexistence of phases for a three-dimensional Ising model*, Teor. Verojatnost. i Primenen. **17** (1972), 619–639. MR0421546

25. M. D. Donsker and S. R. S. Varadhan, *Asymptotic evaluation of certain Markov process expectations for large time. I*, Comm. Pure Appl. Math. **28** (1975), 1–47. MR0386024

26. Richard M. Dudley, *Real analysis and probability*, The Wadsworth & Brooks/Cole Mathematics Series, Wadsworth & Brooks/Cole Advanced Books & Software, Pacific Grove, CA, 1989. MR982264

27. Rick Durrett, *Probability: theory and examples*, fourth ed., Cambridge Series in Statistical and Probabilistic Mathematics, Cambridge University Press, Cambridge, 2010. MR2722836

28. E. B. Dynkin, *Initial and final behavior of the trajectories of Markov processes*, Uspehi Mat. Nauk **26** (1971), no. 4(160), 153–172. MR0298758

29. Uwe Einmahl and James Kuelbs, *Moderate deviation probabilities for open convex sets: nonlogarithmic behavior*, Ann. Probab. **32** (2004), no. 2, 1316–1355. MR2060299

30. Ivar Ekeland and Roger Témam, *Convex analysis and variational problems*, english ed., Classics in Applied Mathematics, vol. 28, Society for Industrial and Applied Mathematics (SIAM), Philadelphia, PA, 1999, Translated from the French. MR1727362

31. Richard S. Ellis, *Large deviations for a general class of random vectors*, Ann. Probab. **12** (1984), no. 1, 1–12. MR723726

32. ______, *Entropy, large deviations, and statistical mechanics*, Grundlehren der Mathematischen Wissenschaften [Fundamental Principles of Mathematical Sciences], vol. 271, Springer-Verlag, New York, 1985. MR793553

33. ______, *Large deviations for the empirical measure of a Markov chain with an application to the multivariate empirical measure*, Ann. Probab. **16** (1988), no. 4, 1496–1508. MR958199

34. Gerald B. Folland, *Real analysis: Modern techniques and their applications*, second ed., Pure and Applied Mathematics (New York), John Wiley & Sons Inc., New York, 1999. MR1681462

35. Hans Föllmer, *Random fields and diffusion processes*, École d'Été de Probabilités de Saint-Flour XV–XVII, 1985–87, Lecture Notes in Math., vol. 1362, Springer, Berlin, 1988, pp. 101–203. MR983373

36. C. M. Fortuin and P. W. Kasteleyn, *On the random-cluster model. I. Introduction and relation to other models*, Physica **57** (1972), 536–564. MR0359655

37. Klaus Fritzsche and Hans Grauert, *From holomorphic functions to complex manifolds*, Graduate Texts in Mathematics, vol. 213, Springer-Verlag, New York, 2002. MR1893803

38. Jürg Fröhlich and Charles-Edouard Pfister, *Spin waves, vortices, and the structure of equilibrium states in the classical XY model*, Comm. Math. Phys. **89** (1983), no. 3, 303–327. MR709469

39. Jürg Fröhlich and Thomas Spencer, *Kosterlitz-Thouless transition in the two-dimensional plane rotator and Coulomb gas*, Phys. Rev. Lett. **46** (1981), no. 15, 1006–1009. MR607429

40. Jürgen Gärtner, *On large deviations from an invariant measure*, Teor. Verojatnost. i Primenen. **22** (1977), no. 1, 27–42. MR0471040

41. Hans-Otto Georgii, *Gibbs measures and phase transitions*, de Gruyter Studies in Mathematics, vol. 9, Walter de Gruyter & Co., Berlin, 1988. MR956646

42. Louis Gordon, Mark F. Schilling, and Michael S. Waterman, *An extreme value theory for long head runs*, Probab. Theory Relat. Fields **72** (1986), no. 2, 279–287. MR836278

43. Robert B. Griffiths, *Peierls proof of spontaneous magnetization in a two-dimensional Ising ferromagnet*, Phys. Rev. (2) **136** (1964), A437–A439. MR0189681

44. Geoffrey Grimmett, *Percolation*, second ed., Grundlehren der Mathematischen Wissenschaften [Fundamental Principles of Mathematical Sciences], vol. 321, Springer-Verlag, Berlin, 1999. MR1707339

45. Olle Häggström, *Random-cluster representations in the study of phase transitions*, Markov Process. Related Fields **4** (1998), no. 3, 275–321. MR1670023

46. Y. Higuchi, *On the absence of non-translation invariant Gibbs states for the two-dimensional Ising model*, Random fields, Vol. I, II (Esztergom, 1979), Colloq. Math. Soc. János Bolyai, vol. 27, North-Holland, Amsterdam, 1981, pp. 517–534. MR712693

47. Roger A. Horn and Charles R. Johnson, *Matrix analysis*, Cambridge University Press, Cambridge, 1990, Corrected reprint of the 1985 original. MR1084815

48. D.A. Huffman, *A method for the construction of minimum-redundancy codes*, Proceedings of the IRE **40** (1952), no. 9, 1098–1101.

49. D. A. Huse and C. L. Henley, *Pinning and roughening of domain wall in Ising systems due to random impurities*, Phys. Rev. Lett. **54** (1985), 2708–2711.

50. Robert B. Israel, *Convexity in the theory of lattice gases*, Princeton University Press, Princeton, N.J., 1979, Princeton Series in Physics, With an introduction by Arthur S. Wightman. MR517873

51. Naresh C. Jain, *Large deviation lower bounds for additive functionals of Markov processes*, Ann. Probab. **18** (1990), no. 3, 1071–1098. MR1062059

52. Yi Wen Jiang and Li Ming Wu, *Large deviations for empirical measures of not necessarily irreducible countable Markov chains with arbitrary initial measures*, Acta Math. Sin. (Engl. Ser.) **21** (2005), no. 6, 1377–1390. MR2190007

53. G. Kassay, *A simple proof for König's minimax theorem*, Acta Math. Hungar. **63** (1994), no. 4, 371–374. MR1261480

54. Davar Khoshnevisan, *Probability*, Graduate Studies in Mathematics, vol. 80, American Mathematical Society, Providence, RI, 2007. MR2296582

55. Heinz König, *Über das von Neumannsche Minimax-Theorem*, Arch. Math. (Basel) **19** (1968), 482–487. MR0240600

56. Rohini Kumar, *Space-time current process for independent random walks in one dimension*, ALEA Lat. Am. J. Probab. Math. Stat. **4** (2008), 307–336. MR2456971

57. O. E. Lanford, III and D. Ruelle, *Observables at infinity and states with short range correlations in statistical mechanics*, Comm. Math. Phys. **13** (1969), 194–215. MR0256687

58. Thomas M. Liggett, *Interacting particle systems*, Grundlehren der Mathematischen Wissenschaften [Fundamental Principles of Mathematical Sciences], vol. 276, Springer-Verlag, New York, 1985. MR776231

59. A. Messager, S. Miracle-Sole, and C. Pfister, *Correlation inequalities and uniqueness of the equilibrium state for the plane rotator ferromagnetic model*, Comm. Math. Phys. **58** (1978), no. 1, 19–29. MR0475552

60. James R. Munkres, *Topology: a first course*, Prentice-Hall, Inc., Englewood Cliffs, N.J., 1975. MR0464128

61. Lars Onsager, *Crystal statistics. I. A two-dimensional model with an order-disorder transition*, Phys. Rev. (2) **65** (1944), 117–149. MR0010315

62. K. R. Parthasarathy, *Probability measures on metric spaces*, AMS Chelsea Publishing, Providence, RI, 2005, Reprint of the 1967 original. MR2169627

63. Robert R. Phelps, *Lectures on Choquet's theorem*, second ed., Lecture Notes in Mathematics, vol. 1757, Springer-Verlag, Berlin, 2001. MR1835574

64. Firas Rassoul-Agha and Timo Seppäläinen, *Quenched point-to-point free energy for random walks in random potentials*, Probab. Theory Related Fields **158** (2014), no. 3-4, 711–750. MR3176363

65. Firas Rassoul-Agha, Timo Seppäläinen, and Atilla Yilmaz, *Quenched free energy and large deviations for random walks in random potentials*, Comm. Pure Appl. Math. **66** (2013), no. 2, 202–244. MR2999296

66. A. Rényi, *Probability theory*, North-Holland Publishing Co., Amsterdam, 1970, Translated by László Vekerdi, North-Holland Series in Applied Mathematics and Mechanics, Vol. 10. MR0315747

67. Sidney Resnick, *Adventures in stochastic processes*, Birkhäuser Boston, Inc., Boston, MA, 1992. MR1181423

68. R. Tyrrell Rockafellar, *Convex analysis*, Princeton Mathematical Series, No. 28, Princeton University Press, Princeton, N.J., 1970. MR0274683

69. Walter Rudin, *Functional analysis*, second ed., International Series in Pure and Applied Mathematics, McGraw-Hill Inc., New York, 1991. MR1157815

70. Erwin Schrödinger, *Statistical thermodynamics*, A course of seminar lectures delivered in January-March 1944, at the School of Theoretical Physics, Dublin Institute for Advanced Studies. Second edition, reprinted, Cambridge University Press, New York, 1962. MR0149891

71. E. Seneta, *Non-negative matrices and Markov chains*, Springer Series in Statistics, Springer, New York, 2006, Revised reprint of the second (1981) edition [Springer-Verlag, New York; MR0719544]. MR2209438

72. Timo Seppäläinen, *Large deviations for lattice systems. I. Parametrized independent fields*, Probab. Theory Related Fields **96** (1993), no. 2, 241–260. MR1227034

73. ———, *Large deviations for lattice systems. II. Nonstationary independent fields*, Probab. Theory Related Fields **97** (1993), no. 1-2, 103–112. MR1240718

74. ———, *Entropy, limit theorems, and variational principles for disordered lattice systems*, Comm. Math. Phys. **171** (1995), no. 2, 233–277. MR1344727

75. ———, *Maximum entropy principles for disordered spins*, Probab. Theory Related Fields **101** (1995), no. 4, 547–576. MR1327225

76. Timo Seppäläinen and J. E. Yukich, *Large deviation principles for Euclidean functionals and other nearly additive processes*, Probab. Theory Related Fields **120** (2001), no. 3, 309–345. MR1843178

77. Barry Simon, *The statistical mechanics of lattice gases. Vol. I*, Princeton Series in Physics, Princeton University Press, Princeton, NJ, 1993. MR1239893

78. V. Strassen, *The existence of probability measures with given marginals*, Ann. Math. Statist. **36** (1965), 423–439. MR0177430

79. S. R. S. Varadhan, *Large deviations and applications*, CBMS-NSF Regional Conference Series in Applied Mathematics, vol. 46, Society for Industrial and Applied Mathematics (SIAM), Philadelphia, PA, 1984. MR758258

80. ———, *Probability theory*, Courant Lecture Notes in Mathematics, vol. 7, New York University Courant Institute of Mathematical Sciences, New York, 2001. MR1852999

81. Ofer Zeitouni, *Random walks in random environments*, Lecture Notes in Mathematics, vol. 1837, Springer-Verlag, Berlin, 2004, Lectures from the 31st Summer School on Probability Theory held in Saint-Flour, July 8–25, 2001, Edited by Jean Picard. MR2071629

82. Ya. B. Zel′dovich, A. A. Ruzmaĭkin, and D. D. Sokoloff, *The almighty chance*, World Scientific Lecture Notes in Physics, vol. 20, World Scientific Publishing Co. Inc., River Edge, NJ, 1990, Translated from the Russian by Anvar Shukurov. MR1141627

Notation index

A°	topological interior of the set A
$\overline{A}$	topological closure of the set A
$\lvert A\rvert$	cardinality of the set A
A^c	complement of the set A
A^ε	points within distance $< \varepsilon$ of the set A
$\operatorname{aff} A$	affine hull of the set A
$A \smallsetminus B$	set difference, equals $A \cap B^c$
$A \triangle B$	symmetric set difference $(A \smallsetminus B) \cup (B \smallsetminus A)$
$a_n \sim b_n$	means $a_n/b_n \to 1$
$a_n \nearrow a$	a_n is a nondecreasing sequence that converges to a
αp	push-forward of measure α by kernel p (page 188)
$\alpha \times p$	joint measure with first marginal α and conditional given by kernel p (page 188)
$\alpha \circ f^{-1}$	push-forward of measure α by mapping f (page 35)
$\mathscr{B}_\mathcal{X}$	Borel σ-algebra on the topological space $\mathcal{X}$
$\mathbb{B}$	space of absolutely summable shift-invariant continuous interaction potentials
β	inverse temperature
$b\mathscr{B}$	space of bounded real-valued $\mathscr{B}$-measurable functions
$\operatorname{bd}(A)$	topological boundary of A, equals $\overline{A} \setminus A^\circ$
$B(x, r)$	open ball of radius r centered at x
$\mathscr{C}(\mathcal{X})$	space of continuous real-valued functions on $\mathcal{X}$

$\mathscr{C}_b(\mathcal{X})$	space of bounded continuous real-valued functions on $\mathcal{X}$
$\mathscr{C}_b^+(\mathcal{X})$	space of functions in $\mathscr{C}_b(\mathcal{X})$ that are strictly positive and bounded away from 0
$\mathscr{C}_{b,\mathrm{loc}}(\Omega)$	space of bounded continuous local functions (page 84)
$\overline{\mathrm{co}}\, A$	closed convex hull of the set A
δ_x	point mass at x: $\delta_x(A) = \mathbb{1}_A(x)$ for a set A
∂f	subdifferential of f (page 56)
$\mathrm{dom}\, f$	effective domain of f (page 167)
$e_1, \dots, e_d$	canonical basis of $\mathbb{R}^d$
$E[X]$, $E(X)$, EX	expectation of X
$E[X, A]$, $E(X, A)$	expectation of X on the event A, same as $E[X\mathbb{1}_A]$
$E^\mu[X]$, $E^\mu(X)$, $E^\mu X$	expectation of X under probability measure μ
$\mathrm{epi}\, f$	epigraph of f (page 20)
P-ess inf	essential infimum relative to P
P-ess sup	essential supremum relative to P
$\mathrm{ex}(K)$	set of extreme points of K
$\mathscr{F}$	σ-algebra on Ω, the product σ-algebra if $\Omega = \mathcal{X}^{\mathbb{Z}^d}$
$\mathscr{F}_V$	σ-algebra generated by coordinates in volume V
f^*	convex conjugate (or dual) of f
f^{**}	convex biconjugate of f
f_{lsc}	lower semicontinuous regularization of f
$\mathscr{G}^\Pi$	set of Gibbs measures consistent with specification Π
$H_\Lambda(\nu \mid \lambda)$	entropy of ν_Λ relative to λ_Λ
$H_n(\nu \mid \lambda)$	$H_{V_n}(\nu \mid \lambda)$
$H(\nu \mid \lambda)$	entropy of ν relative to λ
$h(\nu \mid \Phi)$	specific entropy relative to potential Φ
$h(\nu \mid \lambda)$	specific entropy of ν relative to λ
$\mathcal{H}_\Lambda^{\mathrm{free}}$	Hamiltonian in volume Λ with free boundary conditions
$\mathcal{H}_\Lambda^0$	free Ising Hamiltonian
$\mathcal{H}_\Lambda^{\tau_{\Lambda^c}}$	Hamiltonian in volume Λ with boundary condition τ_{Λ^c}
$\mathcal{I}$	σ-algebra of invariant sets, also used as an arbitrary index set

$i \sim j$	i and j are nearest neighbors in $\mathbb{Z}^d$
$\mathbb{1}_A$, $\mathbb{1}\{A\}$	indicator function of the set A
J	convex conjugate of $\bar{p}$ (page 214), coupling constant in the Ising model (page 102), also Donsker-Varadhan entropy (page 188)
$\overline{\kappa}$	upper rate function (page 26)
$\underline{\kappa}$	lower rate function (page 26)
$\kappa(\omega, f)$	expectation of f under kernel $\kappa(\omega)$
Λ	a (sometimes finite) subset of $\mathbb{Z}^d$
$\Lambda_n \nearrow \Lambda$	Λ_n is a nondecreasing sequence of sets the union of which is Λ
$\lambda^{\otimes\mathbb{N}}$, $\lambda^{\otimes\mathbb{Z}^d}$	i.i.d. product measure with marginal λ
$\mathrm{LDP}(\mu_n, r_n, I)$	$\{\mu_n\}$ satisfies a large deviation principle with normalization r_n and rate function I
$\overline{\lim}$	limsup
$\underline{\lim}$	liminf
L_n	empirical measure $L_n = \frac{1}{n}\sum_{k=1}^{n} \delta_{X_k}$
$L_n(f)$	expectation of f under L_n, equals $n^{-1}\sum_{k=0}^{n-1} f(X_k)$
$\mathcal{M}(\mathcal{X})$	space of finite signed measures on $\mathcal{X}$
$\mathcal{M}_e(\Omega)$	space of ergodic probability measures on Ω
$\mathcal{M}_\theta(\Omega)$	space of shift-invariant probability measures on Ω
$\mathcal{M}_1(\mathcal{X})$	space of probability measures on $\mathcal{X}$
$\mu^{\omega}_{\Lambda,\beta,h,J}$	Ising Gibbs measure in volume Λ with boundary condition ω and parameters (β, h, J)
$\mu^{0}_{\Lambda,\beta,h,J}$	Ising Gibbs measure with free boundary condition
$\mu^{+}_{\Lambda,\beta,h,J}$	Ising Gibbs measure with plus boundary condition $\omega = 1$
$\mu^{-}_{\Lambda,\beta,h,J}$	Ising Gibbs measure with minus boundary condition $\omega = -1$
$\mu \leq \nu$	probability measure ν stochastically dominates μ
$\mathbb{N}$	set of positive integers
∇f	gradient of f
$\nu \ll \lambda$	ν is absolutely continuous relative to λ
ν_Λ	the restriction of $\nu \in \mathcal{M}_1(\mathcal{S}^{\mathbb{Z}^d})$ to $\mathcal{S}^\Lambda$
ν_n	the restriction of $\nu \in \mathcal{M}_1(\mathcal{S}^{\mathbb{Z}^d})$ to $\mathcal{S}^{V_n}$

$\|f\|_\infty$	supremum norm, equals $\sup_x \|f(x)\|$
$\|\mu - \nu\|$	total variation distance of measures (see (7.10))
$\|x\|$	ℓ^2 norm $\|x\| = \sqrt{\|x_1\|^2 + \cdots + \|x_d\|^2}$, $x = (x_1, \ldots, x_d)$
$\|x\|_1$	ℓ^1 norm $\|x\|_1 = \|x_1\| + \cdots + \|x_d\|$, $x = (x_1, \ldots, x_d)$
$\|x\|_\infty$	ℓ^∞ norm $\|x\|_\infty = \max(\|x_1\|, \ldots, \|x_d\|)$, $x = (x_1, \ldots, x_d)$
$a_n = \mathcal{O}(b_n)$	means $\|a_n\| \le C b_n$ for a constant C independent of n
$a_n = o(b_n)$	means $a_n / b_n \to 0$
$\mathrm{osc}_j(f)$	oscillation of f as spin ω_j varies (page 115)
Ω	a general probability space and often $\Omega = \mathcal{S}^{\mathbb{Z}^d}$
Ω_Λ	space $\mathcal{S}^\Lambda$ of spin configurations ω_Λ in volume $\Lambda \subset \mathbb{Z}^d$
ω_Λ	spin configuration in volume Λ, $\omega_\Lambda = (\omega_i)_{i \in \Lambda}$
$\omega^{(n)}$	periodized configuration (page 85)
$\omega \le \sigma$	partial order on configurations: $\omega_i \le \sigma_i \ \forall i \in \mathbb{Z}^d$
Π	specification
$P(\Phi)$	infinite volume pressure corresponding to potential Φ
$p(\cdot)$	pressure function (page 59)
$\bar{p}(\cdot)$	upper pressure (page 213)
$\pi_\Lambda(\tau, \cdot)$, $\pi_\Lambda^\tau(\cdot)$	specification with boundary condition τ_{Λ^c}
$\mathbb{Q}$	set of rational numbers
$\mathbb{R}$	set of real numbers
$\mathrm{ri}\, A$	relative interior of the set A
R_n	level-3 empirical field $R_n(\omega) = \|V_n\|^{-1} \sum_{i \in V_n} \delta_{\theta_i \omega}$
$R_n(f)$, $R_n(\omega, f)$	expectation of f under R_n, equals $\|V_n\|^{-1} \sum_{i \in V_n} f(\theta_i \omega)$
$\widetilde{R}_n$	periodic empirical field $\widetilde{R}_n(\omega) = \frac{1}{\|V_n\|} \sum_{i \in V_n} \delta_{\theta_i \omega^{(n)}}$
$\widetilde{R}_n(f)$, $\widetilde{R}_n(\omega, f)$	expectation of f under $\widetilde{R}_n$
$\sigma_A = 1$	means that $\sigma_i = 1$ for all $i \in A$
$\mathcal{S}$	space, often metric or Polish
S_n	empirical mean $S_n = (X_1 + \cdots + X_n)/n$
$\mathscr{T}$	tail σ-algebra (page 284), topology (page 260)
θ_i	shift mapping on configurations: $(\theta_i \omega)_j = \omega_{i+j}$
$\mathscr{U}_b(\mathcal{X})$	space of bounded uniformly continuous functions on a metric space $\mathcal{X}$
$\mathscr{U}_{b,d}(\mathcal{X})$	space of bounded uniformly continuous functions on a metric space $\mathcal{X}$ with metric d specified

$\mathscr{U}_{b,\mathrm{loc}}(\Omega)$	space of bounded uniformly continuous local functions on Ω
V_n	the cube $\{i \in \mathbb{Z}^d : -n < i_1, \ldots, i_d < n\}$
$\mathcal{X}$, $\mathcal{Y}$	spaces (Hausdorff, metric, vector, Polish, etc)
$\lfloor x \rfloor$	integral part of x, that is, largest integer smaller or equal to x
$\lceil x \rceil$	smallest integer larger or equal to x
$x \wedge y$	$\min(x, y)$
$x \vee y$	$\max(x, y)$
x^+	$x \vee 0$
x^-	$-(x \wedge 0)$
$\mathbb{Z}$	set of integers
$\mathbb{Z}_+$	set of nonnegative integers
$\langle x, y \rangle$, $x \cdot y$	inner product in $\mathbb{R}^d$
$\langle \cdot\,, \cdot \rangle$	bilinear duality between two vector spaces

RWRE Notation:

Ω_+	same as $\Omega \times \mathcal{R}^{\mathbb{N}}$
$\pi_{x,y}(\omega)$	transition probability from x to y in environment ω
$\mathcal{R}$	set of admissible steps
$T_x\omega$	shift by x of environment ω: $(T_x\omega)_y = \omega_{x+y}$
$\mathscr{U}_b(\Omega)$	space of bounded uniformly continuous local functions on Ω
$\hat{x}_n$, $\hat{x}_n(\xi)$	same as $(\lfloor n\xi_1 \rfloor, n - \lfloor n\xi_1 \rfloor)$
ξ_1	first coordinate of $\xi = (\xi_1, 1 - \xi_1)$
z_i	increment $x_{i+1} - x_i$ of a path (x_i)

Author index

General index

A bold page number points to the page where the term is defined or used in an especially significant way. The number in parentheses is the page number on which the cited exercise may be found.

Selected Published Titles in This Series

162 **Firas Rassoul-Agha and Timo Seppäläinen,** A Course on Large Deviations with an Introduction to Gibbs Measures, 2015
160 **Marius Overholt,** A Course in Analytic Number Theory, 2014
158 **Fritz Colonius and Wolfgang Kliemann,** Dynamical Systems and Linear Algebra, 2014
157 **Gerald Teschl,** Mathematical Methods in Quantum Mechanics: With Applications to Schrödinger Operators, Second Edition, 2014
156 **Markus Haase,** Functional Analysis, 2014
155 **Emmanuel Kowalski,** An Introduction to the Representation Theory of Groups, 2014
154 **Wilhelm Schlag,** A Course in Complex Analysis and Riemann Surfaces, 2014
153 **Terence Tao,** Hilbert's Fifth Problem and Related Topics, 2014
152 **Gábor Székelyhidi,** An Introduction to Extremal Kähler Metrics, 2014
151 **Jennifer Schultens,** Introduction to 3-Manifolds, 2014
150 **Joe Diestel and Angela Spalsbury,** The Joys of Haar Measure, 2013
149 **Daniel W. Stroock,** Mathematics of Probability, 2013
148 **Luis Barreira and Yakov Pesin,** Introduction to Smooth Ergodic Theory, 2013
147 **Xingzhi Zhan,** Matrix Theory, 2013
146 **Aaron N. Siegel,** Combinatorial Game Theory, 2013
145 **Charles A. Weibel,** The K-book, 2013
144 **Shun-Jen Cheng and Weiqiang Wang,** Dualities and Representations of Lie Superalgebras, 2012
143 **Alberto Bressan,** Lecture Notes on Functional Analysis, 2013
142 **Terence Tao,** Higher Order Fourier Analysis, 2012
141 **John B. Conway,** A Course in Abstract Analysis, 2012
140 **Gerald Teschl,** Ordinary Differential Equations and Dynamical Systems, 2012
139 **John B. Walsh,** Knowing the Odds, 2012
138 **Maciej Zworski,** Semiclassical Analysis, 2012
137 **Luis Barreira and Claudia Valls,** Ordinary Differential Equations, 2012
136 **Arshak Petrosyan, Henrik Shahgholian, and Nina Uraltseva,** Regularity of Free Boundaries in Obstacle-Type Problems, 2012
135 **Pascal Cherrier and Albert Milani,** Linear and Quasi-linear Evolution Equations in Hilbert Spaces, 2012
134 **Jean-Marie De Koninck and Florian Luca,** Analytic Number Theory, 2012
133 **Jeffrey Rauch,** Hyperbolic Partial Differential Equations and Geometric Optics, 2012
132 **Terence Tao,** Topics in Random Matrix Theory, 2012
131 **Ian M. Musson,** Lie Superalgebras and Enveloping Algebras, 2012
130 **Viviana Ene and Jürgen Herzog,** Gröbner Bases in Commutative Algebra, 2011
129 **Stuart P. Hastings and J. Bryce McLeod,** Classical Methods in Ordinary Differential Equations, 2012
128 **J. M. Landsberg,** Tensors: Geometry and Applications, 2012
127 **Jeffrey Strom,** Modern Classical Homotopy Theory, 2011
126 **Terence Tao,** An Introduction to Measure Theory, 2011
125 **Dror Varolin,** Riemann Surfaces by Way of Complex Analytic Geometry, 2011
124 **David A. Cox, John B. Little, and Henry K. Schenck,** Toric Varieties, 2011
123 **Gregory Eskin,** Lectures on Linear Partial Differential Equations, 2011